College Algebra

Summary Notebook

Kirk Trigsted

University of Idaho

Editorial Director, Mathematics: Christine Hoag
Senior Acquisitions Editor: Adam Jaworski
Sponsoring Editor: Dawn Murrin
Editorial Assistant: Joseph Colella
Art Director: Heather Scott
Cover Designer: Tamara Newnam
Production Manager: Ron Hampton
Senior Math Media Producer: Ceci Fleming
Project Manager, MathXL: Eileen Moore
AV Project Manager: Thomas Benfatti
Senior Marketing Manager: Katherine Greig
Marketing Assistant: Katherine Minton
Market Development Manager: Dona Kenly
Manufacturing Manager: Evelyn Beaton

This book is dedicated to

JAMES D. "PAPA JIM" GARRED
1936–2008

Photograph Credits: Page 1, Shutterstock; Page 56, Shutterstock; Page 89, Digital Vision; Page 159, Getty RF; Page 224, Shutterstock; Page 275, Getty Images Sport; Page 321, Shutterstock

Copyright © 2010 Pearson Education, Inc.
Publishing as Pearson Prentice Hall, Upper Saddle River, NJ 07458.

1 2 3 4 5 6—RRDHBS—11 10 09 08

ISBN-13: 978-0-13-174469-1
ISBN-10: 0-13-174469-0

Contents

Preface

Introduction

This *Summary Notebook* contains the pages of Kirk Trigsted's *College Algebra* eBook in a portable, spiral-bound format. The structure of the *Summary Notebook* helps students organize their notes by providing them with space to summarize the videos and animations. Students can also use it to review the eBook's material anytime, anywhere.

A Note to Students

This textbook was created for you! Unlike a traditional text, I wanted to create content that gives you, the reader, the ability to be an active participant in your learning. This ebook was specifically designed to be read online. The eBook pages have large, readable fonts and were designed without the need to scroll. Throughout the material, I have carefully placed hundreds of hyperlinks to definitions, previous chapters, interactive videos, animations, and other important content. Many of the videos and animations require you to actively participate and interact. Take some time to "click around" and get comfortable with the navigation of the eBook and explore its many features.

Before you attempt each homework assignment, read the appropriate section(s) of the eBook. At the beginning of each section (starting in Chapter One), you will encounter a feature called Things to Know. This feature includes all the prerequisite objectives from previous sections that you will need to successfully learn the material presented in the new section. If you do not think that you have a basic understanding of any of these objectives, click on any of the hyperlinks and rework through those objectives, taking advantage of any videos or animations.

Try testing yourself by working through the corresponding You Try It exercises. Remember, students learn math by doing math! The more time you spend working through the videos, animations, and exercises, the more you will understand. If your instructor assigns homework in MyMathLab or MXL, rework the exercises until you get them right. Be sure to go back and read the eBook at anytime while you are working on your homework. This text caters to your educational needs. I hope you enjoy the experience!

A Note to Instructors

I have taught with MyMathLab for many years, and have experienced first-hand how fewer and fewer students have been referring to their traditional textbooks. As the use of technology plays an ever increasing role in how we are teaching our students, it seems only natural to have a textbook that mirrors the way our students are learning. I am excited to have written a book that was conceived from the ground up to be used as an online interactive tool. Unlike traditional printed textbooks, this book was specifically designed for students to read online while working in MyMathLab. Therefore, I wrote this book entirely from an online perspective, with MyMathLab and its existing functionality specifically in mind. Every hyperlink, video, and animation was strategically integrated within the context of each page to maximize the student learning experience. All of the interactive media was designed so students could actively participate while they learn math. Many of the interactive videos and animations require student interaction, giving specific feedback for incorrect responses.

I am a proponent of students learning terms and definitions. Therefore, I have created hyperlinks throughout the text to the definitions of important mathematical terms. I have also inserted a tremendous amount of "just-in-time review" through-out the text by creating links to prerequisite topics. Students have the ability to reference these review materials with just a click of the mouse. You will see that the exercise sets are concise and nonrepetitive. Since MyMathLab will be used as the main assessment engine, there is no need for a repetitive exercise set in the hardcopy version of the text. Every single exercise is available for you to assign within MyMathLab or MathXL. For the first time, instructors can assign reading! I have created five conceptual reading assessment questions for every section of the book, giving the student specific feedback for both correct and incorrect responses. Each feedback message directs them back to the appropriate pages for review. The students of today live in a technological world where everything is "click-able." My hope is that every student that uses the eBook will have a positive learn-ing experience.

Acknowledgments

First of all, I want to thank my wife, Wendy, for all her loving support, encourage-ment, and sacrifices over the last four years. I could not have completed this project without her! I also could not have finished without the love of my children, Benjamin, Emily, and Gabrielle. Thank you for being so patient when daddy needed to work and for reminding daddy when he needed to take a break.

Writing this textbook has been one of the most difficult and rewarding experi-ences of my life. I truly could not have accomplished my goal without the support and contributions of so many wonderful, talented people. These extraordinary talents deserve every accolade I can give them. From Pearson, I would like to thank Adam Jaworski, my editor and friend, for believing in me; Greg Tobin and Chris Hoag for their continued support; Dawn Murrin for keeping me on track; Becky Anderson for helping to get this project off the ground; Heather Scott for her design brilliance; Dona Kenly and Katherine Greig for their marketing expert-ise; Ceci Fleming and Ruth Berry for taking care of all of the details; Eileen Moore, Alicia Anderson, and the rest of the MathXL development team for an amazing job; Tom Benfatti and the art team for their great eye for detail; and Ron Hampton, our production manager, for all his support.

Along the way, I had the help of so many people from around the world and from many different walks of life. Phoebe Rouse's contribution to this book cannot be measured. Her detailed notebooks, reviews of the reading assessment questions, and our "Thursday conversations" were invaluable throughout the entire writing process. Thank you for everything, Phoebe! I appreciate everyone else who made this book a reality: Crystal Clifton of Progressive Publishing Alternatives helped me in so many ways; Alice Champlin and Anthony T. J. Kilian at Magnitude Entertainment for creating quality videos and animations; Matt Jepsen for making the prospectus CD; Pamela Trim for reading the entire eBook and catching the little things; Donna Densmore for her work in the creation of the PowerPoint slides; and Eloy Lopez for sitting through all the video shoots.

The following list includes all reviewers, focus group attendees, and class testers. Please accept my deepest apologies if I have inadvertently omitted any-one. I am deeply grateful and I genuinely thank you all from the bottom of my heart.

Krishan Agrawal, Virginia State University
Ali M. Ahmad, New Mexico State University–Dona Ana Community College
Theresa Allen, University of Idaho
Patty G. Amick, Greenville Technical College
Alexandra Athanassiou, University of Missouri–Columbia
Richard Avery, Dakota State University

Ben Baker, University of Idaho
Linda Barton, Ball State University
Sam Bazzi, Henry Ford Community College
Molly Beauchman, Yavapai College
Patricia Blus, National-Louis University
Nina Bohrod, Anoka Ramsey Community College
David Bramlet, Jackson State University
Julie Brown, University of Idaho
Connie Buller, Metropolitan Community College
Joe Castillo, Broward Community College
Mariana Coanda, Broward Community College
Earl W. Cook, Chattahoochee Valley Community College
Kemba C. Countryman, Chattahoochee Valley Community College
Momoyo Dahle, Las Positas Community College
Diane Daniels, Mississippi State University
Emmett Dennis, Southern Connecticut State University
Donna Densmore, Bossier Parish Community College
Holly Dickin, University of Idaho
Christina Dwyer, Manatee Community College
Jeanette Eggert, Concordia University
Brett Elliott, Southeastern Oklahoma State University
Nicki Feldman, Pulaski Technical College
Catherine Ferrer, Valencia Community College
Cynthia Francisco, Oklahoma State University
Robert Frank, Westmoreland County Community College
Jim Frost, St. Louis Community College–Meramec
Angelito Garcia, Truman College
Lee Gibson, University of Louisville
Charles B. Green, University of North Carolina–Chapel Hill
Mike Hall, Arkansas State University
Melissa Hardeman, University of Arkansas, Little Rock
Celeste Hernandez, Richland College
Pamela Howard, Boise State University
Jeffrey Hughes, Hinds Community College
Eric Hutchinson, College of Southern Nevada
Robert Indrihovic, Florence-Darlington Tech
Philip Kaatz, Mesalands Community College
Cheryl Kane, University of Nebraska–Lincoln
Susan Knights, Boise State University
Jeff Laub, Central Virginia Community College
Jennifer Legrand, St. Charles Community College
Kurt Lewandowski, Clackamas Community College
Pamela Spurlock Mills, University of Arkansas
Peggy L. Moch, Valdosta State University
Rebecca Morgan, Wayne State University
Rebecca Muller, Southeastern Louisiana University
Veronica Murphy, Saint Leo University
Karla Neal, Louisiana State University
Prince Raphael A. Okojie, Fayetteville State University
Enyinda Onunwor, Stark State University
Shahla Peterman, University of Missouri–St. Louis
Sandra Poinsett, College of Southern Maryland
Ray Purdom, University of North Carolina at Greensboro
Nancy Ressler, Oakton Community College
Joe Rody, Arizona State University
Phoebe Rouse, Louisiana State University
Patricia Rowe, Columbus State Community College

Chie Sakabe, University of Idaho
Jorge Sarmiento, County College of Morris
John Savage, Montana State University–College of Technology
Victoria Seals, Gwinnett Technical College
Rita Sowell, Volunteer State Community College
John Squires, Cleveland State Community College
Pam Stogsdill, Bossier Parish Community College
Eleanor Storey, Front Range Community College
Lalitha Subramanian, Potomac State College of West Virginia University
Mary Ann Teel, University of North Texas
Jamie Thomas, University of Wisconsin–Manitowoc
Keith B. Thompson, Davidson County Community College
Suzanne Topp, Salt Lake Community College
Marcia Vergo, Metropolitan Community College
Kimberly Walters, Mississippi State University
Jacci White, Saint Leo University
Ralph L. Wildy Jr., Georgia Military College
Sherri M. Wilson, Fort Lewis College
Janet Wyatt, Metropolitan Community College–Longview
Ghidei Zedingle, Normandale Community College

Review Chapter

R.1 Real Numbers

OBJECTIVES

1 Understanding the Real Number System

2 Writing Sets Using Set-Builder Notation and Interval Notation

3 Determining the Intersection and Union of Sets and Intervals

4 Understanding Absolute Value

OBJECTIVE 1: UNDERSTANDING THE REAL NUMBER SYSTEM

A **set** is a collection of objects. Each object in the set is called an **element** or a **member** of the set. We typically use braces { } to enclose all elements of a set. Capital letters are sometimes used to name a set. For example, suppose we define sets A and B as

$$A = \{1, 2, 3, 4, 5, 6, 7, 8, 9, 10\} \text{ and } B = \{2, 4, 6, 8, 10\}$$

Notice that each element of set B is also an element of set A. We say that set B is a **subset** of set A and write $A \subset B$. The set of **real numbers** consists of several subsets of numbers. We now describe these subsets.

R-1

Natural Numbers

The set of natural numbers, sometimes referred to as the counting numbers, is defined as $\mathbb{N} = \{1, 2, 3, 4 \ldots\}$. The three dots ($\ldots$), called an ellipsis, are used to indicate that the elements of the set follow the same pattern.

Whole Numbers

The set of whole numbers is composed of the natural numbers combined with the number 0. Thus, $\mathbb{W} = \{0, 1, 2, 3, 4 \ldots\}$. Note that $\mathbb{N} \subset \mathbb{W}$.

Integers

Combining the negatives of the natural numbers with the whole numbers gives the set of integers. The symbol $\mathbb{Z}$ is often used to denote the set of integers; thus, $\mathbb{Z} = \{\ldots, -3, -2, -1, 0, 1, 2, 3, \ldots\}$. Note that $\mathbb{N} \subset \mathbb{W} \subset \mathbb{Z}$.

Rational Numbers

A rational number is a number that can be expressed as a quotient of two integers. The symbol $\mathbb{Q}$ is often used to represent the set of rational numbers. We can use **set-builder notation** to define the set of rational numbers as $\mathbb{Q} = \{\frac{p}{q} \mid p \text{ and } q \text{ are integers}, q \neq 0\}$, which is read as "the set of all elements $\frac{p}{q}$ such that p and q are integers, $q \neq 0$." We revisit set-builder notation later in this section. The set of rational numbers includes all repeating and terminating decimals. Examples of rational numbers include $7, \frac{1}{2}, -\frac{11}{43}, -0.56, .3333\ldots$, and $.\overline{1234}$. (The bar above the decimal .1234 represents a repeating decimal.) Note that $\mathbb{N} \subset \mathbb{W} \subset \mathbb{Z} \subset \mathbb{Q}$.

Irrational Numbers

Numbers that are not rational (numbers that cannot be expressed as a quotient of two integers) are called irrational numbers. We can define the set of irrational numbers as $\mathbb{I} = \{x \mid x \text{ is not a rational number}\}$. Examples of irrational numbers include $\sqrt{7}, \sqrt{17}, \sqrt[3]{29}$, and π.

The set of real numbers, denoted as $\mathbb{R}$, is the set of rational numbers joined with the set of irrational numbers. Figure 1 gives us a visual image of the set of real numbers.

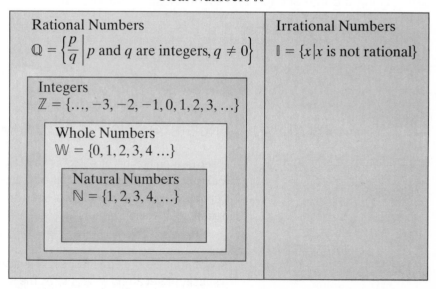

Figure 1 The real numbers

Real numbers can be represented by points located on a number line called the *real axis*. The point corresponding to 0 is called the *origin*. Figure 2 illustrates several real numbers plotted on a number line. In this text, we use a solid circle (●) to represent a point on a number line.

Figure 2

Example 1 Classify Real Numbers

Classify each number in the set $\{-5, -\frac{1}{3}, 0, \sqrt{3}, 1.\overline{4}, 2\pi, 11\}$ as a natural number, whole number, integer, rational number, irrational number, and/or real number.

Solution

The number -5 is an integer, rational number, and real number.

The number $-\frac{1}{3}$ is a rational number and a real number.

The number 0 is a whole number, integer, rational number, and real number.

The number $\sqrt{3}$ is an irrational number and a real number.

The number $1.\overline{4}$ is a rational number and a real number.

The number 2π is an irrational number and a real number.

The number 11 is a natural number, whole number, integer, rational number, and real number.

Work exercises 1–6 in this textbook or in the *MyMathLab*® Study Plan.

OBJECTIVE 2:   **WRITING SETS USING SET-BUILDER NOTATION AND INTERVAL NOTATION**

Set-builder notation can be a convenient way to describe certain sets of numbers. For example, we used set-builder notation to describe the set of **rational numbers** and the set of **irrational numbers**. Suppose that we want to describe the set of all real numbers less than 10. Using set-builder notation, we write this set as $\{x|x < 10\}$. This set is read as "the set of all x such that x is less than 10." Figure 3 illustrates how we can represent this set on a number line.

$$\xleftarrow{\hspace{3cm}} \underset{10}{\overset{}{\diamond}} \xrightarrow{\hspace{1cm}}$$

Figure 3 The set $\{x|x < 10\}$ represented on a number line.

The open circle (○) at the number 10 in Figure 3 represents that 10 is *not* included in the set. If we wanted to include the number 10, we would have used a solid circle (●). The arrow extending to the left indicates that the set goes on indefinitely toward negative infinity. In this text, we often use **interval notation** to describe sets of numbers such as the set displayed in Figure 3. The interval that describes this set is called an open infinite interval and is written as $(-\infty, 10)$. It is extremely important to be able to write intervals of numbers in both set-builder notation and interval notation. Table 1 illustrates the different types of intervals and the corresponding set-builder notation.

Table 1

Type of Interval and Graph	Interval Notation	Set-Builder Notation
Open interval $a \qquad b$	(a, b)	$\{x \mid a < x < b\}$
Closed interval $a \qquad b$	$[a, b]$	$\{x \mid a \le x \le b\}$
Half-open intervals $a \qquad b$ $a \qquad b$	$(a, b]$ $[a, b)$	$\{x \mid a < x \le b\}$ $\{x \mid a \le x < b\}$
Open infinite intervals a b	(a, ∞) $(-\infty, b)$	$\{x \mid x > a\}$ $\{x \mid x < b\}$
Closed infinite intervals a b	$[a, \infty)$ $(-\infty, b]$	$\{x \mid x \ge a\}$ $\{x \mid x \le b\}$

Example 2 Write a Set Using Set-Builder Notation and Interval Notation

Given the set sketched on the number line, a) identify the type of interval, b) write the set using set-builder notation, and c) write the set using interval notation.

$$-\tfrac{6}{5} \qquad 0 \qquad 3$$

Solution

a. This interval is called a half-open interval. The solid circle at 3 indicates that the set includes the number 3. The open circle at $-\frac{6}{5}$ indicates that the set does not include $-\frac{6}{5}$.

b. This set can be described in set-builder notation as $\left\{x \mid -\frac{6}{5} < x \le 3\right\}$.

c. We write this set in interval notation as $\left(-\frac{6}{5}, 3\right]$.

My video summary

▣ Example 3 Write a Set Using Set-Builder Notation and Interval Notation

a. Write the set $\left[-\frac{1}{3}, \infty\right)$ in set-builder notation and graph the set on a number line.

b. Write the set $\left\{x \mid -\frac{7}{2} < x \le \pi\right\}$ in interval notation and graph the set on a number line.

Solution Watch the video to see the solution worked out in detail.

Work exercises 7–18 in this textbook or in the *MyMathLab* Study Plan.

OBJECTIVE 3: DETERMINING THE INTERSECTION AND UNION OF SETS
AND INTERVALS

We are often interested in looking at two or more sets at a time. Joining all elements of a set A with all elements of a set B is known as the **union** of sets and is denoted as $A \cup B$. The set of elements that are *common* to both sets A and B is known as the **intersection** of sets A and B and is written as $A \cap B$. The set containing no elements is called the empty set and is denoted $\varnothing$. Sometimes the intersection of two sets results in the empty set.

Example 4 Find the Intersection and Union of Sets

Let $A = \left\{-5, 0, \frac{1}{3}, 11, 17\right\}$, $B = \{-6, -5, 4, 17\}$, and $C = \left\{-4, 0, \frac{1}{4}\right\}$.

  a. Find $A \cup B$.
  b. Find $A \cap B$.
  c. Find $B \cap C$.

Solution

  a. $A \cup B = \left\{-6, -5, 0, \frac{1}{3}, 4, 11, 17\right\}$ *List all elements in set A or set B.*

  b. $A \cap B = \{-5, 17\}$ *List all elements common to set A and set B.*

  c. $B \cap C = \varnothing$ *There are no elements common to both B and C.*

My video summary

 Example 4 Find the Intersection of Intervals

Find the intersection of the following intervals and graph the set on a number line.
  a. $[0, \infty) \cap (-\infty, 5]$ **b.** $((-\infty, -2) \cup (-2, \infty)) \cap [-4, \infty)$

Solution Watch the video to verify the following solutions.

  a. $[0, \infty) \cap (-\infty, 5] = [0, 5]$

$$0 \qquad 5$$

  b. $((-\infty, -2) \cup (-2, \infty)) \cap [-4, \infty) = [-4, -2) \cup (-2, \infty)$

$$-4 \qquad -2 \qquad 0$$

Work exercises 19–27 in this textbook or in the *MyMathLab* Study Plan.

OBJECTIVE 4: UNDERSTANDING ABSOLUTE VALUE

The absolute value of a real number a is defined as the distance between a and 0 on a number line and is denoted $|a|$. For example, $|-5| = 5$ because the number -5 is 5 units away from 0 on a number line. See Figure 4.

$$|-5| = 5$$
-5 is 5 units away from 0

Figure 4 The absolute value of -5 is 5 because -5 is 5 units away from 0.

DEFINITION Absolute Value

The absolute value of a real number a is defined by

$$|a| = \begin{cases} a & \text{if } a \geq 0 \\ -a & \text{if } a < 0 \end{cases}$$

and represents the distance between a and 0 on a number line.

Because the absolute value of a number a represents the *distance* from a to 0 on a number line, it follows that $|a| \geq 0$ for any real number a. This property and several other properties of absolute value are stated as follows.

Properties of Absolute Value

For all real numbers a and b,

1. $|a| \geq 0$ **2.** $|ab| = |a||b|$ **3.** $\left|\dfrac{a}{b}\right| = \dfrac{|a|}{|b|}, b \neq 0$

4. $|a|^2 = a^2$ **5.** $|a - b| = |b - a|$

Example 5 Evaluate Absolute Value Expressions

Evaluate the following expressions involving absolute value.

a. $\left|-\sqrt{2}\right|$ **b.** $\left|\dfrac{-4}{16}\right|$ **c.** $|2 - 5|$

Solution

a. $\left|-\sqrt{2}\right| = -(-\sqrt{2}) = \sqrt{2}$

b. $\left|\dfrac{-4}{16}\right| = \dfrac{|-4|}{|16|} = \dfrac{-(-4)}{16} = \dfrac{4}{16} = \dfrac{1}{4}$

c. $|2 - 5| = |-3| = -(-3) = 3$ or $|2 - 5| = |5 - 2| = |3| = 3$

Let's take a closer look at the absolute value expression $|2 - 5|$ from Example 5c and interpret its meaning geometrically. The expression $|2 - 5|$ represents the distance between the numbers 2 and 5 on a number line. Notice in Figure 5 that the distance between 2 and 5 on a number line is three units.

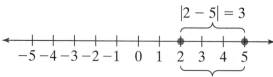

$$|2 - 5| = 3$$

The distance between 2 and 5 is 3 units.   **Figure 5**

Because the distance between 2 and 5 is the same as the distance between 5 and 2, we can conclude that $|2 - 5| = |5 - 2|$. Thus, for any real numbers a and b, $|a - b| = |b - a|$, which is precisely property 5 of absolute values. We now restate property 5 as the distance between two real numbers on a number line.

Distance between Two Real Numbers on a Number Line

For any real numbers a and b, the distance between a and b on a number line is defined by $|a - b|$ or $|b - a|$.

Example 6 Find the Distance between Two Real Numbers

Find the distance between the numbers -5 and 3 using absolute value.

Solution Let $a = -5$ and $b = 3$. The distance between a and b is given by $|a - b|$ or $|b - a|$.

$$|a - b| = |-5 - 3| = |-8| = 8.$$ So, the distance between -5 and 3 is 8 units.

Work exercises 28–35 in this textbook or in the *MyMathLab*® Study Plan.

R.1 Exercises

In Exercises 1–4, classify each number as a natural number, whole number, integer, rational number, irrational number, and/or real number. Each number may belong to more than one set.

1. 259   **2.** $-11{,}401$ **3.** $-\dfrac{43}{7}$ **4.** $-\sqrt{17}$

In Exercises 5 and 6, given each set of real numbers, list the numbers that are (a) natural numbers, (b) whole numbers, c) integers, d) rational numbers, and/or e) irrational numbers.

5. $\left\{-17, -\sqrt{5}, -\dfrac{25}{19}, 0, 0.331, 1, \dfrac{\pi}{2}\right\}$ **6.** $\left\{-11, -3.\overline{2135}, -\dfrac{3}{9}, \dfrac{\sqrt{7}}{2}, 21.1\right\}$

In Exercises 7–10, given the set sketched on the number line, a) identify the type of interval, b) write the set using set-builder notation, and c) write the set using interval notation.

7.

8.

9.

10.

In Exercises 11–14, write each interval in set-builder notation and graph the set on a number line.

11. $\left[-\dfrac{1}{2}, 5\right]$ **12.** $\left(0, \dfrac{5}{2}\right)$ **13.** $(-\infty, 3]$ **14.** $(-1, \infty)$

In Exercises 15–18, write the given set in interval notation and graph the set on a number line.

15. $\left\{x \,\middle|\, -\dfrac{5}{2} \le x \le 1\right\}$ **16.** $\{x \mid 0 \le x < 3\}$ **17.** $\left\{x \,\middle|\, x \ge \dfrac{3}{4}\right\}$ **18.** $\{x \mid x > -4\}$

In Exercises 19–21, sets A and B are given. Find (a) $A \cup B$ and (b) $A \cap B$.

19. $A = \{-1, 0, 1, 2\}$ and $B = \{0, 2, 4\}$

20. $A = \left\{ -\dfrac{3}{4}, -\dfrac{1}{4}, \dfrac{1}{4}, \dfrac{3}{4} \right\}$ and $B = \left\{ -1, \dfrac{1}{4}, 1 \right\}$

21. $A = \{1, 2, 3, 4\}$ and $B = \{0, 5, 6\}$

In Exercises 22–27, find the intersection of the given intervals.

22. $(-\infty, \infty) \cap (-\infty, 3)$ **23.** $(-8, 5] \cap (-12, 3)$

24. $[0, \infty) \cap (-\infty, 4]$ **25.** $((-\infty, 1) \cup (1, \infty)) \cap (-\infty, \infty)$

26. $((-\infty, 0) \cup (0, \infty)) \cap [-3, \infty)$ **27.** $(-\infty, 10) \cap (-5, \infty) \cap (-\infty, 1]$

In Exercises 28–32, evaluate the absolute value expression.

28. $|286|$ **29.** $|-2.2|$ **30.** $|-2 \cdot 3|$ **31.** $-\left| \dfrac{-2 \cdot 6}{4} \right|$ **32.** $|2 - \pi|$

In Exercises 33–35, find the distance between the given two numbers using absolute value.

33. 3 and 14   **34.** -5 and 7   **35.** -24 and -1.5

R.2 The Order of Operations and Algebraic Expressions

OBJECTIVES

1 Understanding the Properties of Real Numbers

2 Using Exponential Notation

3 Using the Order of Operations to Simplify Numeric and Algebraic Expressions

In this section, we learn how to simplify **algebraic expressions**. An algebraic expression consists of one or more terms that include **variables**, constants, and operating symbols such as $+$ or $-$. For example, $4x^3y^4 - 1$ is an algebraic expression consisting of two terms. The letters x and y used in this expression are called **variables** and are used to represent any number. Before we work with algebraic expressions, it is important to understand some basic properties of real numbers. Our understanding of these properties will enhance our ability to simplify numeric and algebraic expressions.

OBJECTIVE 1:   UNDERSTANDING THE PROPERTIES OF REAL NUMBERS

We all know that when adding two real numbers or multiplying two numbers, the order in which we perform the addition or multiplication does not affect the result. For example, $3 + 5 = 5 + 3$ and $(-4)(6) = (6)(-4)$. The operations of addition and multiplication are said to be commutative. The commutative property of addition and multiplication are stated here along with two other important properties of real numbers.

Commutative Property of Addition

If a and b are real numbers, then $a + b = b + a$.

Example: $3 + 5 = 8$ and $5 + 3 = 8$

Commutative Property of Multiplication

If a and b are real numbers, then $ab = ba$.

Example: $(-4)(6) = -24$ and $(6)(-4) = -24$

Associative Property of Addition

If a, b, and c are real numbers, then $(a + b) + c = a + (b + c)$.

Example: $(-7 + 11) + 6 = 4 + 6 = 10$ and $-7 + (11 + 6) = -7 + 17 = 10$

Associative Property of Multiplication

If a, b, and c are real numbers, then $(ab)c = a(bc)$.

Example: $(-8 \cdot -2) \cdot 3 = 16 \cdot 3 = 48$ and $-8 \cdot (-2 \cdot 3) = -8 \cdot (-6) = 48$

Distributive Property

If a, b, and c are real numbers, then $a(b + c) = ab + ac$.

Example: $3(x - 4) = 3 \cdot x + 3 \cdot (-4)$
$$= 3x - 12$$

Example 1 Use the Properties of Real Numbers to Rewrite Expressions

a. Use the commutative property of addition to rewrite $11 + y$ as an equivalent expression.
b. Use the commutative property of multiplication to rewrite $7(y + 4)$ as an equivalent expression.
c. Use the associative property of multiplication to rewrite $8(pq)$ as an equivalent expression.
d. Use the distributive property to multiply $11(a - 4)$.
e. Use the distributive property to rewrite $aw + az$ as an equivalent expression.

Solution

a. By the commutative property of addition, the algebraic expression $11 + y$ is equivalent to $y + 11$.
b. By the commutative property of multiplication, the algebraic expression $7(y + 4)$ is equivalent to $(y + 4) \cdot 7$.
c. By the associative property of multiplication, the algebraic expression $8(pq)$ is equivalent to $(8p)q$.
d. To use the distributive property, we multiply each variable or number inside the parentheses by the factor outside the parentheses.

$$11(a - 4) = 11 \cdot a - 11 \cdot 4 = 11a - 44$$

e. By the distributive property, $aw + az = a(w + z)$.

Work exercises 1–18 in this textbook or in the *MyMathLab*® Study Plan.

OBJECTIVE 2:   USING EXPONENTIAL NOTATION

Numeric and algebraic expressions often have terms involving exponents such as x^3, a^4, or $(-2)^5$. It is important to understand how to simplify expressions involving exponents. We now define exponential notation.

Exponential Notation

If b is a real number and n is a positive integer, then b^n is equivalent to
$$\underbrace{b \cdot b \cdot b \cdots b}_{n \text{ factors of } b}.$$

Note: b is called the base of the exponential expression and n is called the exponent.

In Section R.3, we establish the laws of exponents.

Example 2 Using Exponential Notation

a. Rewrite the expression $5 \cdot 5 \cdot 5$ using exponential notation. Then identify the base and the exponent.
b. Identify the base and the exponent of the expression $(-3)^4$, and then evaluate.
c. Identify the base and the exponent of the expression -2^5, and then evaluate.

Solution

a. We can rewrite $5 \cdot 5 \cdot 5$ as 5^3. The base is 5, and the exponent is 3.
b. The base of the expression $(-3)^4$ is -3, and the exponent is 4. Therefore, $(-3)^4 = (-3) \cdot (-3) \cdot (-3) \cdot (-3) = 81$.
c. The expression -2^5 can be rewritten as $-1 \cdot 2^5$. The base of the exponential expression is 2, and the exponent is 5. Thus, $-2^5 = -1 \cdot 2^5 = -1 \cdot 2 \cdot 2 \cdot 2 \cdot 2 \cdot 2 = -1 \cdot 32 = -32$.

Work exercises 19–25 in this textbook or in the *MyMathLab*® Study Plan.

OBJECTIVE 3:   USING THE ORDER OF OPERATIONS TO SIMPLIFY NUMERIC AND ALGEBRAIC EXPRESSIONS

When we encounter an expression with more than one operating symbol, it is crucial that we simplify the expression in the correct order. For example, suppose we are given the expression $5 - 3 \cdot 7$. In this expression, there are two operating symbols, symbolizing subtraction and multiplication. If we choose to subtract first and then multiply, we get

$$5 - 3 \cdot 7 = 2 \cdot 7 = 14$$

If we multiply first and then subtract the result from 5, we get

$$5 - 3 \cdot 7 = 5 - 21 = -16$$

As you can see, depending on the order in which we perform the operations of subtraction and multiplication, we obtain different results. The correct way to simplify the expression $5 - 3 \cdot 7$ is to perform the multiplication before the subtraction. Therefore, $5 - 3 \cdot 7 = -16$.

When simplifying an expression having more than one operating symbol, we use the following order of operations.

Order of Operations

1. Always start within the innermost grouping symbols. This includes parentheses ( ), brackets [ ], braces { }, and absolute value bars | |.
2. Simplify exponential expressions.
3. Perform multiplication and division working from left to right.
4. Perform addition and subtraction working from left to right.

My interactive video summary

 Example 3 **Use the Order of Operations to Simplify a Numeric Expression**

Simplify each expression.

a. $-2^3 + [3 - 5 \cdot (1 - 3)]$ b. $\dfrac{|2 - 3^3| + 5}{5^2 - 4^2}$

Solution Watch the interactive video to verify that

a. $-2^3 + [3 - 5 \cdot (1 - 3)] = 5$ and b. $\dfrac{|2 - 3^3| + 5}{5^2 - 4^2} = \dfrac{10}{3}$.

Work exercises 26–33 in this textbook or in the *MyMathLab*® Study Plan.

My video summary

Example 4 **Use the Order of Operations to Evaluate an Algebraic Expression**

Evaluate the algebraic expression $-x^3 - 4x$ for $x = -2$.

Solution $-x^3 - 4x$ *Write the original algebraic expression.*

$= -(-2)^3 - 4(-2)$ *Substitute -2 for x.*

$= -(-8) - 4(-2)$ *$(-2)^3 = -8$.*

$= 8 + 8$ *Simplify.*

$= 16$

Work exercises 34–41 in this textbook or in the *MyMathLab*® Study Plan.

When simplifying algebraic expressions, we must remove all grouping symbols and combine **like terms**. Like terms are two or more terms consisting of the same variable (or variables) raised to the same power. For example, $3x^2$ and $5x^2$ are like terms because each term contains the same variable, x, raised to the same power, 2.

Example 5 **Simplify an Algebraic Expression**

Simplify each algebraic expression.

a. $3x^2 - 2 + 5x^2$ b. $-3(4a - 7) + 5(3 - 5a)$

Solution

a. $3x^2 - 2 + 5x^2$ *Write the original algebraic expression.*

$= 3x^2 + 5x^2 - 2$ *Use the commutative property of addition to rearrange the terms.*

$= (3 + 5)x^2 - 2$ *Use the distributive property $ab + ac = (b + c)a$ to combine like terms.*

$= 8x^2 - 2$ *Simplify.*

b. $-3(4a - 7) + 5(3 - 5a)$ Write the original algebraic expression.

$= -3(4a) - 3(-7) + 5(3) + 5(-5a)$ Use the distributive property.

$= -12a + 21 + 15 - 25a$ Simplify.

$= -12a - 25a + 21 + 15$ Use the commutative property of addition to rearrange the terms.

$= (-12 - 25)a + 21 + 15$ Use the distributive property $ab + ac = (b + c)a$ to combine like terms.

$= -37a + 36$ Simplify.

Work exercises 42–51 in this textbook or in the *MyMathLab*® Study Plan.

R.2 Exercises

In Exercises 1–3, use the commutative property of addition to rewrite each expression as an equivalent expression.

1. $13 + m$ **2.** $ab + c$ **3.** $11w + 10t$

In Exercises 4–6, use the commutative property of multiplication to rewrite each expression as an equivalent expression.

4. $3 \cdot z$ **5.** mn **6.** $6(v + 4)$

In Exercises 7–9, use the associative property of addition to rewrite each expression as an equivalent expression.

7. $4 + (a + 11)$ **8.** $(d + a) + 30$ **9.** $(zw + a) + y$

In Exercises 10 and 11, use the associative property of multiplication to rewrite the expression as an equivalent expression.

10. $45 \cdot (y \cdot z)$ **11.** $8(2(y + z))$

In Exercises 12–15, use the distributive property to multiply the given expressions.

12. $12(11a + 4)$ **13.** $(x + y)20$ **14.** $15(a - 10 + 6b)$ **15.** $(h + 2v - 6)5$

In Exercises 16–18, use the distributive property to rewrite each sum or difference as a product.

16. $20c + 12c$ **17.** $9a - 11a$ **18.** $zx - zy$

In Exercises 19–21, write each expression using exponential notation. Then identify the base and the exponent.

19. $4 \cdot 4 \cdot 4$ **20.** $w \cdot w \cdot w \cdot w$ **21.** $(-6y) \cdot (-6y) \cdot (-6y) \cdot (-6y) \cdot (-6y)$

In Exercises 22–25, identify the base and exponent of each expression, and then evaluate.

22. 8^3 **23.** -5^4 **24.** $(-4)^3$ **25.** $(2z)^5$

In Exercises 26–33, simplify each expression using the order of operations.

26. $7 - (4 \cdot 6 - 26)$ **27.** $2 + 8^3 \div (-56) - 7$ **28.** $21^3 \div \dfrac{3^2}{5 - 2} - (-1)$ **29.** $-\dfrac{7}{6}\left(\dfrac{1}{2}\right) + \dfrac{5}{9} \div \dfrac{7}{6}$

30. $10^2 + 20^2 \div 5^2$ **31.** $3 \cdot (18 + 3)^2 - 4 \cdot (8 - 3)^2$ **32.** $\dfrac{7 \cdot 3 - 2^2}{16 - 2^3}$ **33.** $\dfrac{|-5^2 + (-3)^2| + 4 \cdot 5}{6 \div 2 - 3 \cdot 2^2}$

In Exercises 34–41, evaluate each algebraic expression for the indicated variable.

34. $4a \div 16a^2$ for $a = 4$ **35.** $30 \div 3x$ for $x = -2$ **36.** $-z^2 - 6z$ for $z = -3$

37. $\dfrac{6b - 9b^2}{b^2 - 6}$ for $b = 3$ **38.** $54 \div 3^2 \cdot n(n - 4)$ for $n = 7$ **39.** $|8x^2 - 5y|$ for $x = -1$ and $y = 3$

40. $b^2 - 4ac$ for $a = -4, b = 3$, and $c = -2$ **41.** $\dfrac{y_2 - y_1}{x_2 - x_1}$ for $x_1 = 4, x_2 = 6, y_1 = -2$, and $y_2 = 7$

In Exercises 42–51, simplify each algebraic expression.

42. $-7x - 10x$ **43.** $10z^2 - z^2$ **44.** $-7y^5 - 7y^5$ **45.** $7a - 20 - 17a + 50$

46. $3k + 3k^2 + 7k + 7k^2$ **47.** $5(3t - 5) - 6t$ **48.** $6 - 3[4 - (7r - 2)]$

49. $-5(3w - 9) + 2(2w + 7)$ **50.** $56p^2 - 36 - [4(p^2 - 9) + 1]$ **51.** $\dfrac{3}{4}x - \dfrac{4}{3} + \dfrac{7}{2}x + \dfrac{5}{6}$

R.3 The Laws of Exponents; Radicals

OBJECTIVES

1 Simplifying Exponential Expressions Involving Integer Exponents

2 Simplifying Radical Expressions

3 Simplifying Exponential Expressions Involving Rational Exponents

OBJECTIVE 1: SIMPLIFYING EXPONENTIAL EXPRESSIONS INVOLVING INTEGER EXPONENTS

In Section R.2, we introduced exponential notation. In this section, we learn how to simplify exponential expressions of the form $b^{m/n}$, where m/n is a **rational number**. We start by simplifying expressions involving *positive* integer exponents.

Properties of Positive Integer Exponents

Suppose m and n are positive integers and a and b are real numbers, then

1. $b^m b^n = b^{m+n}$

2. $\dfrac{b^m}{b^n} = b^{m-n}$, where $m > n$ and $b \neq 0$

$\dfrac{b^m}{b^n} = 1$, where $m = n$

$\dfrac{b^m}{b^n} = \dfrac{1}{b^{n-m}}$, where $m < n$ and $b \neq 0$

3. $(b^m)^n = b^{mn}$

4. $(ab)^n = a^n b^n$

5. $\left(\dfrac{a}{b}\right)^n = \dfrac{a^n}{b^n}$, where $b \neq 0$

Although these five properties are true for positive integer exponents, they are actually true for *all* integer exponents, including exponents that are negative integers or zero. By property for positive integer exponents m and n, we know that $b^m b^n = b^{m+n}$. If we want this property to hold true for $n = 0$, then $b^m b^0 = b^{m+0} = b^m$ if and only if $b^0 = 1$. This suggests the following rule.

Zero Exponent Rule

If b is a real number such that $b \neq 0$, then $b^0 = 1$.

Similarly, if we let $n = -m$, then $b^m \cdot b^{-m} = b^{m+(-m)} = b^0 = 1$ if and only if b^{-m} is the **reciprocal** of b^m or $b^{-m} = \frac{1}{b^m}$. This is known as the reciprocal rule of exponents.

> **Reciprocal Rule of Exponents**
>
> If b is a real number such that $b \neq 0$ and m is a positive integer, then $b^{-m} = \frac{1}{b^m}$.

We now state the following laws of exponents, which hold true for *all* integer exponents.

> **Laws of Exponents**
>
> Suppose m and n are integers and a and b are real numbers, then
>
> **1.** $b^m b^n = b^{m+n}$ *Product rule of exponents*
>
> **2.** $\dfrac{b^m}{b^n} = b^{m-n}$, where $b \neq 0$ *Quotient rule of exponents*
>
> **3.** $(b^m)^n = b^{mn}$ *Power rule of exponents*
>
> **4.** $(ab)^n = a^n b^n$ *Product to power rule of exponents*
>
> **5.** $\left(\dfrac{a}{b}\right)^n = \dfrac{a^n}{b^n}$, where $b \neq 0$ *Quotient to power rule of exponents*
>
> **6.** $b^0 = 1$ *Zero exponent rule*
>
> **7.** $b^{-m} = \dfrac{1}{b^m}$, where $b \neq 0$ *Reciprocal rule of exponents*

My interactive video summary

▶ Example 1 Simplify Exponential Expressions Using the Laws of Exponents

Simplify each exponential expression. Write your answers using positive exponents. Assume all variables represent positive real numbers.

    a. $-3^{-4} \cdot 3^4$ **b.** $5z^4 \cdot (-9z^{-5})$ **c.** $\left(\dfrac{a^{-3}b^4c^{-6}}{2a^5b^{-4}c}\right)^{-3}$

Solution

a. $-3^{-4} \cdot 3^4 = -3^{-4+4}$ **b.** $5z^4 \cdot (-9z^{-5}) = 5 \cdot (-9) \cdot z^4 \cdot z^{-5}$

$\phantom{-3^{-4} \cdot 3^4} = -3^0$ $= -45 \cdot z^{4-5}$

$\phantom{-3^{-4} \cdot 3^4} = -1$ $= -45 \cdot z^{-1}$

$\phantom{-3^{-4} \cdot 3^4 xxxxxxxxxx} = \dfrac{-45}{z}$

c. $\left(\dfrac{a^{-3}b^4c^{-6}}{2a^5b^{-4}c}\right)^{-3} = \left(\dfrac{b^4 b^4}{2a^5 a^3 c c^6}\right)^{-3}$

$\phantom{\left(\dfrac{a^{-3}b^4c^{-6}}{2a^5b^{-4}c}\right)^{-3}} = \left(\dfrac{b^8}{2a^8 c^7}\right)^{-3}$

$\phantom{\left(\dfrac{a^{-3}b^4c^{-6}}{2a^5b^{-4}c}\right)^{-3}} = \dfrac{b^{-24}}{2^{-3}a^{-24}c^{-21}}$

$\phantom{\left(\dfrac{a^{-3}b^4c^{-6}}{2a^5b^{-4}c}\right)^{-3}} = \dfrac{2^3 a^{24} c^{21}}{b^{24}}$

$\phantom{\left(\dfrac{a^{-3}b^4c^{-6}}{2a^5b^{-4}c}\right)^{-3}} = \dfrac{8a^{24}c^{21}}{b^{24}}$

You may want to watch the interactive video to see these three solutions worked out in detail.

Work exercises 1–30 in this textbook or in the *MyMathLab*® Study Plan.

OBJECTIVE 2:   SIMPLIFYING RADICAL EXPRESSIONS

Before we discuss exponential expressions with rational exponents, we must first define a radical.

A number a is said to be an nth root of b if $a^n = b$. For example, 4 is a square root (2nd root) of 16 because $(4)^2 = 16$. The number -4 is also a square root of 16 because $(-4)^2 = 16$. Similarly, 3 and -3 are 4th roots of 81 because $(3)^4 = 81$ and $(-3)^4 = 81$. It follows that every positive real number has two real nth roots when n is an even integer. This is not the case if n is an odd integer. Every real number has exactly one nth root when n is an odd integer. For example, the only real cube root (3rd root) of -8 is -2 because $(-2)^3 = -8$. We use the symbol $\sqrt[n]{b}$ to mean the *positive* nth root of b, also called the **principal nth root** of a number b. The symbol $\sqrt[n]{}$ is called a **radical sign**. The integer n is called the **index**. The complete expression $\sqrt[n]{b}$ is called a **radical**. We now define the principal nth root.

DEFINITION Principal nth Root

For any integer $n \geq 2$, the **principal nth root** of a number b is defined as

$$\sqrt[n]{b} = a \text{ if and only if } a^n = b$$

If n is even, then $b \geq 0$ and $a \geq 0$.
If n is odd and $b \geq 0$, then $a \geq 0$.
If n is odd and $b < 0$, then $a < 0$.

For a real principal nth root to exist when n is a positive integer, the **radicand** must be greater than or equal to zero. Otherwise, the nth root is not a real number.

Example 2 Simplify Radical Expressions

Simplify each radical expression.

  a. $\sqrt{144}$ b. $\sqrt[3]{\dfrac{1}{27}}$ c. $\sqrt[5]{-32a^5b^5}$ d. $\sqrt[6]{-64}$

Solution

  a. $\sqrt{144} = 12$ because $12^2 = 144$.

  b. $\sqrt[3]{\dfrac{1}{27}} = \dfrac{1}{3}$ because $\left(\dfrac{1}{3}\right)^3 = \dfrac{1}{27}$.

  c. $\sqrt[5]{-32a^5b^5} = -2ab$ because $(-2ab)^5 = -32a^5b^5$.

  d. The radical expression $\sqrt[6]{-64}$ has no real principal 6th root.

Work exercises 31–36 in this textbook or in the *MyMathLab*® Study Plan.

Note that the numeric part of each of the **radicands** from Examples 2a, 2b, and 2c are **perfect nth powers**. To simplify radicals in which the radicand is not a perfect nth power, we must first state some properties of radicals.

Properties of Radicals

If a and b are real numbers and $n \geq 2$ and $m \geq 2$ are integers such that $\sqrt[n]{a}$ and $\sqrt[m]{b}$ are **real numbers***, then

---

$$1. \ \sqrt[n]{ab} = \sqrt[n]{a}\sqrt[n]{b} \qquad 2. \ \sqrt[n]{\dfrac{a}{b}} = \dfrac{\sqrt[n]{a}}{\sqrt[n]{b}} \qquad 3. \ \sqrt[n]{a^n} = a \text{ if } n \text{ is odd}$$

$$4. \ \sqrt[n]{a^n} = |a| \text{ if } n \text{ is even}$$

To simplify radicals in which the **radicand** is not a perfect nth power, we first factor all perfect power factors of the radicand and then use the properties of radicals to simplify.

Example 3 Simplify Radical Expressions

Simplify each radical expression.

a. $\sqrt{108}$ **b.** $\sqrt[3]{-40x^5y^3}$ **c.** $\sqrt[4]{16x^4y^5}$

Solution

a. $\sqrt{108}$ *Write the original expression.*

$= \sqrt{36 \cdot 3}$ *Factor out 36, the largest perfect square factor of 108.*

$= \sqrt{36}\sqrt{3}$ *Rewrite using radical property (1).*

$= 6\sqrt{3}$ *The principal square root of 36 is 6.*

b. $\sqrt[3]{-40x^5y^3}$ *Write the original expression.*

$= \sqrt[3]{-8 \cdot 5 \cdot x^3 \cdot x^2 \cdot y^3}$ *Factor out all perfect cubes.*

$= \sqrt[3]{-8}\sqrt[3]{x^3}\sqrt[3]{y^3}\sqrt[3]{5x^2}$ *Rewrite using radical property (1).*

$= -2xy\sqrt[3]{5x^2}$ $\sqrt[3]{-8} = -2, \ \sqrt[3]{x^3} = x \text{ and } \sqrt[3]{y^3} = y.$

c. $\sqrt[4]{16x^4y^5}$ *Write the original expression.*

$= \sqrt[4]{16 \cdot x^4 \cdot y^4 \cdot y}$ *Factor out the perfect 4th powers.*

$= \sqrt[4]{16}\sqrt[4]{x^4}\sqrt[4]{y^4}\sqrt[4]{y}$ *Rewrite using radical property (1).*

$= 2|x||y|\sqrt[4]{y}$ $\sqrt[4]{16} = 2, \ \sqrt[4]{x^4} = |x| \text{ and } \sqrt[4]{y^4} = |y|.$

$= 2|xy|\sqrt[4]{y}$ *Use the absolute value property* $|a||b| = |ab|.$

We can add or subtract radicals only if the **index** and the **radicand** of each expression are the same.

Example 4 Combine Radical Expressions

Simplify the radical expression. Assume all variables represent positive real numbers.

$$5\sqrt[3]{16x} + 3\sqrt[3]{2x} - \sqrt[3]{2x^4}$$

Solution $5\sqrt[3]{16x} + 3\sqrt[3]{2x} - \sqrt[3]{2x^4}$

$= 5\sqrt[3]{8}\sqrt[3]{2x} + 3\sqrt[3]{2x} - \sqrt[3]{x^3}\sqrt[3]{2x}$ *Factor out all perfect cubes.*

$= 5(2)\sqrt[3]{2x} + 3\sqrt[3]{2x} - x\sqrt[3]{2x}$ $\sqrt[3]{8} = 2 \text{ and } \sqrt[3]{x^3} = x.$

$= 10\sqrt[3]{2x} + 3\sqrt[3]{2x} - x\sqrt[3]{2x}$ *Simplify.*

$= 13\sqrt[3]{2x} - x\sqrt[3]{2x}$ *Combine radicals.*

$= (13 - x)\sqrt[3]{2x}$ *Use the distributive property.*

Determine whether you can simplify the radical expressions from Example 5. Watch the interactive video to see each solution worked out in detail.

My interactive video summary

▶ Example 5 Simplify Radical Expressions

Simplify each radical expression. Assume all variables represent positive real numbers.

a. $\sqrt{192}$

b. $(3\sqrt{2} + \sqrt{5})(2\sqrt{2} - 4\sqrt{5})$

c. $\sqrt[5]{2x^4y^4}\ \sqrt[5]{48xy^2}$

d. $8\sqrt{2x^2} - 3\sqrt{28x} - 2\sqrt{8x^2}$

Solution Watch the interactive video to see each solution worked out in detail.

Work exercises 37–50 in this textbook or in the MyMathLab® Study Plan.

When a radical appears in the denominator of an algebraic expression, we typically multiply the numerator and denominator by a suitable form of 1 in order to remove the radical from the denominator. This process is called **rationalizing the denominator**. Example 6 illustrates this process.

▶ Example 6 Rationalize the Denominator of a Rational Expression

Rationalize the denominator of each expression.

a. $\sqrt{\dfrac{3}{2}}$

b. $\dfrac{\sqrt[3]{375}}{\sqrt[3]{4}}$

c. $\dfrac{7}{\sqrt{5} + \sqrt{2}}$

Solution

a. $\sqrt{\dfrac{3}{2}}$ *Write the original expression.*

$= \dfrac{\sqrt{3}}{\sqrt{2}}$ *Rewrite using radical property (2).*

$= \dfrac{\sqrt{3}}{\sqrt{2}} \cdot \dfrac{\sqrt{2}}{\sqrt{2}}$ *Multiply the expression by* $\dfrac{\sqrt{2}}{\sqrt{2}} = 1.$

$= \dfrac{\sqrt{6}}{2}$ *Simplify. The denominator is now rationalized.*

b. $\dfrac{\sqrt[3]{375}}{\sqrt[3]{4}}$ *Write the original expression.*

$= \dfrac{\sqrt[3]{125}\ \sqrt[3]{3}}{\sqrt[3]{4}}$ *Rewrite the numerator using radical property (1).*

$= \dfrac{5\sqrt[3]{3}}{\sqrt[3]{4}}$ *Simplify* $\sqrt[3]{125} = 5.$

$= \dfrac{5\sqrt[3]{3}}{\sqrt[3]{4}} \cdot \dfrac{\sqrt[3]{2}}{\sqrt[3]{2}}$ *Multiply the expression by* $\dfrac{\sqrt[3]{2}}{\sqrt[3]{2}} = 1.$

$= \dfrac{5\sqrt[3]{6}}{\sqrt[3]{8}}$ *Multiply using radical property (1).*

$= \dfrac{5\sqrt[3]{6}}{2}$ *Simplify* $\sqrt[3]{8} = 2.$ *The denominator is now rationalized.*

c. $\dfrac{7}{\sqrt{5} + \sqrt{2}}$ *Write the original expression.*

$$= \dfrac{7}{\sqrt{5} + \sqrt{2}} \cdot \dfrac{\sqrt{5} - \sqrt{2}}{\sqrt{5} - \sqrt{2}}$$ *Multiply the expression by* $\dfrac{\sqrt{5} - \sqrt{2}}{\sqrt{5} - \sqrt{2}} = 1.$

(The expression $\sqrt{5} - \sqrt{2}$ *is called the radical conjugate* of the expression* $\sqrt{5} + \sqrt{2}$.*)*

$$= \dfrac{7\sqrt{5} - 7\sqrt{2}}{3}$$ *Simplify.*

Work exercises 51–57 in this textbook or in the *MyMathLab*® Study Plan.

OBJECTIVE 3: **SIMPLIFYING EXPONENTIAL EXPRESSIONS INVOLVING RATIONAL EXPONENTS**

We can use our knowledge of radicals to simplify expressions involving rational exponents. We start by defining the expression $b^{1/n}$.

DEFINITION $b^{1/n}$

For any integer $n \geq 2$ and any real number b, we define the expression $b^{1/n}$ as

$$b^{1/n} = \sqrt[n]{b}$$

provided that the expression $\sqrt[n]{b}$ is a **real number***.

Example 7 Evaluate Expressions Involving Rational Exponents

Simplify each expression.

     a. $4^{1/2}$ **b.** $(-27)^{1/3}$ **c.** $\left(\dfrac{1}{64}\right)^{1/6}$

Solution

     a. $4^{1/2} = \sqrt{4} = 2$ **b.** $(-27)^{1/3} = \sqrt[3]{-27} = -3$ **c.** $\left(\dfrac{1}{64}\right)^{1/6} = \sqrt[6]{\dfrac{1}{64}} = \dfrac{1}{2}$

Each exponent from Example 7 was a **rational number** of the form $\frac{1}{n}$. The following definition provides a way to deal with expressions involving rational exponents of the form $\frac{m}{n}$.

DEFINITION $b^{m/n}$

If $\frac{m}{n}$ is a **rational number** with $n \geq 2$ and b is a real number, then we define the expression $b^{m/n}$ as

$$b^{m/n} = \sqrt[n]{b^m} = (\sqrt[n]{b})^m$$

provided that the expression $\sqrt[n]{b}$ is a **real number***.

The laws of exponents that were stated previously for integer exponents hold true for rational exponents as well and are worth stating again.

Laws of Exponents

Suppose s and t are rational numbers and a and b are real numbers, then

1. $b^s b^t = b^{s+t}$ — Product rule of exponents.

2. $\dfrac{b^s}{b^t} = b^{s-t}$, where $b \neq 0$ — Quotient rule of exponents.

3. $(b^s)^t = b^{st}$ — Power rule of exponents.

4. $(ab)^s = a^s b^s$ — Product to power rule of exponents.

5. $\left(\dfrac{a}{b}\right)^s = \dfrac{a^s}{b^s}$, where $b \neq 0$ — Quotient to power rule of exponents.

6. $b^0 = 1$ — Zero exponent rule.

7. $b^{-s} = \dfrac{1}{b^s}$, where $b \neq 0$ — Reciprocal rule of exponents.

When we simplify expressions involving rational exponents, we always simplify until all exponents are positive and until each base appears only once in the expression.

My interactive video summary

🎦 **Example 8   Simplify Expressions Involving Rational Exponents**

Simplify each expression using positive exponents. Assume all variables represent positive real numbers.

a. $(9)^{3/2}$ **b.** $(-32)^{-3/5}$ **c.** $\dfrac{\left(125x^4 y^{-1/4}\right)^{2/3}}{(x^2 y)^{1/3}}$

Solution

a. $(9)^{3/2} = \sqrt{9^3} = (\sqrt{9})^3 = 3^3 = 27$

b. $(-32)^{-3/5} = \sqrt[5]{(-32)^3} = (\sqrt[5]{-32})^3 = (-2)^3 - 8$

c. $\dfrac{\left(125x^4 y^{-1/4}\right)^{2/3}}{(x^2 y)^{1/3}} = \dfrac{125^{2/3} x^{8/3} y^{-1/6}}{x^{2/3} y^{1/3}} = \dfrac{(\sqrt[3]{125})^2 x^{8/3 - 2/3}}{y^{1/3 + 1/6}} = \dfrac{25x^2}{y^{1/2}}$

Watch the interactive video to see each solution worked out in detail.

Work exercises 58–70 in this textbook or in the *MyMathLab*® Study Plan.

R.3 Exercises

In Exercises 1–30, use the laws of exponents to simplify each expression using positive exponents only. Assume all variables represent nonzero real numbers.

1. $3^2 \cdot 3^3$ **2.** $k^4 \cdot k^5$ **3.** $x^5 \cdot x^9 \cdot x^{12}$ **4.** $(a^5 b^4)(a^7 b^2)$

5. $\dfrac{t^9}{t^2}$ **6.** $(r^5)^{-7}$ **7.** $(w^{-8})^{-6}$ **8.** $\dfrac{n}{n^{-7}}$

9. $\dfrac{m^{-6}}{m^{-9}}$ **10.** $(-5y)^0$ **11.** -6^0 **12.** 2^{-2}

13. -4^{-2} **14.** n^{-3} **15.** $5x^{-6}$ **16.** $-7y^{-2}$

17. $\dfrac{1}{t^{-2}}$

18. $\dfrac{5}{y^{-9}}$

19. $\dfrac{a^{-11}}{b^{-4}}$

20. $\dfrac{-5p^{-7}q^4}{w^5}$

21. $b^{11}b^{-4}$

22. $a^{-11}a^{-3}$

23. $3^{-4} \cdot 3$

24. $\dfrac{z^{14}}{z^{-4}}$

25. $(-5x^2y^5)^2$

26. $\dfrac{24x^3y^5}{32x^8y^{-7}}$

27. $\dfrac{(a^{-3}b)^{-2}}{(a^3b^{-1})^2}$

28. $(16a^{-3}bc^{-6})(2ab)^{-5}$

29. $\left(\dfrac{q^3p^4w^5}{q^{-3}p^{-4}w^{-5}}\right)^{-2}$

30. $\dfrac{(3^{-1}x^{-1}y^{-1})^{-2}(3x^{-4}y^3)(9x^{-2}y^3)^0}{(3x^{-3}y^{-5})^2}$

In Exercises 31–50, simplify each radical expression. Assume all variables represent positive real numbers.

31. $\sqrt{169}$

32. $\sqrt{\dfrac{1}{25}}$

33. $\sqrt[3]{216}$

34. $\sqrt[3]{-343}$

35. $\sqrt{49w^2}$

36. $\sqrt[4]{(x+1)^4}$

37. $\sqrt{160}$

38. $\sqrt{363x^2}$

39. $\sqrt[3]{250}$

40. $\sqrt[3]{-27x^4}$

41. $\sqrt[4]{32z^5}$

42. $\sqrt{180x^{12}y^{27}}$

43. $\sqrt[5]{\dfrac{x^{24}y^{13}}{x^9y^8}}$

44. $\sqrt{6}\left(\sqrt{5} - \sqrt{11}\right)$

45. $\left(3\sqrt{2} + 3\sqrt{7}\right)\left(\sqrt{2} - 2\sqrt{7}\right)$

46. $\left(4 + \sqrt{3}\right)^2$

47. $\sqrt[3]{x^2}\left(8\sqrt[3]{4x} - 5\sqrt[3]{x^5}\right)$

48. $11\sqrt{5} + 3\sqrt{5}$

49. $\sqrt{48x} - \sqrt{12x}$

50. $\sqrt[3]{875xy^3} - 3\sqrt[3]{56xy^3} + 2y\sqrt[3]{448x}$

In Exercises 51–57, rationalize the denominator. Assume all variables represent positive real numbers.

51. $\dfrac{\sqrt{5}}{\sqrt{6}}$

52. $\sqrt{\dfrac{1}{7}}$

53. $\sqrt[3]{\dfrac{7}{2}}$

54. $\dfrac{\sqrt[3]{2y^5}}{\sqrt[3]{5x}}$

55. $\dfrac{9}{5 - \sqrt{2}}$

56. $\dfrac{\sqrt{5} - \sqrt{3}}{\sqrt{5} + \sqrt{3}}$

57. $\dfrac{\sqrt{a}}{\sqrt{a} + \sqrt{b}}$

In Exercises 58–70, simplify each exponential expression. Write any negative exponents as positive exponents. Assume all variables represent positive real numbers.

58. $25^{1/2}$

59. $\left(\dfrac{1}{16}\right)^{1/4}$

60. $-81^{1/4}$

61. $16^{3/4}$

62. $8^{-4/3}$

63. $x^{-1/3}$

64. $x^{-3/4} \cdot x^{7/4}$

65. $x^{1/3} \cdot x^{1/4} \cdot x^{-1/2}$

66. $(x^5y^{10})^{1/5}$

67. $\dfrac{(x^2y^4)^{4/7}(xy^2)^{1/7}}{x^{4/7}y^{4/7}}$

68. $(9x^2y^{-5/2})^{1/2}$

69. $(32x^{-5}y^{10})^{-1/5}(xy^{1/5})$

70. $\left(\dfrac{x^{-3/4}y^{1/4}}{x^{-1/4}}\right)^{-8}$

R.4 Polynomials

OBJECTIVES

1 Understanding the Definition of a Polynomial

2 Adding and Subtracting Polynomials

3 Multiplying Polynomials

4 Dividing Polynomials Using Long Division

OBJECTIVE 1:   UNDERSTANDING THE DEFINITION OF A POLYNOMIAL

In Section R.2, we introduced algebraic expressions. An algebraic expression with one term in which the exponents of all variable factors are nonnegative integers is called a **monomial**. Examples include 7, x^5, $-\sqrt{2}x^2y^7$, and πxyz. The **coefficient** of a monomial is the numeric factor. The **degree** of a monomial is the sum of the variable exponents.

Monomial	Degree	Coefficient
7	0	7
x^5	5	1
$-\sqrt{2}x^2y^7$	9	$-\sqrt{2}$
πxyz	3	π

Work exercises 1–5 in this textbook or in the *MyMathLab*®️ Study Plan.

A **polynomial** is the sum or difference of one or more monomials. A polynomial with two terms is called a **binomial**. A polynomial with three terms is called a **trinomial**. The **degree** of a polynomial is the degree of the highest monomial term.

Example 1 Identify Polynomials

Determine whether each algebraic expression is a polynomial. If the expression is a polynomial, state the degree.

a. $8x^3y^2 - 2$ **b.** $5w^5 - 3y^2 - \dfrac{6}{w}$ **c.** $7p^8m^4 - 5p^5m^3 + 2pm$

Solution

a. The expression $8x^3y^2 - 2$ is a polynomial. Because this polynomial has two terms, we say that $8x^3y^2 - 2$ is a binomial. The degree of this binomial is 5.

b. The expression $5w^5 - 3y^2 - \frac{6}{w}$ is not a polynomial because the term $\frac{6}{w}$ is equivalent to $6w^{-1}$. For an expression to be a polynomial, all exponents must be nonnegative integers.

c. The expression $7p^8m^4 - 5p^5m^3 + 2pm$ is a polynomial. Because this polynomial has three terms, we say that $7p^8m^4 - 5p^5m^3 + 2pm$ is a trinomial. The degree of this trinomial is 12.

Work exercises 6–10 in this textbook or in the *MyMathLab*®️ Study Plan.

For the remainder of this section, we learn the arithmetic of polynomials. That is, we learn how to add, subtract, multiply, and divide polynomials. Our discussion focuses mainly on polynomials in one variable.

DEFINITION Polynomial in One Variable

The algebraic expression $a_nx^n + a_{n-1}x^{n-1} + a_{n-2}x^{n-2} + \cdots + a_1x + a_0$ is a polynomial in one variable of degree n, where n is a nonnegative integer. The numbers $a_0, a_1, a_2, \ldots, a_n$ are called the **coefficients** of the polynomial.

We typically write polynomials in descending order with the highest power listed first. A polynomial written in descending order is said to be in **standard form**.

OBJECTIVE 2: ADDING AND SUBTRACTING POLYNOMIALS

To add or subtract polynomials, we simply add or subtract the coefficients of the corresponding terms with like exponents. We always write the final sum or difference in standard form.

My video summary

▦ Example 2 Add and Subtract Polynomials

Add or subtract as indicated and express your answer in standard form.

a. $(-x^5 + 3x^4 - 9x^2 + 7x + 9) + (x^4 - 5x^5 + x^3 - 3x^2 - 8)$

b. $(-x^5 + 3x^4 - 9x^2 + 7x + 9) - (x^4 - 5x^5 + x^3 - 3x^2 - 8)$

Solution

a. To add polynomials, we remove the parentheses then combine the like terms.

$$
\begin{aligned}
&(-x^5 + 3x^4 - 9x^2 + 7x + 9) && \text{Write the original} \\
&+ (x^4 - 5x^5 + x^3 - 3x^2 - 8) && \text{expression.} \\
&= -x^5 + 3x^4 - 9x^2 + 7x + 9 && \text{Remove the} \\
&+ x^4 - 5x^5 + x^3 - 3x^2 - 8 && \text{parentheses.} \\
&= (-x^5 - 5x^5) + (3x^4 + x^4) && \text{Collect like terms.} \\
&+ x^3 + (-9x^2 - 3x^2) + 7x + (9 - 8) \\
&= -6x^5 + 4x^4 + x^3 - 12x^2 + 7x + 1 && \text{Simplify and write in} \\
& && \text{standard form.}
\end{aligned}
$$

b. To subtract polynomials, we remove the parentheses, making sure to change the sign of each term of the second polynomial and then combine the like terms.

$$
\begin{aligned}
&(-x^5 + 3x^4 - 9x^2 + 7x + 9) && \text{Write the original} \\
&- (x^4 - 5x^5 + x^3 - 3x^2 - 8) && \text{expression.} \\
&= -x^5 + 3x^4 - 9x^2 + 7x + 9 && \text{Remove the parentheses} \\
&- x^4 + 5x^5 - x^3 + 3x^2 + 8 && \text{and change the sign of} \\
& && \text{each term of the second} \\
& && \text{polynomial.} \\
&= (-x^5 + 5x^5) + (3x^4 - x^4) - x^3 && \text{Collect like terms.} \\
&+ (-9x^2 + 3x^2) + 7x + (9 + 8) \\
&= 4x^5 + 2x^4 - x^3 - 6x^2 + 7x + 17 && \text{Simplify and write in} \\
& && \text{standard form.}
\end{aligned}
$$

Watch the video to see each solution worked out in detail.

Work exercises 11–15 in this textbook or in the *MyMathLab*® Study Plan.

OBJECTIVE 3: MULTIPLYING POLYNOMIALS

To find the product of two monomials in one variable, we multiply the coefficients together and use the **product rule of exponents** to add any exponents. For example, $(8x^4)(-3x^7) = (8)(-3) \cdot x^4 \cdot x^7 = -24x^{11}$. To multiply a monomial by a polynomial with more than one term, we need to use the **distributive property** multiple times. Thus, if m is a monomial and $p_1, p_2, p_3, \ldots p_n$ are the terms of a polynomial, then

$$m(p_1 + p_2 + p_3 + \cdots + p_n) = mp_1 + mp_2 + mp_3 + \cdots + mp_n$$

Example 3 Multiply a Polynomial by a Monomial

Multiply: $(-2x^5)(3x^3 - 5x^2 + 1)$

Solution $(-2x^5)(3x^3 - 5x^2 + 1) = -2x^5(3x^3) - 2x^5(-5x^2) - 2x^5(1)$

$$= -6x^8 + 10x^7 - 2x^5$$

When we find the product of any two polynomials, we use the **distributive property** to multiply each term of the first polynomial by each term of the second polynomial and then simplify.

My video summary

Example 4 Multiply Polynomials

Find the product and express your answer in standard form.

$$(2x^2 - 3)(3x^3 - x^2 + 1)$$

Solution Watch the video to verify that
$(2x^2 - 3)(3x^3 - x^2 + 1) = 6x^5 - 2x^4 - 9x^3 + 5x^2 - 3.$

Work exercises 16–18 in this textbook or in the *MyMathLab*® Study Plan.

The FOIL Method

When we find the product of two binomials, we can use a technique known as the FOIL method. FOIL is an acronym that stands for **F**irst, **O**utside, **I**nside, and **L**ast. We illustrate the FOIL method by finding the product of the binomials $3x + 5$ and $2x + 7$.

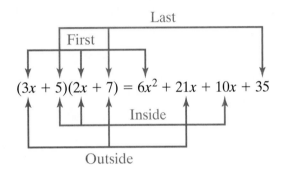

$$(3x + 5)(2x + 7) = 6x^2 + 21x + 10x + 35$$

The product of the **F**irst terms: $(3x)(2x) = 6x^2$
The product of the **O**utside terms: $(3x)(7) = 21x$
The product of the **I**nside terms: $(5)(2x) = 10x$
The product of the **L**ast terms: $(5)(7) = 35$

We now simplify by combining like terms:

$$6x^2 + 21x + 10x + 35 = 6x^2 + 31x + 35.$$

Therefore, $(3x + 5)(2x + 7) = 6x^2 + 31x + 35.$

Work exercises 19–21 in this textbook or in the *MyMathLab*® Study Plan.

Example 5 Multiply Binomials Using the FOIL Method

Find the product: $(x + 4)(x - 4)$

Solution $(x + 4)(x - 4)$	*Write the original expression.*
$= x^2 - 4x + 4x - 16$	*Use the FOIL method.*
$= x^2 - 16$	*Combine like terms.*

R.4 Polynomials   R-23

Notice that the product of the two binomials in Example 4 is also a binomial. (There is no x-term.) The product in Example 5 is a special product known as the **difference of two squares** because the resulting binomial is the difference of two perfect square terms. In general, $(a + b)(a - b) = a^2 - b^2$. This formula is worth remembering. There are other special products that are worth remembering as well. These products include the **square of a binomial** and the **cube of a binomial**. We now list some special products.

Special Products

Let a and b represent any real number or algebraic expression, then

$$(a + b)(a - b) = a^2 - b^2 \qquad \text{Difference of two squares.}$$
$$(a + b)^2 = a^2 + 2ab + b^2 \qquad \text{Squaring a binomial.}$$
$$(a - b)^2 = a^2 - 2ab + b^2$$
$$(a + b)^3 = a^3 + 3a^2b + 3ab^2 + b^3 \qquad \text{Cubing a binomial.}$$
$$(a - b)^3 = a^3 - 3a^2b + 3ab^2 - b^3$$

Determine whether you can use these special product formulas to find the products in Example 5.

Example 6 Use Special Product Formulas

Find each product using a special product formula.

   a. $(2x + 3)(2x - 3)$ **b.** $(3x + 2y)^2$ **c.** $(5 - 3z)^3$

Solution

   a. We can use the difference of two squares formula
   $(a + b)(a - b) = a^2 - b^2$ with $a = 2x$ and $b = 3$ to get
   $(2x + 3)(2x - 3) = (2x)^2 - (3)^2 = 4x^2 - 9$.

   b. We can use the squaring a binomial formula
   $(a + b)^2 = a^2 + 2ab + b^2$ with $a = 3x$ and $b = 2y$ to get
   $(3x + 2y)^2 = (3x)^2 + 2(3x)(2y) + (2y)^2 = 9x^2 + 12xy + 4y^2$.

   c. Using the cubing a binomial formula
   $(a - b)^3 = a^3 - 3a^2b + 3ab^2 - b^3$ with $a = 5$ and $b = 3z$ we get
   $(5 - 3z)^3 = (5)^3 - 3(5)^2(3z) + 3(5)(3z)^2 - (3z)^3 = 125 - 225z + 135z^2 - 27z^3$.

Work exercises 22–28 in this textbook or in the *MyMathLab*® Study Plan.

OBJECTIVE 4:   DIVIDING POLYNOMIALS USING LONG DIVISION

We can long divide two polynomials in much the same way as the long division of two integers.

    For example, suppose we want to divide the polynomial $2x^2 - 7x + 4$ by the polynomial $x + 8$. The polynomial $2x^2 - 7x + 4$ is the **dividend** and $x + 8$ is the **divisor**. We set up this problem using the familiar symbol used for long division with the dividend on the inside of the symbol and the divisor on the outside.

$$x + 8 \overline{)2x^2 - 7x + 4}$$

We now illustrate the long division process.

Step 1: Divide the first term of the dividend, $2x^2$, by the first term of the divisor, x. Write the result, $2x$, above the first term of the dividend.

$$\begin{array}{r} 2x \\ x + 8 \overline{)2x^2 - 7x + 4} \end{array} \qquad \frac{2x^2}{x} = 2x, \text{ or equivalently, } 2x \cdot x = 2x^2.$$

Step 2: Multiply $2x$ by $x + 8$ and line up like terms vertically.

$$\begin{array}{r} 2x \\ x + 8 \overline{)2x^2 - 7x + 4} \\ 2x^2 + 16x \end{array} \qquad 2x(x + 8) = 2x^2 + 16x.$$

Step 3: Subtract $2x^2 + 16x$ from $2x^2 - 7x$. We perform the subtraction by **changing the sign** of each term of the polynomial being subtracted and then combining like terms.

$$\begin{array}{r} 2x \\ x + 8 \overline{)2x^2 - 7x + 4} \\ -2x^2 - 16x \\ \hline -23x \end{array} \qquad \text{Change the signs: } -(2x^2 + 16x) = -2x^2 - 16x.$$

Step 4: Bring down the constant 4.

$$\begin{array}{r} 2x \\ x + 8 \overline{)2x^2 - 7x + 4} \\ -2x^2 - 16x \\ \hline -23x + 4 \end{array}$$

Step 5: Divide $-23x$ by the first term of the divisor, x. Write the result, -23, above the second term of the dividend.

$$\begin{array}{r} 2x - 23 \\ x + 8 \overline{)2x^2 - 7x + 4} \\ -2x^2 - 16x \\ \hline -23x + 4 \end{array} \qquad \frac{-23x}{x} = -23, \text{ or equivalently, } -23 \cdot x = -23x.$$

Step 6: Multiply -23 by $x + 8$ and line up like terms vertically.

$$\begin{array}{r} 2x - 23 \\ x + 8 \overline{)2x^2 - 7x + 4} \\ -2x^2 - 16x \\ \hline -23x + 4 \\ -23x - 184 \end{array} \qquad -23(x + 8) = -23x - 184.$$

Step 7: Subtract $-23x - 184$ from $-23x + 4$.

$$\begin{array}{r} 2x - 23 \\ x + 8 \overline{)2x^2 - 7x + 4} \\ -2x^2 - 16x \\ \hline -23x + 4 \\ +23x + 184 \\ \hline 188 \end{array} \qquad \text{Change the signs. } -(23x - 184) = 23x + 184$$

We know that we are done because 188 is of lower degree (degree zero) than the degree of the divisor (degree one). The polynomial $2x - 23$ is called the **quotient**, and the constant 188 is the **remainder**. We can always check our long division process by verifying that the dividend is equal to the product of the quotient and divisor plus the remainder.

$$\text{dividend} = (\text{divisor})(\text{quotient}) + \text{remainder}$$

$$2x^2 - 7x + 4 \overset{?}{=} (x + 8)(2x - 23) + 188$$

$$2x^2 - 7x + 4 \overset{?}{=} 2x^2 - 7x - 184 + 188$$

$$2x^2 - 7x + 4 = 2x^2 - 7x + 4 \checkmark$$

My video summary

▦ Example 7 Divide Polynomials

Find the quotient and remainder when $3x^4 + x^3 + 7x + 4$ is divided by $x^2 - 1$.

Solution Using the long division symbol, place the dividend on the inside of the symbol and the divisor on the outside of the symbol. We write both polynomials in **standard form**, inserting **coefficients** of 0 for any missing terms.

$$x^2 + 0x - 1 \overline{\smash{)}3x^4 + x^3 + 0x^2 + 7x + 4}$$

The long division process is shown here. Watch the video to see every step of this process.

$$
\begin{array}{r}
3x^2 + x + 3 \\
x^2 + 0x - 1 \overline{\smash{)}3x^4 + x^3 + 0x^2 + 7x + 4} \\
\underline{-(3x^4 + 0x^3 - 3x^2)} \\
x^3 + 3x^2 + 7x \\
\underline{-(x^3 + 0x^2 - x)} \\
3x^2 + 8x + 4 \\
\underline{-(3x^2 + 0x - 3)} \\
8x + 7
\end{array}
$$

The quotient is $3x^2 + x + 3$, and the remainder is $8x + 7$.

Work exercises 29–32 in this textbook or in the *MyMathLab*® Study Plan.

R.4 Exercises

In Exercises 1–5, determine whether the algebraic expression is a monomial. If the expression is a monomial, state the degree and coefficient.

1. $5x^4$ **2.** $\dfrac{13}{x^2}$ **3.** $-x^4y^8$ **4.** $2x^2 - 3y^5$ **5.** 0

In Exercises 6–10, determine whether the algebraic expression is a polynomial. If the expression is a polynomial, state the degree.

6. $5x^3 - \dfrac{13}{x^2}$ **7.** π **8.** $3wx^2 - \sqrt{5}$ **9.** $\dfrac{2x^2 - 3y^5}{x^5 - 1}$

10. $2p^3m^4 - 5p^2m^3 + 2p^4m - 4pm^3 + 2p^4m^3$

In Exercises 11–15, perform the indicated operations. Write each answer in standard form.

11. $(4x + 5) + (-5x + 9)$ **12.** $(x^2 + 11x + 7) + (3x - 13)$

13. $(9x^3 - 12x^2 + 11x + 5) - (3x^2 - 5x + 11)$ **14.** $(4x^5 + 8x^3 + 9x) + (7x^4 - 2x^3 + 8x^2)$

15. $(4x^2 - 4x + 4) + (3x^2 - 5x + 3) - (7x^2 + 8)$

In Exercises 16–18, find the indicated product. Write each answer in standard form.

16. $x(x^3 - x + 7)$ **17.** $-5x^4(7x^6 - 2)$ **18.** $(x + 3)(x^2 + 7x - 6)$

In Exercises 19–21, find the indicated product using the FOIL method. Write each answer in standard form.

19. $(x + 9)(x - 4)$ **20.** $(3x + 2)(x + 5)$ **21.** $(-x - 8)(-3x - 7)$

In Exercises 22–28, find each product using a special product formula. Write each answer in standard form.

22. $(6x + 2)(6x - 2)$ **23.** $(x + 5)^2$ **24.** $(4x - 9)^2$
25. $(8x + y)(8x - y)$ **26.** $(x - 6)^3$ **27.** $(3x + 2)^3$
28. $(x - 4)(x + 4)(x^2 - 16)$

In Exercises 29–32, find the quotient and remainder.

29. $7x^3 - 5x^2 + x + 4$ is divided by $x + 3$ **30.** $9x^4 - 2x^2 + 2x + 9$ is divided by $x^2 + 5$
31. $6x^5 - 5x^2 + 7x + 6$ is divided by $2x^3 - 1$ **32.** $7 - x^2 + x^4$ is divided by $x^2 + x + 9$

R.5 Factoring Polynomials

OBJECTIVES

1 Factoring Out a Greatest Common Factor

2 Factoring by Grouping

3 Factoring Trinomials with a Leading Coefficient Equal to One

4 Factoring Trinomials with a Leading Coefficient Not Equal to One

5 Factoring Using Special Factoring Formulas

In this section, we learn techniques used to factor **polynomial expressions**. Being able to factor polynomials (and other algebraic expressions) is an essential algebraic skill and is used throughout this text to solve equations. Factoring a polynomial is basically the reverse of multiplying polynomials. The first factoring technique that we discuss is to factor out a greatest common factor.

OBJECTIVE 1:   FACTORING OUT A GREATEST COMMON FACTOR

My video summary

Factoring out a greatest common factor is essentially the **distributive property**. Note that the expression $AB + AC$ has two terms, AB and AC. The term AB has two factors, A and B. The term AC also has two factors, A and C. Notice that the terms AB and AC share a common factor of A. This common factor is called the **greatest common factor (GCF)** because it is the largest factor common to both terms. We can thus rewrite the expression $AB + AC$ by "pulling out" (factoring) this greatest common factor of A to get $AB + AC = A(B + C)$. When determining the GCF of a polynomial written as the sum and/or difference of monomials, we must determine the GCF of the numeric factors and then determine the GCF of the variable factors. For example, the polynomial $4x^2 - 6x^3 + 2x$ is composed of three monomials. The number 2 is the largest integer numeric factor common to each monomial term. The largest variable factor common to each monomial is x. Therefore, the GCF is $2x$. Using the distributive property and the GCF of $2x$, we get

$$4x^2 - 6x^3 + 2x$$
$$= 2x \cdot 2x - 2x \cdot 3x^2 + 2x \cdot 1$$
$$= 2x(2x - 3x^2 + 1)$$

Note: We can always verify whether we have factored correctly by multiplying the factors together and checking to see if the product is equal to the original polynomial.

Example 1 Factor Out a Greatest Common Factor

Factor each polynomial by factoring out the GCF.

a. $5w^2y^3 - 15w^4y + 20w^3y^2$

b. $x(x^2 + 6) - 5(x^2 + 6)$

Solution

a. The numeric GCF is 5, and the variable GCF is w^2y. Thus, the GCF is $5w^2y$.

$\quad 5w^2y^3 - 15w^4y + 20w^3y^2$ *Write the original polynomial.*

$\quad = 5w^2y \cdot y^2 - 5w^2y \cdot 3w^2 + 5w^2y \cdot 4wy$ *The GCF is $5w^2y$.*

$\quad = 5w^2y(y^2 - 3w^2 + 4wy)$ *Factor out the GCF.*

b. The GCF is the binomial factor $(x^2 + 6)$.

$$x(x^2 + 6) - 5(x^2 + 6) = (x^2 + 6)(x - 5)$$

Work exercises 1–8 in this textbook or in the *MyMathLab*® Study Plan.

OBJECTIVE 2: **FACTORING BY GROUPING**

My video summary

We often encounter polynomials that, at first, appear to be difficult or impossible to factor. For example, consider the polynomial $x(x^2 + 6) - 5(x^2 + 6)$ seen in Example 1. Using the distributive property, this polynomial can be rewritten as $x^3 + 6x - 5x^2 - 30$. At first glance, it appears that we cannot factor this polynomial because the GCF is 1. However, if we group the terms in pairs and factor the GCF from each pair of terms, we see that we can indeed factor the polynomial.

$\quad (x^3 + 6x) + (-5x^2 - 30)$ *Group the terms in pairs.*

$\quad = x(x^2 + 6) - 5(x^2 + 6)$ *Factor the GCF out of each pair.*

$\quad = (x^2 + 6)(x - 5)$ *Factor out the GCF of $(x^2 + 6)$.*

This technique is called **factoring by grouping** and often works when a polynomial has an even number of terms.

Example 2 Factor Polynomials by Grouping

Factor each polynomial by grouping.

a. $6x^2 - 2x + 9x - 3$

b. $2x^2 + 6xw - xy - 3wy$

Solution

a. $6x^2 - 2x + 9x - 3$

$= (6x^2 - 2x) + (9x - 3)$ *Group the terms in pairs.*

$= 2x(3x - 1) + 3(3x - 1)$ *Factor the GCF out of each pair.*

$= (3x - 1)(2x + 3)$ *Factor out the GCF of $(3x - 1)$.*

b. $2x^2 + 6xw - xy - 3wy$

$= (2x^2 + 6xw) + (-xy - 3wy)$ *Group the terms in pairs.*

$= 2x(x + 3w) - y(x + 3w)$ *Factor the GCF out of each pair.*

$= (x + 3w)(2x - y)$ *Factor out the GCF of $(x + 3w)$.*

Work exercises 9–16 in this textbook or in the MyMathLab® Study Plan.

OBJECTIVE 3:   FACTORING TRINOMIALS WITH A LEADING COEFFICIENT EQUAL TO ONE

My video summary

In Section R.4, the FOIL method for multiplying two binomials was introduced. For example, we can use the FOIL method to multiply the two binomials $(x + a)$ and $(x + b)$.

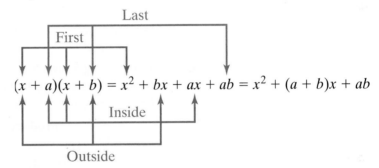

$$(x + a)(x + b) = x^2 + bx + ax + ab = x^2 + (a + b)x + ab$$

The binomial product $(x + a)(x + b)$ is said to be a factored form of the trinomial $x^2 + (a + b)x + ab$. Our goal is to start with a trinomial and try to find two binomials whose product is equal to the given trinomial. If we carefully look at the last two terms of the trinomial $x^2 + (a + b)x + ab$, we see that the middle term, $(a + b)x$, is the *sum* of the products of the inside and outside terms of the two binomials. The third term, ab, is the *product* of the last two terms of each binomial. Thus, when factoring trinomials of the form $x^2 + bx + c$, we must find two numbers whose product is c and whose sum is b.

Example 3 Factor Trinomials with a Leading Coefficient of One

Factor each trinomial.

a. $x^2 + 8x + 15$ **b.** $x^2 - 2x - 24$

c. $x^2 - 12x + 32$ **d.** $x^2 - 2x - 6$

Solution

a. We first note that the constant, 15, is positive and the **coefficient** of the x-term, 8, is positive. The only way for the product *and* the sum of two real numbers to be positive is if *both* numbers are positive. The only two pairs of positive integer factors of 15 are $1 \cdot 15$ and $3 \cdot 5$. Because $3 + 5 = 8$, we conclude that $x^2 + 8x + 15 = (x + 3)(x + 5)$.

b. We are looking for two integers whose product is -24 and whose sum is -2. We list all possible factors until we find the two integers that we are seeking or until we have exhausted all possibilities.

Product	Sum
$(1) \cdot (-24) = -24$	$(1) + (-24) = -23$
$(2) \cdot (-12) = -24$	$(2) + (-12) = -10$
$(3) \cdot (-8) = -24$	$(3) + (-8) = -5$
$(4) \cdot (-6) = -24$	$(4) + (-6) = -2$

We see that 4 and -6 are the correct integers. Thus, $x^2 - 2x - 24 = (x + 4)(x - 6)$.

c. In this case, we are seeking two integers whose product is 32 and whose sum is -12. The only way for the product of two real numbers to be positive and for the sum to be negative is if *both* numbers are negative.

Product	Sum
$(-1) \cdot (-32) = 32$	$(-1) + (-32) = -33$
$(-2) \cdot (-16) = 32$	$(-2) + (-16) = -18$
$(-4) \cdot (-8) = 32$	$(-4) + (-8) = -12$

We see that -4 and -8 are the correct integers. Thus, $x^2 - 12x + 32 = (x - 4)(x - 8)$.

d. We are searching for two integers whose product is -6 and whose sum is -2. The possibilities are listed as follows.

Product	Sum
$(1) \cdot (-6) = -6$	$(1) + (-6) = -5$
$(2) \cdot (-3) = -6$	$(2) + (-3) = -1$
$(3) \cdot (-2) = -6$	$(3) + (-2) = 1$
$(6) \cdot (-1) = -6$	$(6) + (-1) = 5$

We see that there are no two integers whose product is -6 and whose sum is -2. Therefore, we say that the trinomial $x^2 - 2x - 6$ is not factorable with integers. A polynomial that is not factorable with integers is called a **prime polynomial***.

Work exercises 17–27 in this textbook or in the *MyMathLab*® Study Plan.

In Example 3, all trinomials had a leading coefficient of one. We are now ready to explore techniques used to factor trinomials of the form $ax^2 + bx + c$, where $a \neq 1$.

OBJECTIVE 4:   FACTORING TRINOMIALS WITH A LEADING COEFFICIENT NOT EQUAL TO ONE

My video summary

We can try to factor trinomials of the form $ax^2 + bx + c$, where $a \neq 1$ using a **trial and error method**. That is, we basically "guess" until we stumble on the correct two binomial factors. For example, suppose that we want to factor the trinomial $6x^2 - 7x - 3$. The possible first terms of the two binomial factors are x and

$6x$ or $2x$ and $3x$. The only integer factors of the constant, -3, are $(1)\cdot(-3)$ and $(3)\cdot(-1)$. This means that there are eight possible combinations of potential pairs of binomial factors for this trinomial. The possibilities are

1. $(x - 1)(6x + 3)$ **2.** $(x + 1)(6x - 3)$ **3.** $(6x - 1)(x + 3)$

4. $(6x + 1)(x - 3)$ **5.** $(2x - 1)(3x + 3)$ **6.** $(2x + 1)(3x - 3)$

7. $(3x - 1)(2x + 3)$ **8.** $(3x + 1)(2x - 3)$

If we use the FOIL method to multiply each of the eight possibilities, we see that the final two binomial factors on the list are the correct factors because $(3x + 1)(2x - 3) = 6x^2 - 7x - 3$. Although this guessing method works, it can often be a long and arduous process, especially when the leading coefficient and constant coefficient have several factors. Fortunately, there is another more methodical approach to factoring trinomials. We now outline this four-step method using the same trinomial $6x^2 - 7x - 3$ as our example.

Factoring Trinomials of the Form $ax^2 + bx + c, a \neq 0, a \neq 1$

Example: $6x^2 - 7x - 3, a = 6, b = -7, c = -3$

Step 1: **Multiply $a \cdot c$.**

$$(6)\cdot(-3) = -18$$

Step 2: **Find two numbers whose product is ac and whose sum is b.**

The two numbers are -9 and 2 because $(-9)\cdot(2) = -18$ and $(-9) + (2) = -7$.

Step 3: **Rewrite the middle term as two terms using the numbers found in step 2.**

$$6x^2 \underbrace{- 9x + 2x}_{-9x + 2x = -7x} - 3$$

Step 4: **Factor by grouping.**

$$6x^2 - 9x + 2x - 3 = (6x^2 - 9x) + (2x - 3)$$
$$= 3x(2x - 3) + 1(2x - 3)$$
$$= (2x - 3)(3x + 1)$$

As you can see, we obtained the exact same factorization as we did with the trial and error method.

Example 4   Factor a Trinomial with a Leading Coefficient Not Equal to One

Factor $12x^2 - 25x + 12$.

Solution We follow the four steps outlined previously.

Step 1 $a \cdot c = 12 \cdot 12 = 144$

Step 2 $(-9)\cdot(-16) = 144$ and $(-9) + (-16) = -25$

Step 3 $12x^2 - 25x + 12 = 12x^2 - 16x - 9x + 12$

Step 4 $(12x^2 - 16x) + (-9x + 12)$

$$= 4x(3x - 4) - 3(3x - 4)$$
$$= (3x - 4)(4x - 3)$$

Thus, $12x^2 - 25x + 12 = (3x - 4)(4x - 3)$.

Work exercises 28–34 in this textbook or in the *MyMathLab*® Study Plan.

OBJECTIVE 5:   FACTORING USING SPECIAL FACTORING FORMULAS

There are several factoring formulas that are quite useful to remember and worth mentioning at this time. In Section R.4, we saw several special product formulas. We now revisit three of these formulas and restate them as factoring formulas.

Difference of Two Squares

Let a and b represent any real number or algebraic expression, then
$a^2 - b^2 = (a + b)(a - b)$.

Recognizing a polynomial that can be factored using the difference of two squares pattern is fairly straightforward because such a polynomial has two terms, each of which is a perfect square. Following are three examples:

$$z^2 - 4 = (z)^2 - (2)^2 = (z + 2)(z - 2)$$
$$36y^2 - 1 = (6y)^2 - (1)^2 = (6y + 1)(6y - 1)$$
$$z^4 - 16 = (z^2)^2 - (4)^2 = (z^2 + 4)(z^2 - 4) = (z^2 + 4)(z + 2)(z - 2)$$

Note: The sum of two squares, $a^2 + b^2$, is not factorable using integers. A polynomial of this form is a prime polynomial.

In Section R.4, we introduced the following two special product formulas that were stated as the **square of a binomial.**

$$(a + b)^2 = a^2 + 2ab + b^2 \quad \text{and} \quad (a - b)^2 = a^2 - 2ab + b^2$$

The resulting trinomials are called *perfect square trinomials* because they can each be factored as a binomial squared.

Perfect Square Trinomials

Let a and b represent any real number or algebraic expression, then
$$a^2 + 2ab + b^2 = (a + b)^2$$
$$a^2 - 2ab + b^2 = (a - b)^2$$

Note that the first term of a perfect square trinomial is the square of some algebraic expression a. The last term is the square of some algebraic expression b, and the middle term is the sum or difference of twice the product of a and b. Following is an example of a perfect square trinomial:

$$4x^2 - 12xy + 9y^2 = \underbrace{(2x)^2}_{a^2} - \underbrace{2(2x)(3y)}_{2ab} + \underbrace{(3y)^2}_{b^2} = \underbrace{(2x - 3y)^2}_{(a - b)^2}$$

We stated previously that the sum of two squares cannot be factored with integers. However, this is not the case for the sum of two cubes. In fact, the difference of two cubes can be factored as well. We now state these final two special factoring formulas.

Sum and Difference of Two Cubes

Let a and b represent any real number or algebraic expression, then

$$a^3 + b^3 = (a + b)(a^2 - ab + b^2)$$
$$a^3 - b^3 = (a - b)(a^2 + ab + b^2)$$

Note that the sum of two cubes and the difference of two cubes have very similar factoring patterns. The only two differences are the signs of the binomial factor and the sign of the second term of the trinomial factor. Following is an example of a sum of two cubes:

$$8y^3 + 125 = \underbrace{(2y)^3}_{a^3} + \underbrace{(5)^3}_{b^3} = \underbrace{(2y + 5)}_{(a + b)}\underbrace{(4y^2 - 10y + 25)}_{(a^2 - ab + b^2)}$$

Determine whether you can use a special factoring formula to factor one or more of the polynomials stated in Example 5. Do not forget to first try to factor out a GCF if possible. Watch the interactive video to see if you are correct.

My interactive video summary

▶ **Example 5 Use Special Factoring Formulas to Factor Polynomials**

Completely factor each expression.

a. $2x^3 + 5x^2 - 12x$
b. $5x^4 - 45x^2$
c. $8z^3x - 27y + 8z^3y - 27x$

Solution Watch the interactive video to see each solution worked out in detail.

Work exercises 35–50 in this textbook or in the **MyMathLab**® Study Plan.

R.5 **Exercises**

In Exercises 1–8, factor each polynomial by factoring out the GCF.

1. $40x + 4$ **2.** $wx^4 + w$ **3.** $3y^7 + 6xy^8$ **4.** $4x^4 - 20x^3 + 16x^2$

5. $39w^9y^9 - 12w^2y^2 + 12wy + 15w^2y$ **6.** $11(x + 7) + 3a(x + 7)$

7. $8x(z + 1) + (z + 1)$ **8.** $9y(x^2 + 11) - 7(x^2 + 11)$

In Exercises 9–16, factor each polynomial by grouping.

9. $xy + 5x + 3y + 15$ **10.** $ab + 7a - 2b - 14$ **11.** $9xy + 12x + 6y + 8$ **12.** $3xy - 2x - 12y + 8$

13. $5x^2 + 4xy + 15x + 12y$ **14.** $3x^2 + 3xy - 2x - 2y$ **15.** $x^3 + 9x^2 + 4x + 36$

16. $x^3 - x^2 - 7x + 7$

In Exercises 17–50, completely factor each polynomial or state that the polynomial is prime.

17. $x^2 + 13x + 30$ **18.** $y^2 + 5y - 14$ **19.** $w^2 - 17w + 72$ **20.** $x^2 - 2x - 48$

21. $x^2 - 6x + 10$ **22.** $x^2 - 33x - 108$ **23.** $x^2 + 6xy + 8y^2$ **24.** $24 + 2x - x^2$

25. $5x^2 + 10x - 75$ **26.** $4x^2w + 20xw + 24w$ **27.** $5x^2 - 30x - 25$ **28.** $9x^2 + 18x + 5$

29. $12y^2 + 29y + 15$ **30.** $2x^2 + 5x - 25$ **31.** $8x^2 - 6x - 9$ **32.** $10y^2 - 23y + 12$

33. $16x^2 + 36x + 18$ **34.** $12 - x - 20x^2$ **35.** $x^2 - 81$ **36.** $z^2 + 10z + 25$

37. $x^3 + 64$ **38.** $2m^2 - 48m + 288$ **39.** $64x^2 - 49$ **40.** $x^3 - 343$

41. $108t^2 - 147$ **42.** $(x + 8y)^2 - 25$ **43.** $125x^3 + 216$ **44.** $(2x - 5)^3 - 8$

45. $(w + 5)^2 - 100$ **46.** $16x^2 + 40x + 25$ **47.** $x^{14} - x^{11}$ **48.** $6 - 149x^2 - 25x^4$

49. $x^3y^4 - 27y^4$ **50.** $x^2 + 14x + 49 - x^4$

R.6   Rational Expressions

OBJECTIVES

1 Simplifying Rational Expressions

2 Multiplying and Dividing Rational Expressions

3 Adding and Subtracting Rational Expressions

4 Simplifying Complex Rational Expressions

OBJECTIVE 1:   SIMPLIFYING RATIONAL EXPRESSIONS

A **rational number** is the quotient of two integers. A rational expression is the quotient of two **polynomial expressions**.

> **DEFINITION   Rational Expression**
>
> A rational expression is an algebraic expression of the form $\frac{P}{Q}$, where P and Q are polynomials such that $Q \neq 0$.

The following are examples of rational expressions:

$$\frac{x^2 - 4x}{2x - 8}, \qquad \frac{3x^2 - 2x - 8}{3x^2 - 14x - 24}, \qquad \frac{x^2y^3 + 11x^5y^4 - 7xy}{xy^5 - x^2y}$$

Because division by zero is never allowed, we must only consider values of the variable for which the denominator does not equal zero. For example, the rational expression $\frac{x^2 - 4x}{2x - 8}$ is defined for all values of x with the exception of $x = 4$. The value of $x = 4$ produces zero in the denominator and thus cannot be considered as a possible value of the variable. To simplify a rational expression, try factoring the numerator and denominator and cancel any common factors. The restrictions placed on the original expression still hold for the simplified expression. For example, $\frac{x^2 - 4x}{2x - 8} = \frac{x(x - 4)}{2(x - 4)} = \frac{x(x - 4)}{2(x - 4)} = \frac{x}{2}$. Therefore, the rational expression $\frac{x^2 - 4x}{2x - 8}$ is equivalent to $\frac{x}{2}$ for all values of x not equal to 4.

My interactive video summary

 Example 1 Simplify Rational Expressions

Simplify each rational expression.

a. $\dfrac{x^2 + x - 12}{x^2 + 9x + 20}$ **b.** $\dfrac{x^3 + 1}{x + 1}$ **c.** $\dfrac{x^2 - x - 2}{2x - x^2}$

Solution

a. We first factor the rational expression. If you need a quick refresher of how to factor polynomials of the form $x^2 + bx + c$, watch the video that was first introduced in Section R.5. The numerator and denominator of this rational expression factors as follows:

$$\frac{x^2 + x - 12}{x^2 + 9x + 20} = \frac{(x + 4)(x - 3)}{(x + 5)(x + 4)}$$

This rational expression is defined for all values of x such that $x \neq -5$ and $x \neq -4$. This restriction must also apply to any simplified version of the original rational expression.

   We now cancel all common factors within the numerator and denominator.

$$\frac{x^2 + x - 12}{x^2 + 9x + 20} = \frac{(\cancel{x + 4})(x - 3)}{(x + 5)(\cancel{x + 4})} \qquad \textit{Cancel common factors.}$$

Thus, $\dfrac{x^2 + x - 12}{x^2 + 9x + 20} = \dfrac{x - 3}{x + 5}$ for $x \neq -5$ and $x \neq -4$.

b. This rational expression is defined for all values of x not equal to -1. We can factor the numerator using the **sum of two cubes** factoring formula.

$$\frac{x^3 + 1}{x + 1} = \frac{(\cancel{x + 1})(x^2 - x + 1)}{(\cancel{x + 1})} \qquad \textit{Factor the numerator and cancel common factors.}$$

$$= x^2 - x + 1 \text{ for } x \neq -1$$

c. Factor the numerator and denominator.

$$\frac{x^2 - x - 2}{2x - x^2} = \frac{(x - 2)(x + 1)}{x(2 - x)}$$

We see that this rational expression is defined for all values of x such that $x \neq 0$ and $x \neq 2$. At first glance, it does not appear that the numerator and denominator share a common factor. However, note that if we were to multiply the factor $(2 - x)$ by -1, we get $-1 \cdot (2 - x) = -2 + x = x - 2$.

   Therefore, multiply the numerator and denominator by 1 of the form $\frac{-1}{-1}$, and then cancel common factors.

$$\frac{x^2 - x - 2}{2x - x^2} = \frac{(x - 2)(x + 1)}{x(2 - x)}$$

$$= \left(\frac{-1}{-1}\right)\frac{(x - 2)(x + 1)}{x(2 - x)} \qquad \textit{Multiply the rational expression by } \frac{-1}{-1}.$$

$$= -\frac{(x - 2)(x + 1)}{x(x - 2)} \qquad -1 \cdot (2 - x) = (x - 2).$$

$$= -\frac{(\cancel{x - 2})(x + 1)}{x(\cancel{x - 2})} \qquad \textit{Cancel common factors.}$$

$$= -\frac{x + 1}{x} \text{ for } x \neq 0 \text{ and } x \neq 2.$$

Watch the interactive video see the solution to Example 1 worked out in detail.

Work exercises 1–10 in this textbook or in the *MyMathLab*® Study Plan.

Note: From this point forward, we do not make any restrictions on the variables. That is, we assume that all expressions in the denominators of all rational expressions represent nonzero real numbers.

OBJECTIVE 2:   MULTIPLYING AND DIVIDING RATIONAL EXPRESSIONS

We multiply rational expressions similarly to the way we multiply **rational numbers**. If $\frac{A}{B}$ and $\frac{C}{D}$ are rational expressions, then $\frac{A}{B} \cdot \frac{C}{D} = \frac{AC}{BD}$.

Example 2 Multiply Rational Expressions

Multiply $\dfrac{2x^2 + 3x - 2}{3x^2 - 2x - 1} \cdot \dfrac{3x^2 + 4x + 1}{2x^2 + x - 1}$.

Solution Factor the numerator and denominator of each rational expression. Note that each numerator and denominator are polynomials of the form $ax^2 + bx + c$, where $a \neq 1$. If you need to review how to factor polynomials of this form, watch the video that was introduced in Section R.5.

$$\dfrac{2x^2 + 3x - 2}{3x^2 - 2x - 1} \cdot \dfrac{3x^2 + 4x + 1}{2x^2 + x - 1}$$ *Factor the numerators and denominators.*

$$= \dfrac{(2x - 1)(x + 2)}{(x - 1)(3x + 1)} \cdot \dfrac{(3x + 1)(x + 1)}{(2x - 1)(x + 1)}$$

$$= \dfrac{(2x - 1)(x + 2)(3x + 1)(x + 1)}{(x - 1)(3x + 1)(2x - 1)(x + 1)}$$ *Multiply numerators and denominators.*

$$= \dfrac{\cancel{(2x - 1)}(x + 2)\cancel{(3x + 1)}\cancel{(x + 1)}}{(x - 1)\cancel{(3x + 1)}(2x - 1)\cancel{(x + 1)}}$$ *Cancel common factors.*

$$= \dfrac{x + 2}{x - 1}$$

We divide two rational expressions in exactly the same manner in which we divide two **rational numbers**. Recall that if $\frac{a}{b}$ and $\frac{c}{d}$ are rational numbers such that b, c, and d are nonzero, then $\frac{a}{b} \div \frac{c}{d} = \frac{a}{b} \cdot \frac{d}{c} = \frac{ad}{bc}$. The rational numbers $\frac{c}{d}$ and $\frac{d}{c}$ are called **reciprocals**. We follow this same pattern when dividing two rational expressions. To find the quotient of two rational expressions, we find the product of the first expression times the reciprocal of the second rational expression. Thus, if $\frac{A}{B}$ and $\frac{C}{D}$ are rational expressions, then $\frac{A}{B} \div \frac{C}{D} = \frac{A}{B} \cdot \frac{D}{C} = \frac{AD}{BC}$.

My video summary

Example 3 Divide Rational Expressions

Divide and simplify $\dfrac{x^3 - 8}{2x^2 - x - 6} \div \dfrac{x^2 + 2x + 4}{6x^2 + 11x + 3}$.

Solution

$$\dfrac{x^3 - 8}{2x^2 - x - 6} \div \dfrac{x^2 + 2x + 4}{6x^2 + 11x + 3}$$ *Write the original expression.*

$$= \dfrac{x^3 - 8}{2x^2 - x - 6} \cdot \dfrac{6x^2 + 11x + 3}{x^2 + 2x + 4}$$ *Rewrite as a multiplication problem.*

$$= \frac{(x-2)(x^2+2x+4)}{(2x+3)(x-2)} \cdot \frac{(2x+3)(3x+1)}{x^2+2x+4}$$

Factor the numerators and denominators. Use the difference of two cubes factoring formula to factor the first numerator.

$$= \frac{(x-2)(x^2+2x+4)(2x+3)(3x+1)}{(2x+3)(x-2)(x^2+2x+4)}$$

Multiply.

$$= \frac{\cancel{(x-2)}\cancel{(x^2+2x+4)}\cancel{(2x+3)}(3x+1)}{\cancel{(2x+3)}\cancel{(x-2)}\cancel{(x^2+2x+4)}}$$

Cancel common factors.

$$= 3x+1$$

Example 4 Multiply and Divide Rational Expressions

Perform the indicated operations and simplify.

$$\frac{x^2+x-6}{x^2+x-42} \cdot \frac{x^2+12x+35}{x^2-x-2} \div \frac{x+7}{x^2+8x+7}$$

Solution Watch the video to see that the expression simplifies to $\frac{(x+3)(x+5)}{x-6}$.

Work exercises 11–26 in this textbook or in the *MyMathLab*® Study Plan.

OBJECTIVE 3:   ADDING AND SUBTRACTING RATIONAL EXPRESSIONS

Recall that in order to add or subtract rational numbers, we must first rewrite both numbers in a different form using a common denominator. Once both rational numbers have a common denominator, we add or subtract the numerators, writing the result above the common denominator. For example, we subtract the rational number $\frac{1}{6}$ from $\frac{5}{8}$ as follows:

$$\frac{5}{8} - \frac{1}{6} = \left(\frac{5}{8}\right)\left(\frac{3}{3}\right) - \left(\frac{1}{6}\right)\left(\frac{4}{4}\right) = \frac{15}{24} - \frac{4}{24} = \frac{15-4}{24} = \frac{9}{24} = \frac{3}{8}$$

We add and subtract rational expressions in much the same way. We start by determining the **least common denominator (LCD)**. The LCD is the smallest algebraic expression divisible by all denominators.

Example 5 Add and Subtract Rational Expressions

Perform the indicated operations and simplify.

a. $\dfrac{3}{x+1} - \dfrac{2-x}{x+1}$ **b.** $\dfrac{3}{x^2+2x} + \dfrac{x-2}{x^2-x}$

Solution

a. Each rational expression has the same denominator, so we can combine by subtracting the numerators.

$$\frac{3}{x+1} - \frac{2-x}{x+1} \qquad \textit{Write the original expression.}$$

$$= \frac{3-(2-x)}{x+1} \qquad \textit{Subtract numerators.}$$

$$= \frac{3-2+x}{x+1} \qquad \textit{Use the distributive property.}$$

$$= \frac{x+1}{x+1} \qquad \textit{Combine like terms.}$$

$$= \frac{\cancel{x+1}}{\cancel{x+1}} = 1 \qquad \textit{Cancel common factors and simplify.}$$

b. Factor the denominators to determine the LCD.

$$\left.\begin{array}{l} x^2 + 2x = x(x+2) \\ x^2 - \ x = x(x-1) \end{array}\right\} \qquad \text{LCD} = x(x+2)(x-1).$$

$$\frac{3}{x(x+2)} + \frac{x-2}{x(x-1)} \qquad \textit{Write the original expression with the denominators factored.}$$

$$= \frac{3}{x(x+2)}\frac{(x-1)}{(x-1)} + \frac{x-2}{x(x-1)}\frac{(x+2)}{(x+2)} \qquad \textit{Multiply the first rational expression by } \frac{(x-1)}{(x-1)} \textit{ and multiply the second rational expression by } \frac{(x+2)}{(x+2)}.$$

$$= \frac{3(x-1) + (x-2)(x+2)}{x(x+2)(x-1)} \qquad \textit{Add the numerators.}$$

$$= \frac{3x-3+x^2-4}{x(x+2)(x-1)} \qquad \textit{Use the distributive property.}$$

$$= \frac{x^2+3x-7}{x(x+2)(x-1)} \qquad \textit{Combine like terms.}$$

Try working through Example 6 on your own. Watch the interactive video when you are ready to see the solution.

My interactive video summary

▶ Example 6 Add and Subtract Rational Expressions

Perform the indicated operations and simplify.

a. $\dfrac{3}{x-y} - \dfrac{x+5y}{x^2-y^2}$ 　　**b.** $\dfrac{x+4}{3x^2+20x+25} + \dfrac{x}{3x^2+16x+5}$

Solution Watch the interactive video to see each solution worked out in detail.

Work exercises **27–44** in this textbook or in the *MyMathLab* ® Study Plan.

OBJECTIVE 4:　SIMPLIFYING COMPLEX RATIONAL EXPRESSIONS

A complex rational expression (also called a complex fraction) is a fraction that contains a rational expression in the numerator and/or the denominator. The following are examples of complex rational expressions:

$$\frac{4 - \dfrac{5}{x - 1}}{\dfrac{6}{x - 1} - 7} \qquad \frac{\dfrac{1}{x} - \dfrac{3}{x + 4}}{\dfrac{2}{x^2 + 4x} + \dfrac{2}{x}}$$

To simplify a complex rational expression, we want to rewrite the expression using only one fraction bar. There are two methods that can be used to simplify such expressions. We call these Method I and Method II. Method I involves the following three-step process.

Method I for Simplifying Complex Rational Expressions

Step 1 Simplify the rational expression in the numerator.

Step 2 Simplify the rational expression in the denominator.

Step 3 Divide the expression in the numerator by the expression in the denominator. That is, multiply the expression in the numerator by the reciprocal of the expression in the denominator. Simplify if possible.

Example 7 Simplify a Complex Rational Expression Using Method I

Simplify using Method I: $\dfrac{4 - \dfrac{5}{x - 1}}{\dfrac{6}{x - 1} - 7}$

Solution

Step 1 Simplify the numerator.

$$4 - \frac{5}{x - 1} = 4\frac{(x - 1)}{(x - 1)} - \frac{5}{x - 1} = \frac{4(x - 1) - 5}{x - 1}$$

$$= \frac{4x - 4 - 5}{x - 1} = \frac{4x - 9}{x - 1}$$

Step 2 Simplify the denominator.

$$\frac{6}{x - 1} - 7 = \frac{6}{x - 1} - 7\frac{(x - 1)}{(x - 1)} = \frac{6 - 7(x - 1)}{x - 1}$$

$$= \frac{6 - 7x + 7}{x - 1} = \frac{13 - 7x}{x - 1}$$

Step 3 Divide.

$$\frac{4 - \dfrac{5}{x - 1}}{\dfrac{6}{x - 1} - 7} = \frac{\dfrac{4x - 9}{x - 1}}{\dfrac{13 - 7x}{x - 1}} = \frac{4x - 9}{x - 1} \cdot \frac{x - 1}{13 - 7x}$$

$$= \frac{4x - 9}{\cancel{x - 1}} \cdot \frac{\cancel{x - 1}}{13 - 7x} = \frac{4x - 9}{13 - 7x}$$

As stated previously, there is another method that can be used to simplify a complex rational expression. Method II involves multiplying the numerator and denominator of the complex rational expression by the overall LCD. The overall LCD is the LCD of *all* fractions in both the numerator and the denominator.

Method II for Simplifying Complex Rational Expressions

Step 1   Determine the overall LCD.

Step 2   Multiply the numerator and denominator of the complex rational expression by the overall LCD.

Step 3   Simplify.

We now use Method II to simplify the same complex rational expression from Example 7.

Example 8 Simplify a Complex Rational Expression Using Method II

Simplify using Method II: $\dfrac{4 - \dfrac{5}{x - 1}}{\dfrac{6}{x - 1} - 7}$

Solution

Step 1   The overall LCD is $x - 1$.

Step 2   Multiply the numerator and denominator by $x - 1$.

$$\frac{4 - \dfrac{5}{x - 1}}{\dfrac{6}{x - 1} - 7} \cdot \frac{x - 1}{x - 1}$$

Step 3   Simplify.

$$\frac{4 - \dfrac{5}{x - 1}}{\dfrac{6}{x - 1} - 7} \cdot \frac{x - 1}{x - 1} = \frac{4(x - 1) - \dfrac{5(x - 1)}{x - 1}}{\dfrac{6(x - 1)}{x - 1} - 7(x - 1)}$$

$$= \frac{4x - 4 - \dfrac{5\cancel{(x - 1)}}{\cancel{x - 1}}}{\dfrac{6\cancel{(x - 1)}}{\cancel{x - 1}} - 7x + 7}$$

$$= \frac{4x - 4 - 5}{6 - 7x + 7} = \frac{4x - 9}{13 - 7x}$$

As you can see from Examples 7 and 8, it does not matter which method we choose. Pick the method that you are most comfortable with. See if you can simplify the complex rational expression in Example 9. You can watch the interactive video and choose to see the solution using either method.

My interactive video summary

Example 9 Simplify a Complex Rational Expression

Simplify the complex rational expression using Method I or Method II.

$$\frac{\dfrac{1}{x} - \dfrac{3}{x + 4}}{\dfrac{2}{x^2 + 4x} + \dfrac{2}{x}}$$

Solution To see the solution, watch the interactive video and click on Method I or Method II.

Work exercises 45–54 in this textbook or in the MyMathLab® Study Plan.

R.6 Exercises

In Exercises 1–10, simplify each rational expression.

1. $\dfrac{2x - 6x^2}{2x}$

2. $\dfrac{x^2 - 81}{9 + x}$

3. $\dfrac{9a - 27}{5a - 15}$

4. $\dfrac{x^2 + x - 6}{x + 3}$

5. $\dfrac{x - 13}{13 - x}$

6. $\dfrac{x^2 - 4}{-2 - x}$

7. $\dfrac{3x^2 - 2x - 8}{3x^2 - 14x - 24}$

8. $\dfrac{x^3 - 125}{3x - 15}$

9. $\dfrac{4x^2 - 7x - 2}{8x^3 + 2x^2 + 4x + 1}$

10. $\dfrac{36x^2 - 42x + 49}{216x^3 + 343}$

In Exercises 11–26, perform the indicated operations and simplify.

11. $\dfrac{18x + 18}{2x + 6} \cdot \dfrac{x + 3}{6x^2 - 6}$

12. $\dfrac{16x - 12}{35} \cdot \dfrac{10}{3 - 4x}$

13. $\dfrac{15a - 10a^2}{9a^2 + 6a + 1} \cdot \dfrac{6a^2 + 11a + 3}{4a^2 - 9}$

14. $\dfrac{4x^3 - 32}{2x^2 + 4x - 16} \cdot \dfrac{5x + 10}{4x^2 + 8x + 16}$

15. $\dfrac{a^3 + a^2b + a + b}{2a^3 + 2a} \cdot \dfrac{18a^2}{6a^2 - 6b^2}$

16. $\dfrac{x^2 - 6x - 7}{2x^2 - 98} \cdot \dfrac{x^2 + 14x + 49}{3x^2 + 24x + 21}$

17. $\dfrac{x^2 - 4}{9} \cdot \dfrac{x^2 - x - 2}{x^2 - 4x + 4}$

18. $\dfrac{4x}{9} \div \dfrac{8x + 16}{9x + 18}$

19. $\dfrac{a + b}{ab} \div \dfrac{a^2 - b^2}{7a^3b}$

20. $\dfrac{x^2 - 14x + 49}{x^2 - x - 42} \div \dfrac{x^2 - 49}{8}$

21. $\dfrac{x^2 - 6x - 16}{3x^2 - 192} \div \dfrac{x^2 + 10x + 16}{x^2 + 16x + 64}$

22. $\dfrac{3x - x^2}{x^3 - 27} \div \dfrac{x}{x^2 + 3x + 9}$

23. $\dfrac{\dfrac{5x^2 - 4x - 12}{3x^2 - 19x - 14}}{\dfrac{25x^2 + 35x + 6}{3x^2 + 17x + 10}}$

24. $\dfrac{8}{x} \div \dfrac{7xy}{x^4} \cdot \dfrac{5x^2}{x^7}$

25. $\dfrac{12x^2 + 8x + 1}{y^2 - 2y - 3} \cdot \dfrac{y^2 - 7y + 12}{6x^2 + 7x + 1} \div \dfrac{4x^2 - 4x - 3}{3x^2 - 2x - 5}$

26. $\dfrac{4a^2 - 64}{2a^2 - 4a} \div \dfrac{a^3 + 4a^2}{5a^2 - 10a} \cdot \dfrac{6a^3 + 4a^2}{4a^2 - 16a}$

27. $\dfrac{7}{x} + \dfrac{9}{x}$

28. $\dfrac{x + 3}{x - 4} + \dfrac{5x - 2}{x - 4}$

29. $\dfrac{x^2}{3x + 1} - \dfrac{4}{3x + 1}$

30. $\dfrac{6}{49x} + \dfrac{5}{7x^2}$

31. $\dfrac{9}{x - 3} + \dfrac{x}{3 - x}$

32. $\dfrac{8}{x - 1} - \dfrac{6}{x + 8}$

33. $\dfrac{x}{x + 3} + \dfrac{2x - 7}{x - 3}$

34. $\dfrac{x - 2}{x + 4} - \dfrac{x + 1}{x - 4}$

35. $\dfrac{x}{x^2 - 36} + \dfrac{4}{x}$

36. $\dfrac{7}{12x^2y} - \dfrac{13}{4xy}$

37. $\dfrac{10}{a + b} + \dfrac{10}{a - b}$

38. $\dfrac{z + 5}{z - 9} - \dfrac{z - 4}{z + 6}$

39. $\dfrac{9}{x - y} - \dfrac{x + 3y}{x^2 - y^2}$

40. $\dfrac{x - 8}{x^2 + 12x + 27} + \dfrac{x - 9}{x^2 - 9}$

41. $\dfrac{6x}{x^2 + 4x - 12} - \dfrac{x}{x^2 - 36}$

42. $\dfrac{y-8}{2y^2+9y+7}-\dfrac{y+9}{2y^2-3y-35}$

43. $\dfrac{5}{y}-\dfrac{5}{y-3}+\dfrac{16}{(y-3)^2}$

44. $\dfrac{x-3}{x^2+12x+36}+\dfrac{1}{x+6}-\dfrac{2x+3}{2x^2+3x-54}$

In Exercises 45–54, simplify the complex rational expression using Method I or Method II.

45. $\dfrac{7+\dfrac{1}{x}}{7-\dfrac{1}{x}}$

46. $\dfrac{\dfrac{1}{x}+3}{\dfrac{1}{x^2}-9}$

47. $\dfrac{\dfrac{x+2}{x-6}-\dfrac{x+12}{x+5}}{x+82}$

48. $\dfrac{\dfrac{6}{x}+\dfrac{5}{y}}{\dfrac{5}{x}-\dfrac{6}{y}}$

49. $\dfrac{\dfrac{1}{16}-\dfrac{1}{x^2}}{\dfrac{1}{4}+\dfrac{1}{x}}$

50. $\dfrac{\dfrac{8}{x+4}-\dfrac{2}{x+7}}{\dfrac{x+8}{x+4}}$

51. $\dfrac{\dfrac{2}{x+13}+\dfrac{1}{x-5}}{2-\dfrac{x+25}{x+13}}$

52. $\dfrac{\dfrac{x-5}{x+5}+\dfrac{x-5}{x-9}}{1+\dfrac{x+5}{x-9}}$

53. $\dfrac{1+\dfrac{4}{x}-\dfrac{5}{x^2}}{1-\dfrac{2}{x}-\dfrac{35}{x^2}}$

54. $\dfrac{6x^{-1}+6y^{-1}}{xy^{-1}-x^{-1}y}$

CHAPTER ONE

Equations, Inequalities, and Applications

CHAPTER ONE CONTENTS

1.1 Linear Equations

THINGS TO KNOW

Before working through this section, be sure that you are familiar with the following concepts:

VIDEO   ANIMATION   INTERACTIVE

You Try It

1. Factoring Trinomials with a Leading Coefficient Equal to 1 (Section R.5)

You Try It

2. Factoring Trinomials with a Leading Coefficient Not Equal to 1 (Section R.5)

1

OBJECTIVES

1 Recognizing Linear Equations

2 Solving Linear Equations with Integer Coefficients

3 Solving Linear Equations Involving Fractions

4 Solving Linear Equations Involving Decimals

5 Solving Equations That Lead to Linear Equations

OBJECTIVE 1:   RECOGNIZING LINEAR EQUATIONS

In the Review chapter, we manipulated **algebraic expressions**. Remember, algebraic expressions do *not* involve an equal sign. An **equation** indicates that two algebraic expressions are equal. For example, $2x + 6$ and $5x$ are two algebraic expressions. By equating these two expressions with an equal sign, we obtain the equation $2x + 6 = 5x$. This equation is an example of a **linear equation in one variable**. A linear equation in one variable involves only constants and variables that are raised to the first power. Examples of linear equations in one variable include the following:

$$7x - 4 = x - 5, \quad \frac{1}{3}y - \sqrt{2} = 11y, \quad \text{and} \quad 0.4a - 1 = 10$$

DEFINITION Linear Equation in One Variable

A linear equation in one variable is an equation that can be written in the form $ax + b = c$, where $a, b,$ and c are real numbers and $a \neq 0$.

My interactive video summary

 Watch this interactive video to determine if equations are linear or nonlinear, or click here to see an example of how to recognize linear and nonlinear equations.

Work exercises 1–7 in this textbook or in the *MyMathLab*® Study Plan.

OBJECTIVE 2:   SOLVING LINEAR EQUATIONS WITH INTEGER COEFFICIENTS

Let's now begin to solve linear equations. When we solve an equation for x, we are looking for all values of x that when substituted back into the original equation yield a true statement. For example, given the equation $2x - 1 = 5$, we see that if $x = 3$, the equation is a true statement. In other words, by substituting the number 3 into the equation for x, we get a true statement:

$$2x - 1 = 5 \qquad \textit{Write the original equation.}$$
$$2(3) - 1 \overset{?}{=} 5 \qquad \textit{Substitute 3 in for x.}$$
$$6 - 1 \overset{?}{=} 5 \qquad \textit{Multiply.}$$
$$5 = 5 \qquad \textit{True statement!}$$

We start by solving the most basic linear equations, which are linear equations with integer **coefficients**.

Example 1 Solve a Linear Equation with Integer Coefficients

Solve $5(x - 6) - 2x = 3 - (x + 1)$

Solution The goal here is to isolate the variable x on one side of the equation. First, use the **distributive property** to remove the parentheses on the left-hand side:

$5(x - 6) - 2x = 3 - (x + 1)$	*Write the original equation.*
$5x - 30 - 2x = 3 - x - 1$	*Use the distributive property.*
$3x - 30 = 2 - x$	*Combine like terms.*
$3x - 30 + x = 2 - x + x$	*Add x to both sides.*
$4x - 30 = 2$	*Simplify.*
$4x - 30 + 30 = 2 + 30$	*Add 30 to both sides.*
$4x = 32$	*Combine like terms.*
$\dfrac{4x}{4} = \dfrac{32}{4}$	*Divide both sides by 4.*
$= 8$	*Possible solution.*

You should now check by substituting $x = 8$ back into the original equation. Click here to check.

Example 2 Solve a Linear Equation with Integer Coefficients

Solve $6 - 4(x + 4) = 8x - 2(3x + 5)$

Solution To isolate the variable x, we first use the **distributive property** to remove all parentheses and then simplify:

$6 - 4(x + 4) = 8x - 2(3x + 5)$	*Write the original equation.*
$6 - 4x - 16 = 8x - 6x - 10$	*Use the distributive property.*
$-4x - 10 = 2x - 10$	*Combine like terms.*
$-4x - 10 + 4x = 2x - 10 + 4x$	*Add $4x$ to both sides.*
$-10 = 6x - 10$	*Simplify.*
$-10 + 10 = 6x - 10 + 10$	*Add 10 to both sides.*
$0 = 6x$	*Combine like terms.*
$\dfrac{0}{6} = \dfrac{6x}{6}$	*Divide both sides by 6.*
$0 = x$	*Possible solution.*

You should verify that $x = 0$ is the solution by substituting $x = 0$ back into the original equation.

It is important to point out that $x = 0$ can be a valid solution to an equation, as in Example 2. Once we encounter equations in which variables appear in one or more denominators, we have to be extremely careful when checking our answers because division by zero is never allowed! (See Example 6.)

Work exercises 8–11 in this textbook or in the *MyMathLab*® Study Plan.

OBJECTIVE 3:   SOLVING LINEAR EQUATIONS INVOLVING FRACTIONS

To solve a linear equation involving fractions, we want to first transform the equation into a linear equation involving integer coefficients. We do this by multiplying both sides of the equation by the **least common denominator** (LCD).

My video summary

📼 Example 3 Solve a Linear Equation Involving Fractions

Solve $\frac{1}{3}(1 - x) - \frac{x + 1}{2} = -2$

Solution The first thing to do when solving equations involving fractions is to find the LCD. The LCD for this equation is 6. We can eliminate the fractions by multiplying both sides of this equation by 6.

$$6\left(\frac{1}{3}(1 - x) - \frac{x + 1}{2}\right) = 6(-2) \quad \text{Multiply both sides by 6.}$$

$$6\left(\frac{1}{3}(1 - x)\right) - 6\left(\frac{x + 1}{2}\right) = 6(-2) \quad \text{Use the distributive property.}$$

$$2(1 - x) - 3(x + 1) = -12 \quad \text{Multiply.}$$

Notice that the original equation containing fractions has now been "transformed" into a linear equation involving integer coefficients, as in Example 1. You should solve this equation and/or watch the corresponding video to verify that the solution set is $\left\{\frac{11}{5}\right\}$.

Work exercises 12–14 in this textbook or in the *MyMathLab*® Study Plan.

OBJECTIVE 4:   SOLVING LINEAR EQUATIONS INVOLVING DECIMALS

The strategy for solving linear equations involving decimals is similar to the one used to solve linear equations involving fractions. We want to eliminate all decimals. We can eliminate decimals by multiplying both sides of the equation by the appropriate power of 10, such as $10^1 = 10, 10^2 = 100, 10^3 = 1,000$, etc. To determine the appropriate power of 10, look at the constants in the equation and choose the constant that has the greatest number of decimal places. Count those decimal places and then raise 10 to that power. Multiplying by that power of 10 will immediately eliminate all decimal places in the equation.

Example 4 Solve a Linear Equation Involving Decimals

Solve $.1(y - 2) + .03(y - 4) = .02(10)$

📼 **Solution** The constants .03 and .02 in the equation each have *two* decimal places. We can eliminate the decimals in this equation by multiplying both sides of the equation by $10^2 = 100$, thus moving each decimal two places to the right to obtain the new equation $10(y - 2) + 3(y - 4) = 2(10)$. Watch the video to see this problem solved in its entirety. You should verify that the solution set is $\{4\}$.

Work exercises 15–17 in this textbook or in the *MyMathLab*® Study Plan.

OBJECTIVE 5:   SOLVING EQUATIONS THAT LEAD TO LINEAR EQUATIONS

Some equations do not start out as linear equations but often lead to linear equations after simplification. Consider the following three examples.

Example 5 Solve an Equation That Leads to a Linear Equation

Solve $3a^2 - 1 = (a + 1)(3a + 2)$

Solution The left side of this equation is not linear because the variable is raised to the second power. However, after multiplying the two **binomials** on the right-hand side and simplifying, we see that this equation does indeed simplify into a linear equation.

$3a^2 - 1 = (a + 1)(3a + 2)$	*Write the original equation.*
$3a^2 - 1 = 3a^2 + 5a + 2$	*Multiply.*
$3a^2 - 3a^2 - 1 = 3a^2 - 3a^2 + 5a + 2$	*Subtract $3a^2$ from both sides.*
$-1 = 5a + 2$	
$-1 - 2 = 5a + 2 - 2$	*Subtract 2 from both sides.*
$-3 = 5a$	*Combine like terms.*
$\dfrac{-3}{5} = \dfrac{5a}{5}$	*Divide both sides by 5.*
$a = -\dfrac{3}{5}$	

The solution set is $\left\{-\frac{3}{5}\right\}$.

Example 6 Solve an Equation That Leads to a Linear Equation

Solve $\dfrac{2 - x}{x + 2} + 3 = \dfrac{4}{x + 2}$

Solution This equation is nonlinear because there are variables in the denominator. This is an example of a **rational equation**. To solve this rational equation, first multiply both sides of the equation by the LCD, $x + 2$, to eliminate the denominators:

$(x + 2)\left(\dfrac{2 - x}{x + 2} + 3\right) = (x + 2)\left(\dfrac{4}{x + 2}\right)$	*Multiply both sides by $x + 2$.*
$(x + 2)\left(\dfrac{2 - x}{x + 2}\right) + (x + 2)(3) = (x + 2)\left(\dfrac{4}{x + 2}\right)$	*Use the distributive property.*
$2 - x + 3x + 6 = 4$	*Multiply.*
$2x + 8 = 4$	*Combine like terms.*
$2x = -4$	*Subtract 8 from both sides.*
$x = -2$	*Divide both sides by 2.*

If we check by substituting $x = -2$ back into the original equation, we see that we obtain zeros in the denominators. Because division by zero is never permitted, $x = -2$ is not a solution. The solution set to this equation is the empty set or $\varnothing$. ($x = -2$ is called an **extraneous solution**.)

⬦**CAUTION** Because rational equations often have extraneous solutions, *it is imperative* to first determine all values that make any denominator equal to zero. Any solution that makes the denominator equal to zero must be discarded.

🎞 Example 7 Solve an Equation That Leads to a Linear Equation

My video summary

Solve $\dfrac{12}{x^2 + x - 2} - \dfrac{x + 3}{x - 1} = \dfrac{1 - x}{x + 2}$

Solution Watch and work through the video to see the solution.

Work exercises 18–23 in this textbook or in the *MyMathLab*® Study Plan.

1.1 Exercises

In Exercises 1–7, determine whether the given equation is linear or nonlinear.

1. $7x - \dfrac{1}{3} = 5$ **2.** $\dfrac{5}{x} + 4 = 10$ **3.** $\sqrt{2}x - 1 = 0$ **4.** $x^2 + x = 1$

5. $8x - 7 = \pi x - 3$ **6.** $\dfrac{5x - 1}{x + 2} = 3$ **7.** $5 - .2x = .1 - 4x$

In Exercises 8–23, solve each equation:

8. $5x - 8 = 3 - 7(x + 1)$

9. $4(2x + 5) = 10 - 2(x - 5)$

10. $3(3x + 4) = -(x - 3)$

11. $-3(2 - x) + 1 = 4 - (7x - 2)$

12. $\dfrac{1}{2}y - \dfrac{1}{3}(y - 1) = 5y$

13. $\dfrac{a - 3}{6} - \dfrac{3(a - 1)}{10} = \dfrac{2a + 1}{5}$

14. $\dfrac{1}{8}(2p - 1) = \dfrac{7}{3}p - \dfrac{p - 4}{6}$

15. $.12x + .3(x - 4) = .01(2x - 3)$

16. $.002(1 - k) + .01(k - 3) = 1$

17. $-.17x + .01(16x + 5) = .02(3 - x)$

18. $x(3x - 2) + x(5x - 1) = (4x + 1)(2x - 3)$

19. $(x + 3)^3 - 4 = x(x + 4)(x + 5) - 1$

20. $\dfrac{3x}{x - 4} + 1 = \dfrac{12}{x - 4}$

21. $\dfrac{w}{w - 3} - \dfrac{2w}{2w - 1} = \dfrac{w - 3}{2w^2 - 7w + 3}$

22. $\dfrac{3}{x - 1} + \dfrac{4}{x + 1} = \dfrac{8x}{x^2 - 1}$

23. $\dfrac{6}{x^2 - x} - \dfrac{2}{x} = \dfrac{3}{x - 1}$

1.2 Applications of Linear Equations

THINGS TO KNOW

Before working through this section, be sure that you are familiar with the following concepts:

VIDEO   ANIMATION   INTERACTIVE

 You Try It
1. Solving Linear Equations with Integer Coefficients (Section 1.1)

 You Try It
2. Solving Linear Equations Involving Fractions (Section 1.1)

 You Try It
3. Solving Linear Equations Involving Decimals (Section 1.1)

OBJECTIVES

1 Converting Verbal Statements into Mathematical Statements

2 Solving Applications Involving Unknown Numeric Quantities

3 Solving Applications Involving Decimal Equations (Money, Mixture, Interest)

4 Solving Applied Problems Involving Distance, Rate, and Time

5 Solving Applied Working Together Problems

We find algebraic applications in all sorts of subjects, including economics, physics, medicine, and countless other disciplines. In this section, we learn how to solve a variety of useful applications. Before we look at some specific examples, it is important to first practice translating or converting verbal statements into mathematical statements.

OBJECTIVE 1: CONVERTING VERBAL STATEMENTS INTO MATHEMATICAL STATEMENTS

When solving word problems it is important to recognize key words and phrases that translate into algebraic expressions involving addition, subtraction, multiplication, and division.

Example 1 Convert Verbal Statements into Mathematical Statements

Rewrite each statement as an algebraic expression or equation:

   a. 7 more than three times a number
   b. 5 less than twice a number
   c. Three times the quotient of a number and 11
   d. The sum of a number and 9 is 1 less than half of the number.
   e. The product of a number and 4 is 1 more than 8 times the difference of 10 and the number.

Solution

   a. If x is the number, then three times a number is $3x$. Therefore, 7 more than three times a number is equivalent to $3x + 7$ (or $7 + 3x$).
   b. Twice a number is equivalent to $2x$. So, 5 less than twice a number is written as $2x - 5$.

CAUTION The expression $2x - 5$ is not equivalent to $5 - 2x$ because subtraction is not **commutative**; that is, $a - b \neq b - a$.

   c. The quotient of a number and 11 is equivalent to $\frac{x}{11}$. So, three times the quotient of a number and 11 is equivalent to $3\left(\frac{x}{11}\right)$ (or $\frac{3x}{11}$).
   d. Notice the key word "*is*." The word "*is*" often translates into an equal sign, thus indicating that we need to translate the phrase into an **equation**. In this case, the sum of a number and 9 translates to $x + 9$. The statement "1 less than half of the number" translates to $\frac{1}{2}x - 1$. Putting it all together, we get the following equation:

$$\underbrace{x + 9}_{\substack{\text{The sum of} \\ \text{a number} \\ \text{and 9}}} \underset{\text{is}}{=} \underbrace{\frac{1}{2}x - 1}_{\substack{\text{one less than} \\ \text{half the number.}}}$$

   e. This statement translates into the following equation:

$$\underbrace{4x}_{\substack{\text{The product} \\ \text{of a number} \\ \text{and 4}}} \underset{\text{is}}{=} \underbrace{8(10 - x) + 1}_{\substack{\text{one more than} \\ \text{8 times the difference} \\ \text{of 10 and the number.}}}$$

Work exercises 1–5 in this textbook or in the *MyMathLab*® Study Plan.

It is important to have a strategy in place before trying to solve an applied problem. **George Polya**, a Hungarian mathematician who taught at Stanford University for many years, was famous for his book *How to Solve It* (Princeton University Press). The Mathematics Lab at the University of Idaho was named the Polya Mathematics Learning Center in honor of George Polya. His problem-solving model can be summarized into four basic steps.

Polya's Guidelines for Problem Solving

1. Understand the problem.
2. Devise a plan.
3. Carry out the plan.
4. Look back.

Following is a four-step strategy for solving applied problems with Polya's problem-solving guidelines in mind.

Four-Step Strategy for Problem Solving

Step 1 Read the problem several times until you have an understanding of what is being asked. If possible, create diagrams, charts, or tables to assist you in your understanding.   *Understand the problem.*

Step 2 Pick a variable that describes the unknown quantity that is to be found. All other quantities must be written in terms of that variable. Write an equation using the given information and the variable.   *Devise a plan.*

Step 3 Carefully solve the equation.   *Carry out the plan.*

Step 4 Make sure that you have answered the question and then check all answers to make sure they make sense.   *Look back.*

OBJECTIVE 2: **SOLVING APPLICATIONS INVOLVING UNKNOWN NUMERIC QUANTITIES**

Example 2 Number of Touchdowns Thrown

Roger Staubach and Terry Bradshaw were both quarterbacks in the National Football League. In 1973, Staubach threw three touchdown passes more than twice the number of touchdown passes thrown by Bradshaw. If the total number of touchdown passes between Staubach and Bradshaw was 33, how many touchdown passes did each player throw?

Solution

Step 1 After carefully reading the problem, we see that we are trying to figure out how many touchdown passes were thrown by each player.

Step 2 Let B be the number of touchdown passes thrown by Bradshaw. Because Staubach threw 3 more than twice the number of touchdowns thrown by Bradshaw, then $2B + 3 =$ the number of touchdown passes thrown by Staubach.

The sum of the number of touchdown passes is 33, so we can write the following equation:

$$\underbrace{\text{Bradshaw's touchdown passes}}_{B} + \underbrace{\text{Staubach's touchdown passes}}_{(2B+3)} = \underbrace{\text{Total}}_{33}$$

Step 3   Solve:

$B + (2B + 3) = 33$	*Write the equation.*
$3B + 3 = 33$	*Combine like terms.*
$3B = 30$	*Subtract 3 from both sides.*
$B = 10$	*Divide both sides by 3.*

To answer the question, Bradshaw threw 10 touchdown passes in 1973, and Staubach threw $2(B) + 3 = 2(10) + 3 = 23$ touchdown passes in 1973.

Step 4   Check: We see that the total number of touchdown passes is $33(10 + 23)$. The number of touchdown passes thrown by Staubach is 3 more than two times the number of touchdown passes thrown by Bradshaw.

Work exercises 6–9 in this textbook or in the MyMathLab® Study Plan.

OBJECTIVE 3:   SOLVING APPLICATIONS INVOLVING DECIMAL EQUATIONS (MONEY, MIXTURE, INTEREST)

In Example 2, the equation turned out to be a linear equation involving integer coefficients. In the next two examples, we see that the equations turn out to be linear equations involving decimals. You may want to review the technique used to solve **linear equations involving decimals** that was covered in Section 1.1.

My video summary

▦ Example 3   Money Application

Billy has $16.50 in his piggy bank, consisting of nickels, dimes, and quarters. Billy notices that he has 20 fewer quarters than dimes. If the number of nickels is equal to the number of quarters and dimes combined, how many of each coin does Billy have?

Solution

Step 1   We must find out how many nickels, dimes, and quarters Billy has.

Step 2   Let d = number of dimes.

The remaining quantities must also be expressed in terms of the variable d. Because Billy has 20 fewer quarters than dimes, we can express the number of quarters in terms of the number of dimes, or

$$d - 20 = \text{number of quarters}$$

⬥ CAUTION   The expression $d - 20$ is not equivalent to $20 - d$ because subtraction is not **commutative**, that is, $a - b \neq b - a$.

The number of nickels is equal to the sum of the number of dimes and quarters.

$$\underbrace{d}_{\text{number of dimes}} + \underbrace{(d - 20)}_{\text{number of quarters}} = \text{number of nickels}$$

Nickels are worth $.05, dimes are worth $.10, quarters are worth $.25, and the total amount is $16.50. So, we get the following equation:

$$\underbrace{.05(d + (d - 20))}_{\text{value of the nickels}} + \underbrace{.10d}_{\text{value of the dimes}} + \underbrace{.25(d - 20)}_{\text{value of the quarters}} = \underbrace{16.50}_{\text{Total}}$$

Watch the video to see the rest of the solution, or solve the equation yourself to verify that Billy has 80 nickels, 50 dimes, and 30 quarters.

Work exercises 10 and 11 in this textbook or in the *MyMathLab*® Study Plan.

My video summary

Example 4 Mixture Application

How many milliliters of a 70% acid solution must be mixed with 30 mL of a 40% acid solution to obtain a mixture that is 50% acid?

Solution

Step 1 We are mixing two solutions together to obtain a third solution.

Step 2 The unknown quantity in this problem is the amount (in mL) of a 70% acid solution. So, let x = amount of a 70% acid solution. A diagram may help set up the required equation. See Figure 1.

40% acid solution	70% acid solution	50% acid solution (after mixing)
30 mL	x mL	$(30 + x)$ mL

Figure 1

Because we are mixing 30 mL of one solution with x mL of another, the resulting total quantity will be $(x + 30)$ mL, as in Figure 1. To set up an equation, notice that the number of milliliters of pure acid in the 40% solution plus the number of milliliters of pure acid in the 70% solution must equal the number of milliliters of pure acid in the 50% solution. The equation is as follows:

$$\underbrace{.40(30)}_{\substack{\text{amount of pure}\\\text{acid in the 40\%}\\\text{container}}} + \underbrace{.70x}_{\substack{\text{amount of pure}\\\text{acid in the 70\%}\\\text{container}}} = \underbrace{.50(30 + x)}_{\substack{\text{amount of pure}\\\text{acid in the 50\%}\\\text{container}}}$$

Step 3 Solving this linear equation involving decimals, we get $x = 15$ mL. (Watch the video to see the problem solved in its entirety.) Therefore, we must mix 15 mL of 70% solution to reach the desired mixture.

Work exercises 12–15 in this textbook or in the *MyMathLab*® Study Plan.

My video summary

▦ Example 5 Interest Application

Kristen inherited $20,000 from her Aunt Dawn Ann, with the stipulation that she invest part of the money in an account paying 4.5% **simple interest** and the rest in an account paying 6% simple interest locked in for 3 years. If at the end of 1 year, the total interest earned was $982.50, how much was invested at each rate?

Solution Watch this video to see that Kristen invested $14,500 at 4.5% and $5,500 at $6%.

Work exercises 16 and 17 in this textbook or in the *MyMathLab*® Study Plan.

In Example 6, we use the formula distance = rate × time. It is important to draw a picture of the situation, and then create a table to organize the information. Carefully work through Example 6 by watching the corresponding video.

OBJECTIVE 4: SOLVING APPLIED PROBLEMS INVOLVING DISTANCE, RATE, AND TIME

My video summary

▦ Example 6 Distance, Rate, and Time

Rick left his house on his scooter at 9:00 AM to go fishing. He rode his scooter at an average speed of 10 mph. At 9:15 AM, his girlfriend Deb (who did not find Rick at home) pedaled after Rick on her new 10-speed bicycle at a rate of 15 mph. If Deb caught up with Rick at precisely the time they both reached the fishing hole, how far is it from Rick's house to the fishing hole? At what time did Rick and Deb arrive at the fishing hole?

Solution It is 7.5 mi to the fishing hole from Rick's house. Deb and Rick arrived at the fishing hole at 9:45 AM. Watch the video to see this problem worked out in detail.

Example 7 Distance, Rate, and Time

An airplane that can maintain an average velocity of 320 mph in still air is transporting smokejumpers to a forest fire. On takeoff from the airport, it encounters a headwind and takes 34 minutes to reach the jump site. The return trip from the jump site takes 30 minutes. What is the speed of the wind? How far is it from the airport to the fire?

Solution

Step 1 We are asked to find the speed of the wind, and then to find the distance from the airport to the forest fire.

Step 2 Let w = speed of the wind. Knowing that distance = rate × time, we can create the following table.

The plane flies 320 mph in still air (with no wind). Therefore, when the plane encounters a headwind, the plane's rate will decrease by the amount of the wind. So, the net rate (speed) of the plane into a headwind can be described as $320 - w$. Similarly, the net speed of the plane flying with the tailwind is $320 + w$. We know that the time, in hours, is $\frac{34}{60}$ hours into the headwind and $\frac{30}{60} = \frac{1}{2}$ hour with the tailwind.

	Rate	Time (hours)	Distance
Headwind	$320 - w$	$\frac{34}{60}$	$\frac{34}{60}(320 - w)$
Tailwind	$320 + w$	$\frac{1}{2}$	$\frac{1}{2}(320 + w)$

Because both distances are the same, we can equate the two distances:

$$\frac{34}{60}(320 - w) = \frac{1}{2}(320 + w)$$

distance traveled
by the plane into
the headwind

distance traveled
by the plane with
the tailwind

Step 3 Solve:

$\frac{34}{60}(320 - w) = \frac{1}{2}(320 + w)$	*Rewrite the equation.*
$34(320 - w) = 30(320 + w)$	*Multiply both sides by 60.*
$10,880 - 34w = 9,600 + 30w$	*Use the distributive property.*
$1,280 = 64w$	*Combine like terms.*
$20 = w$	*Divide by 64.*

Step 4 To answer the question, the wind is blowing at a speed of 20 mph. We are also asked to find the distance from the airport to the fire. We can substitute $w = 20$ into either distance expression. Substituting $w = 20$ into the distance traveled with the tailwind expression, we see that the distance from the airport to the fire is as follows:

$$\frac{1}{2}(320 + 20) = \frac{1}{2}(340) = 170 \text{ mi}$$

Work exercises 18–21 in this textbook or in the *MyMathLab*® Study Plan.

OBJECTIVE 5: SOLVING APPLIED WORKING TOGETHER PROBLEMS

Example 8 Working Together Application

Brad and Michelle decide to paint the entire upstairs of their new house. Brad can do the job by himself in 8 hours. If it took them 3 hours to paint the upstairs together, how long would it have taken Michelle to paint it by herself?

Solution

Step 1 We are asked to find the time that it would take Michelle to do the job by herself.

Step 2 Let t = time it takes for Michelle to complete the job; thus, she can complete $\frac{1}{t}$ of the job in an hour. Brad can do this job in 8 hours by himself, so he can complete $\frac{1}{8}$ of the job per hour. Working together it took them 3 hours to complete the job. We can fill out the following table:

	Time needed to complete the job in hours	Portion of job completed in 1 hour (rate)
Brad	8	$\frac{1}{8}$
Michelle	t	$\frac{1}{t}$
Together	3	$\frac{1}{3}$

Step 3 There are two methods that we can use to solve this problem:

Method 1: From the table, we see that $\frac{1}{3}$ = portion of the job completed in 1 hour together. Adding Brad and Michelle's rate together, $\left(\frac{1}{8} + \frac{1}{t}\right)$, we get another expression describing the portion of the job completed together in 1 hour. Thus, equating these rates we obtain the following equation:

$$\left(\frac{1}{8} + \frac{1}{t}\right) = \frac{1}{3}$$

To solve for t, we first multiply by the LCD of $24t$ to clear the fractions:

$$\left(\frac{1}{8} + \frac{1}{t}\right) = \frac{1}{3} \qquad \textit{Rewrite the equation.}$$

$$24t\left(\frac{1}{8} + \frac{1}{t}\right) = \left(\frac{1}{3}\right)24t \qquad \textit{Multiply both sides by 24t.}$$

$$24t \cdot \frac{1}{8} + 24t \cdot \frac{1}{t} = \frac{1}{3} \cdot 24t \qquad \textit{Use the distributive property.}$$

$$3t + 24 = 8t \qquad \textit{Simplify.}$$

$$24 = 5t \qquad \textit{Subtract 3t from both sides.}$$

$$\frac{24}{5} = t \qquad \textit{Divide both sides by 5.}$$

So, $t = \frac{24}{5}$ hours or $4\frac{1}{5}$ hours = 4 hours and 12 minutes ($\frac{1}{5}$ hour $\times \frac{60 \text{ minutes}}{1 \text{ hour}}$ = 12 minutes). Therefore, it would take Michelle 4 hours and 12 minutes to paint the upstairs by herself.

Method 2: Another method is based on the fact that (time working on job) $\times$ (rate to complete job) = fraction of job completed.

In this case, it took Brad and Michelle 3 hours to complete the entire job together. Thus, the fraction of the job completed is equal to 1. The rate to complete the job together was $\left(\frac{1}{8} + \frac{1}{t}\right)$. Therefore, we can set up the following equation:

$$\underbrace{3}_{\substack{\text{time} \\ \text{to complete} \\ \text{the job}}} \cdot \underbrace{\left(\frac{1}{8} + \frac{1}{t}\right)}_{\substack{\text{rate to} \\ \text{complete} \\ \text{job}}} = \underbrace{1}_{\substack{\text{complete} \\ \text{job}}}$$

You should verify that $t = \frac{24}{5}$.

Step 4 The solution makes sense because Michelle's time must be less than Brad's time. Also, we can substitute $t = \frac{24}{5}$ into the method 1 or method 2 equation to see that a true statement results.

Work exercises 22 and 23 in this textbook or in the _MyMathLab_® Study Plan.

Example 9 Working Together Application

Jim and Earl were replacing the transmission on Earl's old convertible. Earl could replace the transmission by himself in 8 hours, whereas it would take Jim 6 hours to do the same job. They worked together for 2 hours, but then Jim had to go to his job at the grocery store. How long did it take Earl to finish replacing the transmission by himself?

Solution Using Method 2 as in Example 8, it would take Earl 3 hours and 20 minutes to finish replacing the transmission. Watch this video to see the entire solution.

Work exercise 24 in this textbook or in the _MyMathLab_® Study Plan.

1.2 **Exercises**

In Exercises 1–5, write the corresponding algebraic expression or equation for each verbal statement. Let n represent the unknown number.

1. 10 more than twice a number

2. 5 less than three times a number and 6

3. The quotient of three and four times a number

4. The product of a number and 2 is 3 less than the quotient of a number and 4

5. Three less than four times the quotient of a number and 7 is equal to 8 less than twice the number

In Exercises 6–24, solve the problems algebraically. Clearly define all variables, write an appropriate equation, and solve.

6. One number is 5 more than twice the other number. If the sum of the two numbers is 26, find the two numbers.

7. The perimeter of a rectangular garden is 32 feet. The length of the garden is 4 feet less than three times the width. Find the dimensions of the garden.

8. The sum of three consecutive even integers is 42. Find the integers.

9. Together, Steve and Tom sold 121 raffle tickets for their school. Steve sold 1 more than twice as many raffle tickets as Tom. How many raffle tickets did each boy sell?

10. Emily has $18.75, consisting of only dimes, quarters, and silver dollars. She has seven times as many dimes as silver dollars and 6 more quarters than dimes. How many of each coin does she have?

11. Shanika has invested in three different stocks. Stock A is currently worth $4.50 per share. Stock B is worth $8.00 per share, while stock C is worth $10.00 per share. Shanika purchased half as many shares of stock B as stock A and half as many shares of stock C as stock B. If her total investments are currently worth $704, how many shares of each stock does she own?

12. A coffee shop owner decides to blend a gourmet brand of coffee with a cheaper brand. The gourmet coffee usually sells for $9.00/lb. The cheaper brand sells for $5.00 lb. How much of each type should he mix in order to have 30 lb of coffee that is worth $6.50/lb?

13. How much of an 80% orange juice drink must be mixed with 20 gallons of a 20% orange juice drink to obtain a mixture that is 50% orange juice?

14. How many liters of pure water should be mixed with a 5-L solution of 80% acid to produce a mixture that is 70% water?

15. Ben recently moved from southern California to northern Alaska. After consulting with a local mechanic, Ben realized that his car radiator fluid should consist of 70% antifreeze. Currently, his 4.5-L radiator is full, with a concentration of 40% antifreeze. How much coolant should be drained and filled with pure antifreeze to reach the desired concentration?

16. A woman has $28,000 to invest in two investments that pay simple interest. One investment pays 4% simple annual interest, and the other pays 5% simple annual interest. How much would she have to invest in each investment if the total interest earned is to be $1,320?

17. Mark works strictly on commission of his gross sales from selling two different products for his company. Last month, his gross sales were $82,000. If he earns 6% commission on product A and 5% commission on product B, what were his gross sales for each product if he earned $4,310 in total commission?

18. A freight train leaves the train station 2 hours before a passenger train. The two trains are traveling in the same direction on parallel tracks. If the rate of the passenger train is 20 mph faster than the freight train, how fast is each train traveling if the passenger train passes the freight train in 3 hours?

19. Manuel traveled 31 hours nonstop to Mexico, a total of 2,300 miles. He took a train part of the way, which averaged 80 mph, and then took a bus the remaining distance, which averaged 60 mph. How long was Manuel on the train?

20. A homing pigeon left his coup to deliver a message. On takeoff, the pigeon encountered a tailwind of 10 mph and made the delivery in 15 minutes. The return trip into the wind took 30 minutes. How fast did the pigeon fly in still air?

21. A grain barge travels on a river from point A to point B loading and unloading grain. The barge travels at a rate of 5 mph relative to the water. The river flows downstream at a rate of 1 mph. If the trip upstream takes 2 hours longer than the trip downstream, how far is it from point A to point B?

22. Tommy can paint a fence in 5 hours. If his friend Huck helps, they can paint the fence in 1 hour. How fast could Huck paint the fence by himself?

23. It takes Joan three times longer than Jane to file the reports. Together, they can file the reports in 15 minutes. How long would it take each woman to file the reports by herself?

24. Two pumps were required to pump the water out of Lakeview in New Orleans after Hurricane Katrina. Pump A, the larger of the two pumps, can pump the water out in 24 hours, whereas it would take pump B 72 hours. Both pumps were working for the first 8 hours until pump A broke down. How long did it take pump B to pump out the remaining water?

1.3 **Complex Numbers**

THINGS TO KNOW

Before working through this section, be sure that you are familiar with the following concept:

VIDEO   ANIMATION   INTERACTIVE

You Try It

1.   Simplifying Radicals
     (Section R.3)

OBJECTIVES

1 Simplifying Powers of i

2 Adding and Subtracting Complex Numbers

3 Multiplying Complex Numbers

4 Finding the Quotient of Complex Numbers

5 Simplifying Radicals with Negative Radicands

My video summary

🎞 THE IMAGINARY UNIT

It is easy to verify that $x = -1$ is the solution to the equation $x + 1 = 0$. Now consider the equation $x^2 + 1 = 0$. Is $x = -1$ a solution to this equation? To check, let's substitute $x = -1$ into the equation $x^2 + 1 = 0$ to see if we get a true statement:

$$x^2 + 1 = 0 \quad \text{Write the original equation.}$$
$$(-1)^2 + 1 \overset{?}{=} 0 \quad \text{Substitute } x = -1.$$
$$1 + 1 \neq 0 \quad \text{The left-hand side does not equal the right-hand side.}$$

In fact, the equation $x^2 + 1 = 0$ has *no real solution* because any **real number** squared added to one is greater than zero. Does this equation have a solution? If so, what is the solution? To answer this question, suppose that the number i is a solution to the equation $x^2 + 1 = 0$. In other words, $i^2 + 1 = 0$ or $i^2 = -1$. This number i, which is a solution to the equation $x^2 + 1 = 0$, is known as the **imaginary unit**.

> **DEFINITION Imaginary Unit**
>
> The imaginary unit, i, is defined by $i = \sqrt{-1}$, or equivalently, $i^2 = -1$.

OBJECTIVE 1:   SIMPLIFYING POWERS OF i

Given the previous definition, we have the following:

$$i = \sqrt{-1} \qquad\qquad i^5 = i^4 \cdot i = (1) \cdot i = i$$
$$i^2 = \left(\sqrt{-1}\right)^2 = -1 \qquad\qquad i^6 = i^4 \cdot i^2 = (1) \cdot (-1) = -1$$
$$i^3 = i^2 \cdot i = (-1) \cdot i = -i \qquad\qquad i^7 = i^4 \cdot i^3 = (1) \cdot (-i) = -i$$
$$i^4 = i^2 \cdot i^2 = (-1) \cdot (-1) = 1 \qquad i^8 = i^4 \cdot i^4 = (1) \cdot (1) = 1$$
$$\vdots$$

You can see the cyclic nature of i. After every fourth power, the cycle repeats itself. In fact, every integer power of i can be written as $i, -1, -i$, or 1.

Example 1 Simplify Powers of i

Simplify each of the following:

   a. i^{43} **b.** i^{100} **c.** i^{-21}

Solution Because $i^4 = 1$, we factor out the largest fourth power of i.

   a. $i^{43} = (i^4)^{10} \cdot i^3 = 1^{10} \cdot i^3 = i^3 = -i$

   b. $i^{100} = (i^4)^{25} = 1^{25} = 1$

   c. $i^{-21} = \dfrac{1}{i^{21}} = \dfrac{1}{(i^4)^5 \cdot i} = \dfrac{1}{(1)^5 \cdot i} = \dfrac{1}{i}$

It appears that we are done and $i^{-21} = \frac{1}{i}$. However, if we multiply this expression by $\frac{i}{i}$, we get the following:

$$i^{-21} = \frac{1}{i} \cdot \frac{i}{i} = \frac{i}{i^2} = \frac{i}{-1} = -i$$

Therefore, $i^{-21} = -i$.

Work exercises 1–5 in this textbook or in the *MyMathLab*® Study Plan.

COMPLEX NUMBERS

My video summary

 Now that we have defined the imaginary unit, it is time to introduce a new type of number known as a **complex number.**

DEFINITION Complex Number

A complex number is a number that can be written in the form $a + bi$, where a and b are real numbers. We often use the variable z to denote a complex number.

If $z = a + bi$ is a complex number, then a is called the **real part**, and b is called the **imaginary part**. For example, $z = \sqrt{3} - 2i$ is a complex number, where $a = \sqrt{3}$ is the real part and $b = -2$ is the imaginary part. The number 5 is also a complex number because 5 can be written as $5 + 0i$. Therefore, every real number is also a complex number.

OBJECTIVE 2:   ADDING AND SUBTRACTING COMPLEX NUMBERS

My video summary

To add or subtract complex numbers, simply combine the real parts and combine the imaginary parts.

Example 2 Adding and Subtracting Complex Numbers

Perform the indicated operations:

a. $(7 - 5i) + (-2 + i)$ **b.** $(7 - 5i) - (-2 + i)$

Solution

a. $(7 - 5i) + (-2 + i)$ *Write the original expression.*
$= 7 - 5i - 2 + i$ *Remove parentheses.*
$= 5 - 4i$ *Combine the real parts and the imaginary parts.*

b. $(7 - 5i) - (-2 + i)$ *Write the original expression.*
$= 7 - 5i + 2 - i$ *Remove the parentheses $[-(-2 + i) = 2 - i]$.*
$= 9 - 6i$ *Combine the real parts and the imaginary parts.*

Work exercises 6–9 in this textbook or in the MyMathLab® Study Plan.

OBJECTIVE 3:   MULTIPLYING COMPLEX NUMBERS

My video summary

When multiplying two complex numbers, treat the problem as if it were the multiplication of two binomials. Just remember that $i^2 = -1$.

Example 3 Multiplying Complex Numbers

Multiply $(4 - 3i)(7 + 5i)$

Solution Treating this multiplication as if it were the multiplication of two binomials, we get the following:

$$(4 - 3i)(7 + 5i) = \begin{aligned} &4(7) + 4(5i) \\ &- (3i)(7) - (3i)(5i) \end{aligned}$$ *Use the distributive property.*
$$= 28 + 20i - 21i - 15i^2$$ *Multiply.*
$$= 28 - i - 15(-1)$$ $20i - 21i = -i$ *and* $i^2 = -1$.
$$= 28 - i + 15$$ $-15(-1) = 15$.
$$= 43 - i$$ *Combine the real parts.*

My video summary

Example 4 Squaring a Complex Number

Simplify $(\sqrt{3} - 5i)^2$

Solution You should verify that the answer is $-22 - 10i\sqrt{3}$. Watch the video to see the worked out solution.

We now define the **complex conjugate**.

DEFINITION Complex Conjugate

The complex conjugate of a complex number $z = a + bi$ is denoted as $\bar{z} = \overline{a + bi} = a - bi$.

For example, the complex conjugate of $z = -2 - 7i$ is $\bar{z} = -2 - (-7i) = -2 + 7i$.

Example 5 Multiplying a Complex Number by Its Complex Conjugate

Multiply the complex number $z = -2 - 7i$ by its complex conjugate $\bar{z} = -2 + 7i$.

Solution $z\bar{z} = (-2 - 7i)(-2 + 7i)$ *Multiply.*

$\qquad = 4 - 14i + 14i - 49i^2$

$\qquad = 4 - 49(-1)$ *Simplify and $i^2 = -1$.*

$\qquad = 4 + 49 = 53$ *Simplify.*

Thus, $(-2 - 7i)(-2 + 7i) = 53$.

Notice in Example 5 that the product of $z = -2 - 7i$ and its complex conjugate $\bar{z} = -2 + 7i$ is a **real number**. In fact, every complex number when multiplied by its complex conjugate will yield a real number. The following theorem is worth remembering:

Theorem

The product of a complex number $z = a + bi$ and its conjugate $\bar{z} = a - bi$ is the real number $a^2 + b^2$. In symbols, $z\bar{z} = a^2 + b^2$.

Click here to view the proof of this theorem.

Work exercises 10–19 in this textbook or in the *MyMathLab*® Study Plan.

OBJECTIVE 4:   FINDING THE QUOTIENT OF COMPLEX NUMBERS

My video summary

The goal when dividing two complex numbers is to eliminate the imaginary part from the denominator and express the quotient in the standard form of $a + bi$. We can do this by multiplying the numerator and denominator by the complex conjugate of the denominator.

Example 6 The Quotient of Complex Numbers

Write the quotient in the form $a + bi$:

$$\frac{1 - 3i}{5 - 2i}$$

Solution $\dfrac{1 - 3i}{5 - 2i} \cdot \dfrac{5 + 2i}{5 + 2i}$ *Multiply numerator and denominator by the complex conjugate. (Note: $\frac{5+2i}{5+2i} = 1$.)*

$\qquad = \dfrac{5 - 13i - 6i^2}{(5)^2 + (2)^2}$ *Multiply the numerators and the denominators. Note: $(a + bi)(a - bi) = a^2 + b^2$.*

$\qquad = \dfrac{5 - 13i - 6(-1)}{29}$ *Simplify and $i^2 = -1$.*

$\qquad = \dfrac{11 - 13i}{29}$ *Simplify.*

$\qquad = \dfrac{11}{29} - \dfrac{13}{29}i$ *Write in the form $a + bi$.*

Work exercises 20–23 in this textbook or in the *MyMathLab*® Study Plan.

OBJECTIVE 5: SIMPLIFYING RADICALS WITH NEGATIVE RADICANDS

In Section 1.4, we solve equations that often lead to solutions with a negative number inside of a radical. Therefore, it is important to learn how to simplify a radical that contains a negative radicand such as $\sqrt{-49}$. We simplify such radicals using the following property, remembering that $\sqrt{-1} = i$.

PROPERTY
If M is a positive real number, then $\sqrt{-M} = \sqrt{-1} \cdot \sqrt{M} = i\sqrt{M}$.

Example 7 Simplify a Square Root with a Negative Radicand

Simplify $\sqrt{-108}$

Solution $\sqrt{-108} = \sqrt{-1} \cdot \sqrt{108} = i\sqrt{108} = 6i\sqrt{3}$

Thus, $\sqrt{-108} = 6i\sqrt{3}$. To review simplifying radicals from Section R.3, click here.

⬦CAUTION⬦ The property $\sqrt{a} \cdot \sqrt{b} = \sqrt{ab}$ is only true for real numbers $\sqrt{a}$ and $\sqrt{b}$. The expressions $\sqrt{a}$ and $\sqrt{b}$ are real numbers when both a and b are greater than or equal to zero. This property does not apply for nonreal numbers. Attempting to use this property with square roots of negative numbers will lead to false results:

$$\sqrt{-3} \cdot \sqrt{-12} = \sqrt{(-3) \cdot (-12)} = \sqrt{36} = 6 \quad \textit{False!}$$
$$\sqrt{-3} \cdot \sqrt{-12} = i\sqrt{3} \cdot i\sqrt{12} = i^2\sqrt{36} = -6 \quad \textit{True!}$$

Example 8 Simplifying Expressions That Contain Negative Radicands

Simplify the following expressions:

a. $\sqrt{-8} + \sqrt{-18}$ b. $\sqrt{-8} \cdot \sqrt{-18}$

c. $\dfrac{6 + \sqrt{(6)^2 - 4(2)(5)}}{2}$ d. $\dfrac{4 \pm \sqrt{-12}}{4}$

Solution

a. $\sqrt{-8} + \sqrt{-18} = 2i\sqrt{2} + 3i\sqrt{2} = 5i\sqrt{2}$

b. $\sqrt{-8} \cdot \sqrt{-18} = \left(2i\sqrt{2}\right) \cdot \left(3i\sqrt{2}\right) = 6i^2(2) = -12$

$$or$$

$$\sqrt{-8} \cdot \sqrt{-18} = \left(i\sqrt{8}\right) \cdot \left(i\sqrt{18}\right) = i^2\sqrt{144} = -12$$

c. $\dfrac{6 + \sqrt{(6)^2 - 4(2)(5)}}{2} = \dfrac{-6 + \sqrt{36 - 40}}{2} = \dfrac{-6 + \sqrt{-4}}{2}$

$$= \dfrac{-6 + 2i}{2} = \dfrac{\cancel{2}(-3 + i)}{\cancel{2}} = -3 + i$$

d. $\dfrac{4 \pm \sqrt{-12}}{4} = \dfrac{4 \pm 2i\sqrt{3}}{4} = \dfrac{2(2 \pm i\sqrt{3})}{4} = \dfrac{\cancel{2}(2 \pm i\sqrt{3})}{\cancel{2} \cdot 2}$

$$= \dfrac{2 \pm i\sqrt{3}}{2}$$

Work exercises 24–31 in this textbook or in the *MyMathLab*® Study Plan.

1.3 Exercises

In Exercises 1–5, write each power of i as i, -1, $-i$, or 1.

1. i^{17} **2.** $i^5 \cdot i^9$ **3.** i^{-6} **4.** i^{-59} **5.** $\dfrac{i^{24}}{i^{-23}}$

In Exercises 6–9, find the desired sum or difference. Write each answer in the form $a + bi$.

6. $(3 - 2i) + (-7 + 9i)$ **7.** $(3 - 2i) - (-7 + 9i)$ **8.** $i - (1 + i)$ **9.** $(\sqrt{2} - 3i) + (2\sqrt{2} + 3i)$

In Exercises 10–14, perform the desired operations. Write each answer in the form $a + bi$.

10. $2i(4 - 3i)$ **11.** $-i(1 - i)$ **12.** $(-2 - i)(3 - 4i)$ **13.** $(6 - 2i)^2$ **14.** $(\sqrt{2} - i)^2$

In Exercises 15–19, find the product of the given complex number and its conjugate.
Write each answer in the form $a + bi$.

15. $5 - 2i$ **16.** $1 - i$ **17.** $\dfrac{1}{2} - 3i$ **18.** $\sqrt{5} + i$ **19.** $\dfrac{\sqrt{3}}{2} + \dfrac{1}{2}i$

In Exercises 20–23, write each quotient in the form $a + bi$.

20. $\dfrac{2 - i}{3 + 4i}$ **21.** $\dfrac{1}{2 - i}$ **22.** $\dfrac{3i}{2 + 2i}$ **23.** $\dfrac{5 + i}{5 - i}$

In Exercises 24–31, write each expression in the form $a + bi$.

24. $\sqrt{-36} - \sqrt{49}$ **25.** $\sqrt{-1} + 3 - \sqrt{-64}$ **26.** $\sqrt{-2} \cdot \sqrt{-18}$ **27.** $(\sqrt{-8})^2$

28. $(i\sqrt{-4})^2$ **29.** $\dfrac{-4 - \sqrt{-20}}{2}$ **30.** $\dfrac{-3 - \sqrt{-81}}{6}$ **31.** $\dfrac{4 + \sqrt{-8}}{4}$

1.4 Quadratic Equations

THINGS TO KNOW

Before working through this section, be sure that you are
familiar with the following concepts:

VIDEO   ANIMATION   INTERACTIVE

 1. Simplifying Radicals (Section R.3)
You Try It

 2. Simplifying Radicals with Negative Radicands
You Try It   (Section 1.3)

 3. Factoring Trinomials with a Leading Coefficient
You Try It   Equal to 1 (Section R.5)

 4. Factoring Trinomials with a Leading Coefficient
You Try It   Not Equal to 1 (Section R.5)

OBJECTIVES

1 Solving Quadratic Equations by Factoring and the Zero Product Property

2 Solving Quadratic Equations Using the Square Root Property

3 Solving Quadratic Equations by Completing the Square

4 Solving Quadratic Equations Using the Quadratic Formula

5 Using the Discriminant to Determine the Type of Solutions of a Quadratic Equation

In Section 1.1, we studied linear equations of the form $ax + b = c, a \neq 0$. These equations are also known as first-order **polynomial** equations. In this section, we learn how to solve second-order polynomial equations. Second-order polynomial equations are called **quadratic equations.**

DEFINITION Quadratic Equation in One Variable

A **quadratic equation in one variable** is an equation that can be written in the form $ax^2 + bx + c = 0$, $a \neq 0$. Quadratic equations in this form are said to be in *standard form.*

OBJECTIVE 1: SOLVING QUADRATIC EQUATIONS BY FACTORING AND THE ZERO PRODUCT PROPERTY

My video summary

Some quadratic equations can be easily solved by factoring and by using the following important property.

PROPERTY Zero Product Property
If $AB = 0$, then $A = 0$ or $B = 0$.

The zero product property says that if two factors multiplied together are equal to zero, then at least one of the factors must be zero. Example 1 shows how this property is used to solve certain quadratic equations.

Example 1 Use Factoring and the Zero Product Property to Solve a Quadratic Equation

Solve $6x^2 - 17x = -12$

Solution First, rewrite the equation in standard form by adding the constant 12 to both sides.

$$6x^2 - 17x = -12 \qquad \text{\textit{Write the original equation.}}$$

$$6x^2 - 17x + 12 = 0 \qquad \text{\textit{Add 12 to both sides.}}$$

$$(3x - 4)(2x - 3) = 0 \qquad \text{\textit{Factor the left-hand side. Click here to review the factoring process.}}$$

$$3x - 4 = 0 \text{ or } 2x - 3 = 0 \qquad \text{\textit{Use the zero product property.}}$$

$$3x = 4 \text{ or } \qquad 2x = 3$$

$$x = \frac{4}{3} \text{ or } \qquad x = \frac{3}{2} \qquad \text{\textit{Solve each equation for x.}}$$

The solution is $\left\{\frac{4}{3}, \frac{3}{2}\right\}$.

Work exercises 1–6 in this textbook or in the *MyMathLab*® Study Plan.

OBJECTIVE 2: SOLVING QUADRATIC EQUATIONS USING THE SQUARE ROOT PROPERTY

My video summary

Consider the quadratic equation $x^2 - 9 = 0$. The left-hand side is a **difference of two squares**, so we can easily solve by factoring.

$$x^2 - 9 = 0 \quad \text{Write the original equation.}$$
$$(x - 3)(x + 3) = 0 \quad \text{Factor using difference of squares formula.}$$
$$x - 3 = 0 \text{ or } x + 3 = 0 \quad \text{Use the zero product property.}$$
$$x = 3 \text{ or } \quad x = -3 \quad \text{Solve each equation for } x.$$

The solution is $x = \pm 3$. In fact, any quadratic equation of the form $x^2 - c = 0$ where $c > 0$ can be solved by factoring:

$$x^2 - c = 0$$
$$(x - \sqrt{c})(x + \sqrt{c}) = 0 \quad \text{Factor using difference of squares formula.}$$
$$x - \sqrt{c} = 0 \quad \text{or } x + \sqrt{c} = 0 \quad \text{Use the zero product property.}$$
$$x = \sqrt{c} \text{ or } x = -\sqrt{c}$$

Although quadratic equations of this form can be solved by factoring, they can be more readily solved by using the following square root property.

PROPERTY Square Root Property
The solution to the quadratic equation $x^2 - c = 0$, or equivalently $x^2 = c$, is $x = \pm\sqrt{c}$.

Example 2 Solve Quadratic Equations Using the Square Root Property

Solve   **a.** $x^2 - 16 = 0$ **b.** $2x^2 + 72 = 0$ **c.** $(x - 1)^2 = 7$

Solution

a. $x^2 - 16 = 0$ *Write the original equation.*
$$x^2 = 16 \quad \text{Add 16 to both sides.}$$
$$x = \pm 4 \quad \text{Use the square root property.}$$

b. $2x^2 + 72 = 0$ *Write the original equation.*
$$2x^2 = -72 \quad \text{Subtract 72 from both sides.}$$
$$x^2 = -36 \quad \text{Divide by 2.}$$
$$\sqrt{x^2} = \pm\sqrt{-36} \quad \text{Use the square root property.}$$
$$x = \pm 6i \quad \text{Simplify the radical with a negative radicand.}$$

c. $(x - 1)^2 = 7$
$$x - 1 = \pm\sqrt{7} \quad \text{Use the square root property.}$$
$$x = 1 \pm \sqrt{7} \quad \text{Add 1 to both sides.}$$

Work exercises 7–11 in this textbook or in the *MyMathLab*® Study Plan.

In Example 2, we see that the quadratic equation $(x - 1)^2 = 7$ can be solved using the square root property. The left side of the equation is a perfect square. In fact, every quadratic equation can be written in the form $(x - h)^2 = k$ using a method known as **completing the square**.

OBJECTIVE 3:   SOLVING QUADRATIC EQUATIONS BY COMPLETING THE SQUARE

Consider the following perfect square trinomials:

$$x^2 + 2x + 1 = (x + 1)^2 \qquad x^2 - 6x + 9 = (x - 3)^2 \qquad x^2 - 7x + \frac{49}{4} = \left(x - \frac{7}{2}\right)^2$$

$$\left(\frac{1}{2} \cdot 2\right)^2 = 1 \qquad\qquad \left(\frac{1}{2} \cdot (-6)\right)^2 = 9 \qquad\qquad \left(\frac{1}{2} \cdot (-7)\right)^2 = \frac{49}{4}$$

In each perfect square trinomial, notice the relationship between the coefficient of the linear term (x-term) and the constant term. The constant term of a perfect square trinomial is equal to the square of $\frac{1}{2}$ the linear coefficient.

Example 3 Completing the Square

What number must be added to each binomial to make it a perfect square trinomial?

  a. $x^2 - 12x$ **b.** $x^2 + 5x$ **c.** $x^2 - \dfrac{3}{2}x$

Solution

  a. The linear coefficient is -12, so we must add $\left(\frac{1}{2}(-12)\right)^2 = (-6)^2 = 36$ to complete the square. Thus, the expression $x^2 - 12x + 36$ is a perfect square trinomial and $x^2 - 12x + 36 = (x - 6)^2$.

  b. $\left(\frac{1}{2} \cdot 5\right)^2 = \left(\frac{5}{2}\right)^2 = \frac{25}{4}$ must be added to complete the square:

$$x^2 + 5x + \tfrac{25}{4} = \left(x + \tfrac{5}{2}\right)^2.$$

  c. $\left(\frac{1}{2}\left(-\frac{3}{2}\right)\right)^2 = \left(-\frac{3}{4}\right)^2 = \frac{9}{16}$ must be added to complete the square:

$$x^2 - \tfrac{3}{2}x + \tfrac{9}{16} = \left(x - \tfrac{3}{4}\right)^2.$$

Work exercises 12–15 in this textbook or in the *MyMathLab*® Study Plan.

To solve a quadratic equation of the form $ax^2 + bx + c = 0, a \neq 0$, by completing the square, follow these steps:

My video summary

> **Steps for Solving $ax^2 + bx + c = 0, a \neq 0$, by Completing the Square**
>
> 1. If $a \neq 1$, divide the equation by a.
> 2. Move all constants to the right-hand side.
> 3. Take half the coefficient of the x-term, square it, and add it to both sides of the equation.
> 4. The left-hand side is now a perfect square. Rewrite it as a binomial squared.
> 5. Use the square root property to solve for x.

Example 4 Solve by Completing the Square

Solve $3x^2 - 18x + 19 = 0$ by completing the square.

Solution

  Step 1 Divide the equation by 3:

$$\frac{3x^2}{3} - \frac{18x}{3} + \frac{19}{3} = \frac{0}{3}$$

$$x^2 - 6x + \frac{19}{3} = 0$$

Step 2 Move all constants to the right-hand side:

$$x^2 - 6x = -\frac{19}{3}$$

Step 3 Take half the coefficient of the x-term, square it, and add it to both sides of the equation:

$$x^2 - 6x + 9 = -\frac{19}{3} + 9 \qquad \left(\frac{1}{2} \cdot (-6)\right)^2 = (-3)^2 = 9$$

$$x^2 - 6x + 9 = \frac{8}{3} \qquad \textit{Simplify.}$$

Step 4 The left-hand side is now a perfect square. Rewrite it as a binomial squared.

$$(x - 3)^2 = \frac{8}{3}$$

Step 5 Use the square root property to solve for x.

$$x - 3 = \pm\sqrt{\frac{8}{3}} \qquad\qquad \textit{Use the square root property.}$$

$$x - 3 = \pm\frac{\sqrt{8}}{\sqrt{3}} \qquad\qquad \sqrt{\frac{a}{b}} = \frac{\sqrt{a}}{\sqrt{b}}$$

$$x - 3 = \pm\frac{\sqrt{8}}{\sqrt{3}} \cdot \frac{\sqrt{3}}{\sqrt{3}} \qquad\qquad \textit{Rationalize the denominator.}$$

$$x - 3 = \pm\frac{\sqrt{24}}{3} \qquad\qquad \textit{Simplify.}$$

$$x - 3 = \pm\frac{2\sqrt{6}}{3} \qquad\qquad \sqrt{24} = 2\sqrt{6}$$

$$x = 3 \pm \frac{2\sqrt{6}}{3} \text{ or } x = \frac{9 \pm 2\sqrt{6}}{3} \qquad \textit{Add 3 to both sides and simplify.}$$

Example 5 Solve by Completing the Square

Solve $2x^2 - 10x - 6 = 0$ by completing the square.

Solution

Step 1 Divide the equation by 2:

$$\frac{2x^2}{2} - \frac{10x}{2} - \frac{6}{2} = \frac{0}{2}$$

$$x^2 - 5x - 3 = 0$$

Step 2 Move all constants to the right-hand side:

$$x^2 - 5x = 3$$

Step 3 Take half the coefficient of the x-term, square it, and add it to both sides of the equation:

$$x^2 - 5x + \frac{25}{4} = 3 + \frac{25}{4} \qquad \left(\frac{1}{2} \cdot (-5)\right)^2 = \left(-\frac{5}{2}\right)^2 = \frac{25}{4}$$

$$x^2 - 5x + \frac{25}{4} = \frac{37}{4} \qquad \textit{Simplify.}$$

Step 4 The left-hand side is now a perfect square. Rewrite it as a binomial squared.

$$\left(x - \frac{5}{2}\right)^2 = \frac{37}{4}$$

Step 5 Use the square root property to solve for x.

$$x - \frac{5}{2} = \pm\sqrt{\frac{37}{4}} \qquad \text{Use the square root property.}$$

$$x - \frac{5}{2} = \pm\frac{\sqrt{37}}{\sqrt{4}} \qquad \sqrt{\frac{a}{b}} = \frac{\sqrt{a}}{\sqrt{b}}$$

$$x - \frac{5}{2} = \pm\frac{\sqrt{37}}{2} \qquad \text{Simplify.}$$

$$x = \frac{5}{2} \pm \frac{\sqrt{37}}{2} = \frac{5 \pm \sqrt{37}}{2} \qquad \text{Add } \frac{5}{2} \text{ to both sides and simplify.}$$

Work exercises 16–20 in this textbook or in the *MyMathLab*® Study Plan.

OBJECTIVE 4: **SOLVING QUADRATIC EQUATIONS USING THE QUADRATIC FORMULA**

In Example 5, we see how to solve a quadratic equation by completing the square. In fact, the method of completing the square can be used to solve *any* quadratic equation. However, this method is often long and tedious. If we solve the general quadratic equation $ax^2 + bx + c = 0, a \neq 0$, by the method of completing the square, we can obtain the following quadratic formula. The quadratic formula can be used to solve any quadratic equation and is often less tedious than the method of completing the square. Be sure to work through the derivation of the quadratic formula to see exactly how this formula is derived.

Quadratic Formula

The solution to the quadratic equation $ax^2 + bx + c = 0, a \neq 0$, is given by the following formula:

$$x = \frac{-b \pm \sqrt{b^2 - 4ac}}{2a}$$

Example 6 Solve a Quadratic Equation Using the Quadratic Formula

Solve $3x^2 + 2x - 2 = 0$ using the quadratic formula.

Solution This quadratic equation is in standard form with $a = 3$, $b = 2$, and $c = -2$. Substitute these values into the quadratic formula.

$$x = \frac{-2 \pm \sqrt{(2)^2 - 4(3)(-2)}}{2(3)} \qquad \text{Use the quadratic formula.}$$

$$= \frac{-2 \pm \sqrt{28}}{6} \qquad \text{Simplify.}$$

$$= \frac{-2 \pm 2\sqrt{7}}{6} \qquad \sqrt{28} = 2\sqrt{7}$$

$$= \frac{2(-1 \pm \sqrt{7})}{6} \qquad \text{Factor.}$$

$$= \frac{-1 \pm \sqrt{7}}{3} \qquad \text{Simplify.}$$

Example 7   Solve a Quadratic Equation Using the Quadratic Formula

Solve $4x^2 - x + 6 = 0$ using the quadratic formula.

Solution This quadratic equation is in standard form with $a = 4$, $b = -1$, and $c = 6$. Substitute these values into the quadratic formula.

$$x = \frac{-(-1) \pm \sqrt{(-1)^2 - 4(4)(6)}}{2(4)} \qquad \textit{Use the quadratic formula.}$$

$$= \frac{1 \pm \sqrt{-95}}{8} \qquad \textit{Simplify.}$$

$$= \frac{1 \pm i\sqrt{95}}{8} \qquad \textit{Simplify the radical with a negative radicand.}$$

Work exercises 21–26 in this textbook or in the *MyMathLab*® Study Plan.

OBJECTIVE 5:   USING THE DISCRIMINANT TO DETERMINE THE TYPE OF SOLUTIONS OF A QUADRATIC EQUATION

My video summary

In Example 7, the quadratic equation $4x^2 - x + 6 = 0$ had two nonreal solutions. The solutions were nonreal because the expression $b^2 - 4ac$ under the radical was a negative number. Given a quadratic equation of the form $ax^2 + bx + c = 0$, the expression $b^2 - 4ac$ is called the **discriminant**. Knowing the value of the discriminant can help us determine the number and nature of the solutions to a quadratic equation.

Discriminant

Given a quadratic equation $ax^2 + bx + c = 0$, $a \neq 0$, the expression $D = b^2 - 4ac$ is called the **discriminant**.

If $D > 0$, then the quadratic equation has two real solutions.
If $D < 0$, then the quadratic equation has two nonreal solutions.
If $D = 0$, then the quadratic equation has exactly one real solution.

Example 8   Use the Discriminant

Use the discriminant to determine the number and nature of the solutions to each of the following quadratic equations:

a. $3x^2 + 2x + 2 = 0$ **b.** $4x^2 + 1 = 4x$

Solution

a. The equation $3x^2 + 2x + 2 = 0$ is in standard form with $a = 3$, $b = 2$, and $c = 2$. $D = 2^2 - 4(3)(2) = 4 - 24 = -20 < 0$. Because the discriminant is less than zero, there are two nonreal solutions.

b. The equation $4x^2 + 1 = 4x$ is not in standard form. To get the equation into standard form, subtract $4x$ from both sides: $4x^2 - 4x + 1 = 0$. $D = (-4)^2 - 4(4)(1) = 16 - 16 = 0$. Because the discriminant is zero, there is exactly one real solution.

Work exercises 27–30 in this textbook or in the *MyMathLab*® Study Plan.

1.4 Exercises

In Exercises 1–6, solve each equation by factoring.

1. $x^2 - 8x = 0$ **2.** $x^2 - x = 6$ **3.** $x^2 - 10x + 24 = 0$

4. $3x^2 + 8x - 3 = 0$ **5.** $8m^2 - 15 = 14m$ **6.** $28z^2 - 13z - 6 = 0$

In Exercises 7–11, solve each equation using the square root property.

7. $x^2 - 64 = 0$ **8.** $x^2 + 64 = 0$ **9.** $3x^2 = 72$

10. $(x + 2)^2 - 9 = 0$ **11.** $(2x + 1)^2 + 4 = 0$

In Exercises 12–15, decide what number must be added to each binomial to make a perfect square trinomial.

12. $x^2 - 8x$ **13.** $x^2 + 10x$ **14.** $x^2 - 7x$ **15.** $x^2 + \dfrac{5}{3}x$

In Exercises 16–20, solve each quadratic equation by completing the square.

16. $x^2 - 8x - 2 = 0$ **17.** $x^2 + 7x + 14 = 0$ **18.** $2x^2 + 6x + 8 = 0$

19. $3x^2 + 24x - 7 = 0$ **20.** $3x^2 + 5x + 12 = 0$

In Exercises 21–26, solve each quadratic equation using the quadratic formula.

21. $3x^2 + 8x - 3 = 0$ **22.** $x^2 - 8x - 2 = 0$ **23.** $4x^2 - x + 8 = 0$

24. $3x^2 - 4x - 1 = 0$ **25.** $9x^2 - 6x = -1$ **26.** $5x^2 + 3x + 1 = 0$

In Exercises 27–30, use the discriminant to determine the number and nature of the solutions to each quadratic equation. Do not solve the equations.

27. $x^2 + 2x + 1 = 0$ **28.** $4x^2 + 4x + 1 = 0$ **29.** $2x^2 + x - 5 = 0$ **30.** $3x^2 + \sqrt{12}x + 4 = 0$

1.5 Applications of Quadratic Equations

THINGS TO KNOW

Before working through this section, be sure that you are familiar with the following concepts:

VIDEO   ANIMATION   INTERACTIVE

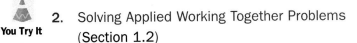

You Try It   **1.** Solving Applied Problems Involving Distance, Rate, and Time
(Section 1.2)

You Try It   **2.** Solving Applied Working Together Problems
(Section 1.2)

   3. Solving Quadratic Equations by Factoring and the Zero Product
You Try It   Property (Section 1.4)

   4. Solving Quadratic Equations by Completing the Square
You Try It   (Section 1.4)

You Try It   **5.** Solving Quadratic Equations Using the Quadratic Formula
(Section 1.4)

OBJECTIVES

1 Solving Applications Involving Unknown Numeric Quantities

2 Using the Projectile Motion Model

3 Solving Geometric Applications

4 Solving Applications Involving Distance, Rate, and Time

5 Solving Working Together Applications

In Section 1.2, we learn how to solve applied problems involving **linear equations**. In this section, we learn to solve applications that involve **quadratic equations**. We follow the same four-step strategy for solving applied problems that is discussed in Section 1.2. The four steps are outlined as follows.

Four-Step Strategy for Problem Solving

Step 1 Read the problem several times until you have an understanding of what is being asked. If possible, create diagrams, charts, or tables to assist you in your understanding. } *Understand the problem.*

Step 2 Pick a variable that describes the unknown quantity that is to be found. All other quantities must be written in terms of that variable. Write an equation using the given information and the variable. } *Devise a plan.*

Step 3 Carefully solve the equation. } *Carry out the plan.*

Step 4 Make sure that you have answered the question and then check all answers to make sure they make sense. } *Look back.*

OBJECTIVE 1:   SOLVING APPLICATIONS INVOLVING UNKNOWN NUMERIC QUANTITIES

Example 1 Find Two Numbers

The product of a number and 1 more than twice the number is 36. Find the two numbers.

Solution

Step 1 We are looking for two numbers.

Step 2 Let $x =$ the first number, and then $2x + 1 =$ the other number. Because their product is 36, we get the equation $x(2x + 1) = 36$, which simplifies to the quadratic equation $2x^2 + x - 36 = 0$.

Step 3 Solve:

$$2x^2 + x - 36 = 0 \qquad \text{\textit{Write the equation.}}$$
$$(2x + 9)(x - 4) = 0 \qquad \text{\textit{Factor.}}$$
$$2x + 9 = 0 \text{ or } x - 4 = 0 \qquad \text{\textit{Use the zero product property.}}$$
$$x = -\frac{9}{2} \text{ or } x = 4 \qquad \text{\textit{Solve.}}$$

Step 4 If $x = -\frac{9}{2}$, then the other number is $2x + 1 = 2\left(-\frac{9}{2}\right) + 1 = -8$.

If $x = 4$, then the other number is $2x + 1 = 2(4) + 1 = 9$.

We can see that there are two possibilities for the two numbers, $-\frac{9}{2}$ and -8 or 4 and 9. In either case, the second number is one more than twice the first, and their products are 36.

Work exercises 1–4 in this textbook or in the *MyMathLab*® Study Plan.

OBJECTIVE 2:   USING THE PROJECTILE MOTION MODEL

An object launched, thrown, or shot vertically into the air with an initial velocity of v_0 meters/second (m/s) from an initial height of h_0 meters above the ground can be modeled by the equation $h = -4.9t^2 + v_0 t + h_0$. The variable h is the height (in meters) of the projectile t seconds after its departure. This equation can be used to model the height of a toy rocket at any time after liftoff, as in Example 2.

Example 2 Launch a Toy Rocket

A toy rocket is launched at an initial velocity of 14.7 m/s from a 49-m-tall platform. The height h of the object at any time t seconds after launch is given by the equation $h = -4.9t^2 + 14.7t + 49$. When will the rocket hit the ground?

Solution

Step 1 We are looking for the time when h is equal to zero.

Step 2 Set h equal to 0, and solve for t: $0 = -4.9t^2 + 14.7t + 49$.

Step 3 Solve:

$$
\begin{aligned}
0 &= -4.9t^2 + 14.7t + 49 &&\textit{Write the original equation.} \\
0 &= t^2 - 3t - 10 &&\textit{Divide both sides by } -4.9. \\
0 &= (t - 5)(t + 2) &&\textit{Factor.} \\
t - 5 &= 0 \text{ or } t + 2 = 0 &&\textit{Use the zero product property.} \\
t &= 5 \text{ or } t = -2 &&\textit{Solve.}
\end{aligned}
$$

Step 4 Because t represents the time (in seconds) after launch, the value $t = -2$ seconds does not make sense. Therefore, the rocket will hit the ground 5 seconds after launch.

Work exercises 5 and 6 in this textbook or in the *MyMathLab*® Study Plan.

OBJECTIVE 3:   SOLVING GEOMETRIC APPLICATIONS

My interactive video summary

▣ Example 3 Find the Dimensions of a Rectangle

The length of a rectangle is 6 in. less than four times the width. Find the dimensions of the rectangle if the area of the rectangle is 54 in.^{2}.

Solution Work through the interactive video to verify that the dimensions of the rectangle are 12 in. by 4.5 in.

Work exercises 7 and 8 in this textbook or in the *MyMathLab*® Study Plan.

1.5 Applications of Quadratic Equations   **29**

Example 4 Find the Width of a High-Definition Television

Jimmy bought a new 40-in. high-definition television. If the length of Jimmy's television is 8 in. longer than the width, find the width of the television.

Solution

Step 1   We are trying to find the width of the television.

Step 2   Let w = width of the television, then $w + 8$ = length of the television. The size of a television is the length of the diagonal, which in this case is 40 in.

We can create an equation using the Pythagorean theorem $a^2 + b^2 = c^2$.

$$w^2 + (w + 8)^2 = 40^2 \quad \text{Use the Pythagorean theorem.}$$

Step 3   Solve:

$$
\begin{aligned}
w^2 + w^2 + 16w + 64 &= 1{,}600 && \text{Square the binomial.} \\
2w^2 + 16w - 1{,}536 &= 0 && \text{Combine like terms.} \\
w^2 + 8w - 768 &= 0 && \text{Divide by 2.} \\
(w - 24)(w + 32) &= 0 && \text{Factor.} \\
w - 24 = 0 \ \text{ or } \ w + 32 &= 0 && \text{Use the zero product property.} \\
w = 24 \ \text{or} \ w &= -32 && \text{Solve for } w.
\end{aligned}
$$

Step 4   Because w represents the width of the television, we can disregard the negative solution. Thus, the width of the television is 24 in.

Work exercise 9 in this textbook or in the MyMathLab® Study Plan.

OBJECTIVE 4:   SOLVING APPLICATIONS INVOLVING DISTANCE, RATE, AND TIME

My video summary

▣ Example 5 Find the Speed of an Airplane

Kevin flew his new Cessna 0-2A airplane from Jonesburg to Mountainview, a distance of 2,560 miles. The average speed for the return trip was 64 mph faster than the average outbound speed. If the total flying time for the round trip was 18 hours, what was the plane's average speed on the outbound trip from Jonesburg to Mountainview?

Solution

Step 1   We are asked to find the average speed of the plane from Jonesburg to Mountainview.

Step 2   Let r = speed of plane on the outbound trip from Jonesburg to Mountainview, and then $r + 64$ = speed of plane on the return trip from Mountainview to Jonesburg.

Step 3   Because distance = rate × time, we know that time = $\frac{\text{distance}}{\text{rate}}$. We can organize our data in the following table.

	Distance	Rate	Time = Distance/Rate
Outbound to Mountainview	2,560	r	$\dfrac{2{,}560}{r}$
Return to Jonesburg	2,560	$r + 64$	$\dfrac{2{,}560}{r + 64}$

The total flying time was 18 hours, so we can set up the following equation:

$$\underbrace{\frac{2{,}560}{r}}_{\substack{\text{Outbound} \\ \text{Time}}} + \underbrace{\frac{2{,}560}{r + 64}}_{\substack{\text{Return} \\ \text{Time}}} = \underbrace{18}_{\substack{\text{Total Flight} \\ \text{Time}}}$$

Solving this equation for r, we get $r = 256$ or $r = -\frac{320}{9} \approx -35.56$. Watch the video to see the solution worked out in detail.

Step 4   Because the rate cannot be negative, the average outbound speed to Mountainview was 256 mph.

Work exercises 10–12 in this textbook or in the *MyMathLab*® Study Plan.

OBJECTIVE 5:   SOLVING WORKING TOGETHER APPLICATIONS

My video summary

▦ **Example 6** Work Together to Create Monthly Sales Reports

Dawn can finish the monthly sales reports in 2 hours less time than it takes Adam. Working together, they were able to finish the sales reports in 8 hours. How long does it take each person to finish the monthly sales reports alone? (Round to the nearest minute.)

Solution

Step 1   We are asked to find the time that it takes each person to finish the monthly sales reports.

Step 2   Let t = Adam's time to do the job alone and $t - 2$ = Dawn's time to do the job alone. We can now set up a table with the given data.

	Time needed to complete the job in hours	Portion of job completed in 1 hour (rate)
Adam	t	$\dfrac{1}{t}$
Dawn	$t - 2$	$\dfrac{1}{t - 2}$
Together	8	$\dfrac{1}{8}$

We can now use techniques discussed in Section 1.2 to set up the following equation:

$$\underbrace{\frac{1}{t}}_{\textit{Adam's Rate}} + \underbrace{\frac{1}{t - 2}}_{\textit{Dawn's Rate}} = \underbrace{\frac{1}{8}}_{\textit{Rate Together}}$$

Step 3   Solving for t, we see that Adam can do the job by himself in approximately 17 hours and 4 minutes, while it would take Dawn approximately 15 hours and 4 minutes to do the same job. To see this problem worked out in detail, watch this video.

Step 4   You should verify that Dawn's time is exactly 2 hours less than Adam's time and that the sum of their rates is equal to $\frac{1}{8}$.

Work exercises 13–15 in this textbook or in the *MyMathLab*® Study Plan.

1.5 Exercises

1. The product of some negative number and 5 less than three times that number is 12. Find the number.

2. The square of a number plus the number is 132. What is the number?

3. The sum of the square of a number and the square of 5 more than a number is 169. What is the number?

4. Three consecutive odd integers are such that the square of the third integer is 15 more than the sum of the squares of the first two. Find the integers.

5. A toy rocket is launched from a 2.8-m-high platform in such a way that its height, h (in meters), after t seconds is given by the equation $h = -4.9t^2 + 18.9t + 2.8$. How long will it take for the rocket to hit the ground?

6. Benjamin threw a rock straight up from a cliff that was 24 ft above the water. If the height of the rock h, in feet, after t seconds is given by the equation $h = -16t^2 + 20t + 24$, how long will it take for the rock to hit the water?

7. The width of a rectangle is 1 cm less than three times the width. If the area of the rectangle is 30 cm^2, find the dimensions of the rectangle.

8. The length of a rectangle is 1 in. less than twice the width. If the diagonal is 2 in. more than the length, find the dimensions of the rectangle.

9. A 35- by 20-ft rectangular swimming pool is surrounded by a walkway of uniform width. If the total area of the walkway is 434 ft^2, how wide is the walkway?

10. A boat traveled downstream a distance of 45 mi and then came right back. If the speed of the current is 5 mph and if the total trip took 3 hours and 45 minutes, find the average speed of the boat relative to the water.

11. Meg rowed her boat upstream a distance of 9 mi and then rowed back to the starting point. The total time of the trip was 10 hours. If the rate of the current was 4 mph, find the average speed of the boat relative to the water.

12. Imogene's car traveled 280 mi averaging a certain speed. If the car had gone 5 mph faster, the trip would have taken 1 hour less. Find the average speed.

13. Twin brothers, Billy and Bobby, can mow their grandparent's lawn together in 56 minutes. Billy could mow the lawn by himself in 15 minutes less time than it would take Bobby. How long would it take Bobby to mow the lawn by himself?

14. Jeff and Kirk can build a 75-ft. retaining wall together in 12 hours. Because Jeff has more experience, he could build the wall by himself 4 hours quicker than Kirk. How long would it take Kirk (to the nearest minute) to build the wall by himself?

15. Homer and Mike were replacing the boards on Mike's old deck. Mike can do the job alone in 1 hour less time than Homer. They worked together for 3 hours until Homer had to go home. Mike finished the job working by himself in an additional 4 hours. How long would it have taken Homer to fix the deck by himself? (*Hint:* Use method 2, as discussed in Section 1.2.)

1.6 Other Types of Equations

THINGS TO KNOW

Before working through this section, be sure that you are familiar with the following concepts:

VIDEO   ANIMATION   INTERACTIVE

 1. Factoring Trinomials with a Leading Coefficient Equal to 1 (Section R.5)

 2. Factoring Trinomials with a Leading Coefficient Not Equal to 1 (Section R.5)

 3. Factoring Polynomials by Grouping (Section R.5)

 4. Solving Quadratic Equations by Factoring and the Zero Product Property (Section 1.4)

5. Solving Quadratic Equations Using the Quadratic Formula (Section 1.4)

OBJECTIVES

1 Solving Higher-Order Polynomial Equations

2 Solving Equations That Are Quadratic in Form (Disguised Quadratics)

3 Solving Equations Involving Radicals

OBJECTIVE 1:   SOLVING HIGHER-ORDER POLYNOMIAL EQUATIONS

My video summary

▶ So far in this text, we have learned methods for solving **linear equations** and **quadratic equations**. Linear equations and quadratic equations in one variable are both examples of **polynomial equations** of first and second degree, respectively. In this section, we first start by looking at certain higher-order polynomial equations that can be solved using special factoring techniques.

One useful technique is to factor out an expression included in each term.

Example 1 Remove the Common Factor

Find all solutions of the equation $3x^3 - 2x = -5x^2$.

Solution First, add $5x^2$ to both sides to set the equation equal to 0.

$$3x^3 + 5x^2 - 2x = 0$$

Now, try to factor the left-hand side.

$$x(3x^2 + 5x - 2) = 0 \quad \text{Factor out the greatest common factor, } x.$$
$$x(3x - 1)(x + 2) = 0 \quad \text{Factor the quadratic.}$$
$$x = 0 \text{ or } 3x - 1 = 0 \text{ or } x + 2 = 0 \quad \text{Use the zero product property.}$$
$$x = 0 \text{ or } \qquad x = \frac{1}{3} \text{ or } \qquad x = -2 \quad \text{Solve for } x.$$

The solution set is $\left\{-2, 0, \frac{1}{3}\right\}$.

CAUTION   In the equation $3x^3 + 5x^2 - 2x = 0$, do *not* divide both sides by x. This would produce the equation $3x^2 + 5x - 2 = 0$, which has only $\frac{1}{3}$ and -2 as solutions. The solution $x = 0$ would be "lost." In addition, because $x = 0$ is a solution of the original equation, dividing by x would mean dividing by 0, which of course is undefined and produces incorrect results.

Sometimes polynomials can be solved by grouping terms and factoring (especially when the polynomial has four terms). You may want to review how to **factor by grouping**, which was discussed in Section R.5.

Example 2 Factor by Grouping to Solve a Polynomial Equation

Find all solutions of the equation $2x^3 - x^2 + 8x - 4 = 0$.

Solution Arrange the terms of the polynomial in descending order, and group the terms of the polynomial in pairs.

$$2x^3 - x^2 + 8x - 4 = 0 \qquad \text{\textit{Write the original equation.}}$$
$$(2x^3 - x^2) + (8x - 4) = 0 \qquad \text{\textit{Group the terms in pairs.}}$$
$$x^2(2x - 1) + 4(2x - 1) = 0 \qquad \text{\textit{Factor the G.C.F from each pair.}}$$
$$(2x - 1)(x^2 + 4) = 0 \qquad \text{\textit{Factor } 2x - 1 \text{ from the left-hand side.}}$$
$$2x - 1 = 0 \quad \text{or} \quad x^2 + 4 = 0 \qquad \text{\textit{Use the zero product property.}}$$
$$2x = 1 \quad \text{or} \qquad x^2 = -4 \qquad \text{\textit{Solve for } x.}$$
$$x = \frac{1}{2} \quad \text{or} \ x = \pm\sqrt{-4} = \pm 2i \qquad \text{\textit{Simplify the radical with a negative radicand.}}$$

The solution set is $\left\{-2i, 2i, \frac{1}{2}\right\}$.

Work exercises 1–6 in this textbook or in the MyMathLab® Study Plan.

OBJECTIVE 2:   **SOLVING EQUATIONS THAT ARE QUADRATIC IN FORM (DISGUISED QUADRATICS)**

Quadratic equations of the form $ax^2 + bx + c = 0, a \neq 0$, are relatively straight-forward to solve because we know several methods for solving quadratics. Sometimes equations that are not quadratic can be made into a quadratic equation by using a substitution. Equations of this type are said to be *quadratic in form* or *disguised quadratics*. These equations typically have the form $au^2 + bu + c = 0, a \neq 0$, after an appropriate substitution. Following are a few examples.

Original Equation	Make an Appropriate Substitution			New Equation Is a Quadratic
$2x^4 - 11x^2 + 12 = 0$	$\xrightarrow[\text{proper substitution}]{\text{Determine the}}$	Let $u = x^2,$	then $u^2 = x^4.$	$\xrightarrow[\text{a quadratic}]{\text{New equation is}}$ $2u^2 - 11u + 12 = 0$
$\left(\dfrac{1}{x-2}\right)^2 + \dfrac{2}{x-2} - 15 = 0$	$\xrightarrow[\text{proper substitution}]{\text{Determine the}}$	Let $u = \dfrac{1}{x-2},$	then $u^2 = \left(\dfrac{1}{x-2}\right)^2.$	$\xrightarrow[\text{a quadratic}]{\text{New equation is}}$ $u^2 + 2u - 15 = 0$
$x^{\frac{2}{3}} - 9x^{\frac{1}{3}} + 8 = 0$	$\xrightarrow[\text{proper substitution}]{\text{Determine the}}$	Let $u = x^{\frac{1}{3}},$	then $u^2 = \left(x^{\frac{1}{3}}\right)^2 = x^{\frac{2}{3}}.$	$\xrightarrow[\text{a quadratic}]{\text{New equation is}}$ $u^2 - 9u + 8 = 0$
$3x^{-2} - 5x^{-1} - 2 = 0$	$\xrightarrow[\text{proper substitution}]{\text{Determine the}}$	Let $u = x^{-1} = \dfrac{1}{x},$	then $u^2 = (x^{-1})^2 = x^{-2}.$	$\xrightarrow[\text{a quadratic}]{\text{New equation is}}$ $3u^2 - 5u - 2 = 0$

We now solve each of these equations listed in Example 3:

My interactive video summary

▶ **Example 3 Solve Equations That Are Quadratic in Form**

Solve each of the following:

a. $2x^4 - 11x^2 + 12 = 0$

b. $\left(\dfrac{1}{x-2}\right)^2 + \dfrac{2}{x-2} - 15 = 0$

c. $x^{\frac{2}{3}} - 9x^{\frac{1}{3}} + 8 = 0$

d. $3x^{-2} - 5x^{-1} - 2 = 0$

Solution

    a. $2x^4 - 11x^2 + 12 = 0$ *Write the original equation.*

    Let $u = x^2$ then $u^2 = x^4$.   *Determine the proper substitution.*

    $2u^2 - 11u + 12 = 0$ *Substitute u for x^2 and u^2 for x^4 into the original equation.*

    $(2u - 3)(u - 4) = 0$ *Factor.*

    $2u - 3 = 0$ or $u - 4 = 0$ *Use the zero product property.*

    $u = \dfrac{3}{2}$ or $u = 4$ *Solve for u.*

Don't stop yet! We want to solve for x. Because we know that $u = x^2$, we get

$$x^2 = \frac{3}{2} \text{ or } x^2 = 4, \text{ so } x = \pm\sqrt{\frac{3}{2}} = \pm\frac{\sqrt{6}}{2} \text{ or } x = \pm 2$$

Try to find the solutions to (b) through (d). Watch the interactive video to see each worked out solution.

Work exercises 7–13 in this textbook or in the *MyMathLab*® Study Plan.

OBJECTIVE 3:   SOLVING EQUATIONS INVOLVING RADICALS

A radical equation is an equation that involves a variable inside a square root, cube root, or any higher root. To solve these equations, we must try to isolate the radical and then raise each side of the equation to the appropriate power to eliminate the radical. Consider the following examples:

Example 4 Solve an Equation Involving a Square Root

Solve $\sqrt{x - 1} - 2 = x - 9$

Solution First, we isolate the radical on the left side of the equation, adding 2 to both sides.

$$\sqrt{x - 1} = x - 7 \qquad \text{*Add 2 to both sides.*}$$

To eliminate the radical, square both sides of the equation.

    $(\sqrt{x - 1})^2 = (x - 7)^2$ *Square both sides of the equation.*

    $x - 1 = x^2 - 14x + 49$ *Expand the right-hand side:*
                                  $(x - 7)^2 = (x - 7)(x - 7) = x^2 - 14x + 49.$

    $0 = x^2 - 15x + 50$ *Move all terms to the right-hand side.*

    $0 = (x - 5)(x - 10)$ *Factor.*

Setting each factor equal to zero gives us $x = 5$ or $x = 10$ as possible solutions. It is tempting to stop here. When solving equations involving radicals it is important to always check each potential solution by substituting back into the original equation.

Check $x = 10$: $\sqrt{x - 1} - 2 = x - 9$ Check $x = 5$: $\sqrt{x - 1} - 2 = x - 9$

                  $\sqrt{10 - 1} - 2 \overset{?}{=} 10 - 9$ $\sqrt{5 - 1} - 2 \overset{?}{=} 5 - 9$

                        $\sqrt{9} - 2 \overset{?}{=} 1$ $\sqrt{4} - 2 \overset{?}{=} -4$

                        $3 - 2 \overset{?}{=} 1$ $2 - 2 \overset{?}{=} -4$

                        $1 = 1$ ✓ $0 \neq -4$ ✕

We see that $x = 10$ checks, but $x = 5$ does not. Therefore, the solution set to this equation is $\{10\}$. $x = 5$ is called an **extraneous solution**.

CAUTION Because the squaring operation can make a false statement true ($-2 \neq 2$, but $(-2)^2 = (2)^2$), it is essential to always check your answers after solving an equation in which this operation was performed. The squaring operation is usually used when solving an equation involving square roots.

CAUTION Be careful when squaring an expression of the form $(a + b)^2$ or $(a - b)^2$. Remember, $(a + b)^2 = (a + b)(a + b) = a^2 + 2ab + b^2$ and $(a - b)^2 = (a - b)(a - b) = a^2 - 2ab + b^2$.

My video summary

Example 5 Solve an Equation Involving Two Square Roots

Solve $\sqrt{2x + 3} + \sqrt{x - 2} = 4$

Solution When solving equations containing multiple radicals, first isolate one of the radicals and square both sides:

$$\sqrt{2x + 3} = 4 - \sqrt{x - 2}$$
$$(\sqrt{2x + 3})^2 = (4 - \sqrt{x - 2})^2$$

Watch the video to see how to obtain the solution of $x = 3$.

Example 6 Solve an Equation Involving a Cube Root

Solve $\sqrt[3]{1 - 4x} + 3 = 0$

Solution To eliminate the radical, we first isolate the radical and then cube both sides.

$$\sqrt[3]{1 - 4x} = -3$$
$$(\sqrt[3]{1 - 4x})^3 = (-3)^3$$
$$1 - 4x = -27$$
$$-4x = -28$$
$$x = 7$$

Check $\sqrt[3]{1 - 4(7)} + 3 = \sqrt[3]{-27} + 3 = -3 + 3 = 0$ ✓

Work exercises 14–20 in this textbook or in the *MyMathLab*® Study Plan.

1.6 Exercises

In Exercises 1–6, find all solutions.

1. $x^3 + x = 0$
2. $2x^3 + 4x^2 - 30x = 0$
3. $x^2(x^2 - 1) - 9(x^2 - 1) = 0$
4. $x^3 - x^2 + 3x = 3$
5. $2x^3 + 3x^2 = 8x + 12$
6. $x^3 - x^2 - x = -1$

In Exercises 7–13, solve the equation after making an appropriate substitution.

7. $x^4 - 6x^2 + 8 = 0$
8. $(13x - 1)^2 - 2(13x - 1) - 3 = 0$
9. $2x^{2/3} - 5x^{1/3} + 2 = 0$
10. $x^6 - x^3 = 6$
11. $\sqrt{x} - 3\sqrt[4]{x} - 4 = 0$
12. $3\left(\dfrac{1}{x - 1}\right)^2 - \dfrac{5}{x - 1} - 2 = 0$
13. $2x^{-2} - 3x^{-1} - 2 = 0$

In Exercises 14–20, solve each radical equation.

14. $\sqrt{2x - 1} = 3$

15. $\sqrt{32 - 4x} - x = 0$

16. $x - 1 = \sqrt{2x + 1}$

17. $\sqrt{3x + 1} - \sqrt{x + 4} = 1$

18. $\sqrt{6 + 5x} + \sqrt{3x + 4} = 2$

19. $\sqrt[3]{2x - 1} = -2$

20. $\sqrt[4]{x^2 - 6x} - 2 = 0$

1.7 **Linear Inequalities**

THINGS TO KNOW

Before working through this section, be sure that you are familiar with the following concepts:

VIDEO   ANIMATION   INTERACTIVE

1. Describing Intervals of Real Numbers
    (Section R.1)

2. Understanding the Intersection and Union
    of Sets (Section R.1)

OBJECTIVES

1 Solving Linear Inequalities

2 Solving Three-Part Inequalities

3 Solving Compound Inequalities

4 Solving Linear Inequality Word Problems

Unlike equations that usually have a finite number of solutions (or no solution at all), inequalities often have infinitely many solutions. For instance, the inequality $2x - 3 \leq 5$ has infinitely many solutions because there are infinite values of x for which the inequality is true. Take, for example, the numbers $-1, \frac{7}{2}$, and 4. Substituting each of these values into the inequality $2x - 3 \leq 5$ yields a true statement:

Inequality $2x - 3 \leq 5$
$x = -1 \quad 2(-1) - 3 \overset{?}{\leq} 5$
$-5 \leq 5 \quad$ *True statement!*
$x = \frac{7}{2} \quad 2\left(\frac{7}{2}\right) - 3 \overset{?}{\leq} 5$
$4 \leq 5 \quad$ *True statement!*
$x = 4 \quad 2(4) - 3 \overset{?}{\leq} 5$
$5 \leq 5 \quad$ *True statement!*

In fact, any real number x that is less than or equal to 4 makes this inequality a true statement. We certainly cannot list all solutions to this inequality individually. Thus, we typically describe the solutions to an inequality in one of three ways:

1. Graph the solution on a number line.

2. Write the solution in set-builder notation.

3. Write the solution in interval notation.

Figure 2 shows the solution to the inequality in these three ways. In this section, you learn how to solve such inequalities. You may want to review the terms set-builder notation and interval notation, which are described in Section R.1.

Solution graphed on a number line:

Solution in set-builder notation: $\{x \mid x \le 4\}$

Solution in interval notation: $(-\infty, 4]$

Figure 2 Solution to the inequality $2x - 3 \le 5$

OBJECTIVE 1: SOLVING LINEAR INEQUALITIES

The inequality $2x - 3 \le 5$ is called a **linear inequality**. In this section, you learn techniques necessary to solve linear inequalities.

DEFINITION

A **linear inequality** is an inequality that can be written in the form $ax + b < c$, where $a, b,$ and c are real numbers and $a \ne 0$.

Note: The inequality symbol "$<$" in the previous definition and the following properties can be replaced with $>$, $\le$, or $\ge$.

Before we solve a linear inequality, it is important to introduce some inequality properties:

Properties of Inequalities

Let $a, b,$ and c be real numbers:

Property	In Words	Example
1. If $a < b$, then $a + c < b + c$	The same number may be added to both sides of an inequality.	$-3 < 7$ $-3 + 4 < 7 + 4$ $1 < 11$
2. If $a < b$, then $a - c < b - c$	The same number may be subtracted from both sides of an inequality.	$9 \ge 2$ $9 - 6 \ge 2 - 6$ $3 \ge -4$
3. For $c > 0$, if $a < b$, then $ac < bc$	Multiplying both sides of an inequality by a *positive* number *does not reverse the direction* of the inequality.	$3 > 2$ $(3)(5) > (2)(5)$ $15 > 10$
4. For $c < 0$, if $a < b$, then $ac > bc$	Multiplying both sides of an inequality by a *negative* number *reverses the direction* of the inequality.	$3 > 2$ $(3)(-5) < (2)(-5)$ $-15 < -10$
5. For $c > 0$, if $a < b$, then $\dfrac{a}{c} < \dfrac{b}{c}$	Dividing both sides of an inequality by a *positive* number *does not reverse the direction* of the inequality.	$6 > 4$ $\dfrac{6}{2} > \dfrac{4}{2}$ $3 > 2$
6. For $c < 0$, if $a < b$, then $\dfrac{a}{c} > \dfrac{b}{c}$	Dividing both sides of an inequality by a *negative* number *reverses the direction* of the inequality.	$6 > 4$ $\dfrac{6}{-2} < \dfrac{4}{-2}$ $-3 < -2$

Example 1 Solve a Linear Inequality

Solve the inequality $-9x - 3 \geq 7 - 4x$. Graph the solution set on a number line, and express the answer in interval notation.

Solution The technique to use when solving linear inequalities is to isolate the variable on one side.

$$
\begin{aligned}
-9x - 3 &\geq 7 - 4x \\
-9x + 4x - 3 &\geq 7 - 4x + 4x && \text{\textit{Add } 4x \text{ \textit{to both sides.}}} \\
-5x - 3 &\geq 7 && \text{\textit{Simplify.}} \\
-5x - 3 + 3 &\geq 7 + 3 && \text{\textit{Add 3 to both sides.}} \\
-5x &\geq 10 && \text{\textit{Simplify.}} \\
\frac{-5x}{-5} &\leq \frac{10}{-5} && \text{\textit{Divide by } -5\text{. Reverse the direction}} \\
&&& \text{\textit{of the inequality!}} \\
x &\leq -2 && \text{\textit{Simplify.}}
\end{aligned}
$$

The graph of the solution is depicted in Figure 3. The solution set in interval notation is $(-\infty, -2]$.

Solution graphed on number line:

Solution in interval notation: $(-\infty, -2]$.

Figure 3 Solution to the inequality $-9x - 3 \geq 7 - 4x$

> **CAUTION**
> Remember to reverse the direction of the inequality when multiplying or dividing a linear inequality by a negative number.

My video summary

▣ Example 2 Solve a Linear Inequality

Solve the inequality $2 - 5(x - 2) < 4(3 - 2x) + 7$. Express the answer in set-builder notation.

Solution The solution in set-builder notation is $\left\{ x \mid x < \frac{7}{3} \right\}$. Watch the video to see the solution worked out in its entirety.

Work exercises 1–8 in this textbook or in the *MyMathLab*® Study Plan.

OBJECTIVE 2:   SOLVING THREE-PART INEQUALITIES

The weight of a normal adult llama is between 250 lb and 450 lb. This fact can be expressed by the following *three-part inequality* $250 < W < 450$, where W is the weight of a normal adult llama. Notice that the variable W is between two inequality symbols. The technique to use when solving three-part inequalities is to simplify until the variable is "sandwiched" in the middle, as in our example, $250 < W < 450$.

Example 3 Solve a Three-Part Inequality

Solve the inequality $-2 \le \frac{2 - 4x}{3} < 5$. Graph the solution set on a number line, and write the solution in set-builder notation.

Solution We start by multiplying all three parts by 3 to eliminate the fraction.

$$-2 \le \frac{2 - 4x}{3} < 5 \qquad \textit{Write the original three-part inequality.}$$

$$3 \cdot (-2) \le 3 \cdot \frac{2 - 4x}{3} < 3 \cdot 5 \qquad \textit{Multiply by 3.}$$

$$-6 \le 2 - 4x < 15 \qquad \textit{Simplify.}$$

$$-6 - 2 \le 2 - 2 - 4x < 15 - 2 \qquad \textit{Subtract 2.}$$

$$-8 \le -4x < 13 \qquad \textit{Simplify.}$$

$$\frac{-8}{-4} \ge \frac{-4x}{-4} > \frac{13}{-4} \qquad \textit{Divide by } -4, \textit{ and reverse the direction of the inequalities.}$$

$$2 \ge x > -\frac{13}{4} \qquad \textit{Simplify.}$$

Although this answer is adequate, it is good practice to rewrite the inequality so that the variable is in the middle and the smaller of the two outside numbers is on the left.

$$-\frac{13}{4} < x \le 2 \qquad \textit{Rewrite the inequality.}$$

We can write the solution in set-builder notation as $\left\{ x \mid -\frac{13}{4} < x \le 2 \right\}$. The graph of the solution is displayed in Figure 4.

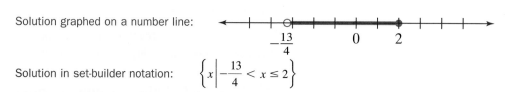

Solution graphed on a number line:

Solution in set-builder notation: $\left\{ x \mid -\frac{13}{4} < x \le 2 \right\}$

Figure 4 Solution to the three-part inequality $-2 \le \frac{2 - 4x}{3} < 5$

Work exercises 9–14 in this textbook or in the *MyMathLab*® Study Plan.

OBJECTIVE 3:   SOLVING COMPOUND INEQUALITIES

Before we solve compound inequalities, it might be a good idea to review the concepts of intersection and union that were introduced in the Review chapter. Click here to review the intersection and union of sets.

   Compound inequalities are two inequalities that are joined together by the words *and* or *or*. A number is a solution to a compound inequality involving *and*, if it is a solution to *both* inequalities. A number is a solution to a compound inequality involving *or*, if it is a solution to *either* inequality.

Example 4 Solve a Compound Inequality with *and*

Solve $2x - 7 < -1$ and $3x + 5 \geq 3$. Graph the solution set, and write the solution in interval notation.

Solution To solve a compound inequality involving the word *and*, first solve each inequality independently and then find the intersection of the two solution sets.

$$2x - 7 < -1 \text{ and } 3x + 5 \geq 3$$
$$2x < 6 \quad \text{and} \quad 3x \geq -2$$
$$x < 3 \quad \text{and} \quad x \geq -\frac{2}{3}$$

We can now graph each solution on a number line:

The solution to this compound inequality is the *intersection* of these two sets. In other words, we are looking for all values of x common to both solutions. The values in common are all points to the right of and including $-2/3$ and to the left of and not including 3, as depicted in Figure 5.

Solution graphed on a number line:

Solution in interval notation: $\left[-\frac{2}{3}, 3\right)$

Figure 5 Solution to the compound inequality $2x - 7 < -1$ and $3x + 5 \geq 3$

Example 5 Solve a Compound Inequality with *or*

Solve $1 - 3x \geq 7$ or $3x + 4 > 7$. Graph the solution set, and write the solution in interval notation.

Solution Solving each inequality independently, we get $x \leq -2$ or $x > 1$. (Click here to see this worked out solution.) We now graph each solution on a number line:

$(-\infty, -2]$

$(1, \infty)$

The solution to this compound inequality will be the *union* of these two sets. To express the solution in interval notation, we join the two intervals with the *union* symbol $\cup$, as shown in Figure 6.

Solution graphed on a number line:

$$-2 \quad 0 \quad 1$$

Solution in interval notation: $(-\infty, -2] \cup (1, \infty)$

Figure 6 Solution to the compound inequality $1 - 3x \geq 7$ and $3x + 4 > 7$

Example 6 Example of an Inequality with No Solution

Solve $3x - 1 < -7$ and $4x + 1 > 9$.

Solution We solve each inequality independently:

$$3x - 1 < -7 \text{ and } 4x + 1 > 9$$
$$3x < -6 \text{ and } 4x > 8$$
$$x < -2 \text{ and } x > 2$$

$\{x \mid x < -2\}$

$\{x \mid x > 2\}$

These graphs do not intersect.

The solution is the intersection of all points to the left of -2 and all points to the right of 2. Because these two sets *do not intersect*, there is no solution to this inequality. We say that the solution set is empty, and use the symbol $\varnothing$ to represent the empty set.

Work exercises 15–19 in this textbook or in the *MyMathLab*® Study Plan.

OBJECTIVE 4:   SOLVING LINEAR INEQUALITY WORD PROBLEMS

Example 7 Solve a Linear Inequality Word Problem

Suppose you rented a forklift to move a pallet with 70-lb blocks stacked on it. The forklift can carry a maximum of 2,500 lb. If the pallet weighs 50 lb by itself with no blocks, how many blocks can be stacked on a pallet and lifted by the forklift?

Solution Let x be the number of blocks that can be stacked and lifted by the forklift. Each block weighs 70 lb, so $70x$ is the total weight in blocks that can be moved. The pallet by itself weighs 50 lb, so we can set up the following inequality:

$$\underbrace{50}_{\text{50 lb for the pallet}} + \underbrace{70x}_{\text{70 lb per block}} \underbrace{\leq}_{\substack{\text{is less than} \\ \text{or equal to}}} \underbrace{2{,}500}_{\text{2,500 lb}}$$

Solving this inequality, we get $x \leq 35$. So, the forklift can lift no more than 35 blocks at a time. Click here to see the worked out solution.

Work exercises 20–23 in this textbook or in the *MyMathLab*® Study Plan.

Example 8 Solve a Three-Part Inequality Word Problem

The perimeter of a rectangular fence is to be at least 80 feet and no more than 140 feet. If the width of the fence is 12 feet, what is the range of values for the length of the fence?

Solution Let l be the length of the rectangle, as shown in Figure 7. The perimeter, P, of the rectangle is $P = 2(12) + 2l = 24 + 2l$.

l **Figure 7**

The perimeter must be greater than or equal to 80 and less than or equal to 140 or $80 \leq P \leq 140$. Because the perimeter is $24 + 2l$, we get the following three-part inequality $80 \leq 24 + 2l \leq 140$.

Now, solve this three-part inequality:

$$80 \leq 24 + 2l \leq 140 \qquad \text{\textit{Rewrite the three-part inequality.}}$$
$$80 - 24 \leq 24 - 24 + 2l \leq 140 - 24 \qquad \text{\textit{Subtract 24.}}$$
$$56 \leq 2l \leq 116 \qquad \text{\textit{Simplify.}}$$
$$\frac{56}{2} \leq \frac{2l}{2} \leq \frac{116}{2} \qquad \text{\textit{Divide by 2.}}$$
$$28 \leq l \leq 58 \qquad \text{\textit{Simplify.}}$$

Therefore, the length of the fence must be between 28 feet and 58 feet inclusive.

Work exercises 24 and 25 in this textbook or in the *MyMathLab*® **Study Plan.**

1.7 Exercises

In Exercises 1–4, solve each linear inequality. Express each solution in set-builder notation.

1. $4x + 2 > -8$ **2.** $2(y + 1) + 3 < -3(y - 2)$

3. $11 - 2a \leq 4a + 35$ **4.** $\frac{1}{2}w - 1 > 4 - \frac{2}{3}w$

In Exercises 5–8, solve each linear inequality. Graph the solution set on a number line, and express each solution using interval notation.

5. $-5x - 3 \leq 2x + 4$ **6.** $3(t - 1) - 5t \geq t + 6$

7. $-2a - 3 > 9 + 3a$ **8.** $\frac{4n - 1}{3} < 3 - n$

In Exercises 9–14, solve each three-part linear inequality. Graph the solution set on a number line, and express each solution using interval notation.

9. $-2 \leq 2m - 5 \leq 2$ **10.** $-3 < 1 - 2W < 4$ **11.** $\frac{1}{4} < \frac{t + 1}{12} \leq 1$

12. $-1 < \frac{3 - x}{2} < 5$ **13.** $-\frac{5}{6} \leq x + 1 < \frac{1}{2}$ **14.** $0 \leq -6 - (a + 3) \leq 1$

In Exercises 15–19, solve each compound inequality. Express each solution using interval notation or state that there is no solution.

15. $2x - 3 \le 5$ and $5x + 1 > 6$

16. $4k + 2 \le -10$ or $3k - 4 > 8$

17. $\dfrac{n - 1}{3} \ge 1$ and $\dfrac{4n - 2}{2} \le 9$

18. $3b - 5 \le 1$ and $1 - b < -3$

19. $2y - 1 \le 7$ or $3y - 5 > 1$

In Exercises 20–25, set up the appropriate inequality and solve.

20. Derek can spend no more than $300 to pay for a limousine. The limousine rental company charges $200 to rent a limo, plus $20 for each hour of use. What is the maximum amount of time that he can rent the limo?

21. Jason and Sadie went to Cancun for a week. They wanted to rent a car, so they looked at two different rental car companies. Company A charges $40 per day plus 5 cents per mile. Company B charges $200 per week plus 30 cents per mile. How many miles must they drive in order for the price of company A to be less than the price of company B?

22. Your scores on your first three algebra exams were 91, 92, and 85. What must you score on exam four so that the average of the four test scores is at least 90.

23. The heights of the two starting guards and two starting forwards on a basketball team are 6′2″, 5′11″, 6′7″, and 6′9″. How tall must the starting center be in order for the starting players to have an average height of at least 6′6″? (*Hint*: First convert all heights to inches.)

24. The perimeter of a rectangular fence is to be at least 100 feet and no more than 180 feet. If the width of the fence is 22 feet, what is the range of values for the length of the fence?

25. The perimeter of a rectangular fence is to be at least 120 feet and no more than 180 feet. If the length of the fence is to be twice the width, what is the range of values for the width of the fence?

1.8   Absolute Value Equations and Inequalities

THINGS TO KNOW

Before working through this section, be sure that you are familiar with the following concepts:

VIDEO   ANIMATION   INTERACTIVE

You Try It

1.   Solving Three-Part Inequalities (Section 1.7)

You Try It

2.   Solving Compound Inequalities (Section 1.7)

OBJECTIVES

1 Solving an Absolute Value Equation

2 Solving an Absolute Value "Less Than" Inequality

3 Solving an Absolute Value "Greater Than" Inequality

My video summary

In Section R.1, we see that the absolute value of a number a, written as $|a|$, represents the **distance** from a number a to zero on the number line. Consider the equation $|x| = 5$. To solve for x, we must find all values of x that are five units away from zero on the number line. The two numbers that are five units away from zero on the number line are $x = -5$ and $x = 5$, as shown in Figure 8.

If $|x| = 5$, then $x = -5$ or $x = 5$.
The solution set is $\{-5, 5\}$.

Figure 8 Solution to $|x| = 5$

The solution to the inequality $|x| < 5$ consists of all values of x whose distance from zero is less than five units on the number line. See Figure 9.

If $|x| < 5$, then $-5 < x < 5$.
The solution set is $\{x \,|\, -5 < x < 5\}$ in set-builder notation or $(-5, 5)$ in interval notation.

Figure 9 Solution to $|x| < 5$

Figure 10 shows the solution to the inequality $|x| > 5$. Notice that we are now looking for all values of x that are more than five units away from zero. The solution is the set of all values of x greater than 5 combined with the set of all values of x less than -5.

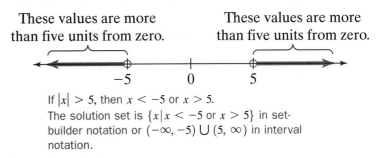

If $|x| > 5$, then $x < -5$ or $x > 5$.
The solution set is $\{x \,|\, x < -5 \text{ or } x > 5\}$ in set-builder notation or $(-\infty, -5) \cup (5, \infty)$ in interval notation.

Figure 10 Solution to $|x| < 5$

Table 1 illustrates how to solve absolute value equations and inequalities. Each absolute value equation and inequality in Table 1 is in standard form. When solving an absolute value equation or inequality, it is necessary to first rewrite it in standard form.

1.8 Absolute Value Equations and Inequalities   45

Table 1 ABSOLUTE VALUE EQUATIONS AND INEQUALITY PROPERTIES

Let u be an algebraic expression and c be a real number such that $c > 0$, then

1. $|u| = c$ is equivalent to $u = -c$ or $u = c$.

2. $|u| < c$ is equivalent to $-c < u < c$.

3. $|u| > c$ is equivalent to $u < -c$ or $u > c$.

OBJECTIVE 1:   SOLVING AN ABSOLUTE VALUE EQUATION

My video summary

▤ Example 1 Solve an Absolute Value Equation

Solve $|1 - 3x| = 4$

Solution This equation is in standard form. Therefore, by property 1, the equation $|1 - 3x| = 4$ is equivalent to the following two equations:

$$1 - 3x = 4 \quad \text{or } 1 - 3x = -4 \qquad \textit{Use property 1.}$$
$$-3x = 3 \quad \text{or} \quad -3x = -5 \qquad \textit{Subtract 1 from both sides.}$$
$$x = -1 \text{ or} \quad x = \frac{5}{3} \qquad \textit{Divide both sides by } -3.$$

The solution set is $\left\{-1, \frac{5}{3}\right\}$. See Figure 11.

Figure 11 Graph of the solution to $|1 - 3x| = 4$

Work exercises 1–6 in this textbook or in the **_MyMathLab_**® Study Plan.

OBJECTIVE 2:   SOLVING AN ABSOLUTE VALUE "LESS THAN" INEQUALITY

▤ Example 2 Solve an Absolute Value *Less Than* Inequality

Solve $|4x - 3| + 2 \leq 7$

Solution This inequality is not in standard form. Rewrite this inequality in standard form and then use property 2 to form a three-part inequality.

$$|4x - 3| + 2 \leq 7 \qquad \textit{Write the original inequality.}$$
$$|4x - 3| \leq 5 \qquad \textit{Subtract 2 from both sides.}$$
$$-5 \leq 4x - 3 \leq 5 \qquad \textit{Use property 2, and rewrite as a three-part inequality.}$$
$$-2 \leq 4x \leq 8 \qquad \textit{Add 3 to all three parts and simplify.}$$
$$-\frac{1}{2} \leq x \leq 2 \qquad \textit{Divide all three parts by 4 and simplify.}$$

The solution is $\left\{x \mid -\frac{1}{2} \leq x \leq 2\right\}$ in set-builder notation or $\left[-\frac{1}{2}, 2\right]$ in interval notation. See Figure 12.

Figure 12 Graph of the solution to $|4x - 3| + 2 \leq 5$

Work exercises 7–12 in this textbook or in the MyMathLab® Study Plan.

OBJECTIVE 3:   SOLVING AN ABSOLUTE VALUE "GREATER THAN" INEQUALITY

My video summary

 Example 3 Solve an Absolute Value *Greater Than* Inequality

Solve $|5x + 1| > 3$

Solution This inequality is in standard form. Therefore, we can use property 3 to rewrite the inequality as a compound inequality and solve.

$$|5x + 1| > 3 \qquad \text{Write the original inequality.}$$

$$5x + 1 < -3 \text{ or } 5x + 1 > 3 \qquad \text{Use property 3, and rewrite as compound inequality.}$$

$$5x < -4 \text{ or } \qquad 5x > 2 \qquad \text{Subtract 1 from both sides.}$$

$$x < -\frac{4}{5} \text{ or } \qquad x > \frac{2}{5} \qquad \text{Divide both sides by 5.}$$

The solution is $\left\{ x \mid x < -\frac{4}{5} \text{ or } x > \frac{2}{5} \right\}$ in set-builder notation or $\left(-\infty, -\frac{4}{5}\right) \cup \left(\frac{2}{5}, \infty\right)$ in interval notation. See Figure 13.

Figure 13 Graph of the solution to $|5x + 1| > 3$

CAUTION   In Example 3, $|5x + 1| > 3$ is *not* equivalent to $-3 > 5x + 1 > 3$. In addition, a common error in this type of problem is to write $5x + 1 > -3$ for the first inequality, instead of $5x + 1 < -3$. Think carefully about the meaning of the inequality before writing it.

Work exercises 13–18 in this textbook or in the MyMathLab® Study Plan.

In Examples 1–3, (and in Table 1), the absolute value equation or inequality is written in standard form, and the constant on the right-hand side is *positive*. Example 4 illustrates how to solve absolute value equations and inequalities written in standard form when the right-hand side is negative or zero.

Example 4 Absolute Value Equations and Inequalities Involving Zero or Negative Constants

Solve each of the following:

a. $|3x - 2| = 0$ b. $|x + 6| = -4$ c. $|7x + 5| \leq 0$

d. $|3 - 4x| < -6$ e. $|8x - 3| > 0$ f. $|1 - 9x| \geq -5$

1.8 Absolute Value Equations and Inequalities   47

Solution

a. The expression $|3x - 2|$ will equal zero only when $3x - 2 = 0$ or when $x = \frac{2}{3}$. The solution set is $\left\{\frac{2}{3}\right\}$.

b. The absolute value of an expression can never be less than zero. Therefore, there is no solution to this equation. The solution set is $\varnothing$.

c. The expression $|7x + 5|$ can never be less than zero. However, $|7x + 5|$ is equal to zero when $7x + 5 = 0$ or when $x = -\frac{5}{7}$. The solution set is $\left\{-\frac{5}{7}\right\}$.

d. The absolute value of an expression can never be less than zero. Therefore, there is no solution to this inequality. The solution set is $\varnothing$.

e. The expression $|8x - 3|$ is equal to zero when $x = \frac{3}{8}$ and is greater than zero for all other values of x. Therefore, the solution set in set-builder notation is $\left\{x \mid x \neq \frac{3}{8}\right\}$ or $\left(-\infty, \frac{3}{8}\right) \cup \left(\frac{3}{8}, \infty\right)$.

f. The absolute value of an expression is always nonnegative. Therefore, $|1 - 9x| \geq -5$ is true for all values of x. The solution is all real numbers or $(-\infty, \infty)$.

1.8 Exercises

In Exercises 1–6, solve each equation.

1. $|x + 1| = 1$ **2.** $|6x + 7| - 8 = 3$ **3.** $\left|\dfrac{3x + 1}{5}\right| = \dfrac{2}{3}$

4. $|x^2 - 2x| = 1$ **5.** $|7x + 3| = 0$ **6.** $|4x - 1| = -5$

In Exercises 7–12, solve each inequality. Express each solution using interval notation.

7. $|x - 2| < 3$ **8.** $|5x - 1| + 7 \leq 9$ **9.** $|3 - 4x| < 11$

10. $\left|\dfrac{2x - 5}{3}\right| \leq \dfrac{1}{5}$ **11.** $|3x - 7| \leq 0$ **12.** $|x - 1| < -2$

In Exercises 13–18, solve each inequality. Express each solution using interval notation.

13. $|x + 4| > 5$ **14.** $|7x - 1| - 5 \geq 8$ **15.** $\left|\dfrac{2x - 6}{7}\right| \geq \dfrac{3}{14}$

16. $|3 - 5x| \geq 2$ **17.** $|7x - 5| > 0$ **18.** $|x + 4| \geq -1$

1.9 Polynomial and Rational Inequalities

THINGS TO KNOW

Before working through this section, be sure that you are familiar with the following concepts:

VIDEO   ANIMATION   INTERACTIVE

You Try It

1. Factoring Trinomials with a Leading Coefficient Equal to 1 (Section R.5)

You Try It

2. Factoring Trinomials with a Leading Coefficient Not Equal to 1 (Section R.5)

You Try It

3. Factoring Polynomials by Grouping (Section R.5)

You Try It

4. Solving Higher-Order Polynomial Equations (Section 1.6)

OBJECTIVES

1 Solving Polynomial Inequalities

2 Solving Rational Inequalities

In Section 1.7, we learned how to solve linear inequalities. In this section, we learn how to solve **polynomial inequalities**.

OBJECTIVE 1:   SOLVING POLYNOMIAL INEQUALITIES

My video summary

⊞ **Example 1** Solve a Polynomial Inequality

Solve $x^3 - 3x^2 + 2x \geq 0$

Solution First, factor the left-hand side to get $x(x - 1)(x - 2) \geq 0$. Second, find all real values of x that make the expression on the left *equal to 0*. These values are called **boundary points**. To find the boundary points, set the factored polynomial on the left equal to 0 and solve for x.

$$x(x - 1)(x - 2) = 0 \qquad \text{Set the factored polynomial equal to 0.}$$

$$x = 0 \text{ or } x - 1 = 0 \text{ or } x - 2 = 0 \qquad \text{Use the zero product property.}$$

The boundary points are $x = 0$, $x = 1$, and $x = 2$.

Now plot each boundary point on a number line. Because the expression $x(x - 1)(x - 2)$ is equal to 0 at our three boundary points, we use a solid circle ● at each of the boundary points to indicate that the inequality $x(x - 1)(x - 2) \geq 0$ is satisfied at these points. (*Note:* If the inequality would have been a strict inequality such as $>$ or $<$, we would have used an open circle ○ to represent that the boundary points were *not* part of the solution. See Example 2). Notice in Figure 14 that we have naturally divided the number line into four intervals.

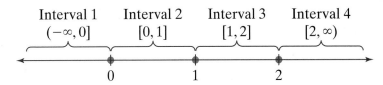

Figure 14

The expression $x(x - 1)(x - 2)$ is equal to zero *only* at the three boundary points 0, 1, and 2. This means that in any of the four intervals shown in Figure 14, the expression $x(x - 1)(x - 2)$ must be either *positive* or *negative* throughout the entire interval. To check whether this expression is positive or negative on each interval, pick a number from each interval called a **test value**. Substitute this test value into the expression $x(x - 1)(x - 2)$, and check to see if it yields a positive or negative value. Possible test values are plotted below in Figure 15.

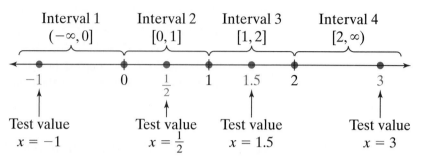

Figure 15

Interval	Test Value	Substitute Test Value into $x(x-1)(x+2)$	Comment
1. $(-\infty, 0]$	$x = -1$	$(-1)(-1-1)(-1-2) \Rightarrow (-)(-)(-) = -$	Expression is negative on $(-\infty, 0]$
2. $[0, 1]$	$x = \dfrac{1}{2}$	$\left(\dfrac{1}{2}\right)\left(\dfrac{1}{2}-1\right)\left(\dfrac{1}{2}-2\right) \Rightarrow (+)(-)(-) = +$	Expression is positive on $[0, 1]$
3. $[1, 2]$	$x = 1.5$	$(1.5)(1.5-1)(1.5-2) \Rightarrow (+)(+)(-) = -$	Expression is negative on $[1, 2]$
4. $[2, \infty)$	$x = 3$	$(3)(3-1)(3-2) \quad\; \Rightarrow (+)(+)(+) = +$	Expression is positive on $[2, \infty)$

If the expression $x(x-1)(x-2)$ is positive on an interval, we place a "+" above the interval on the number line. If the expression $x(x-1)(x-2)$ is negative, we place a "−" above the interval. See Figure 16.

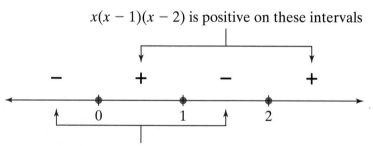

$x(x-1)(x-2)$ is positive on these intervals

$x(x-1)(x-2)$ is negative on these intervals

Figure 16

From the number line in Figure 3, we see that $x(x-1)(x-2)$ is greater than or equal to zero on the interval $[0, 1] \cup [2, \infty)$ which is precisely the solution to the original inequality $x^3 - 3x^2 + 2x \geq 0$.

Example 1 illustrates the following steps for solving a polynomial inequality that can be factored.

Steps for Solving Polynomial Inequalities

Step 1 Move all terms to one side of the inequality leaving zero on the other side.

Step 2 Factor the nonzero side of the inequality.

Step 3 Find all boundary points by setting the factored polynomial equal to zero.

Step 4 Plot the boundary points on a number line. If the inequality is $\le$ or $\ge$, then use a solid circle ●. If the inequality is $<$ or $>$, then use an open circle ○.

Step 5 Now that the number line is divided into intervals, pick a test value from each interval.

Step 6 Substitute the test value into the polynomial, and determine whether the expression is positive or negative on the interval.

Step 7 Determine the intervals that satisfy the inequality.

Example 2 Solve a Polynomial Inequality

Solve $x^2 + 5x < 3 - x^2$

Solution

Step 1 Move all terms to one side of the inequality leaving zero on the other side.

$$2x^2 + 5x - 3 < 0$$

Step 2 Factor the nonzero side of the inequality.

$$(2x - 1)(x + 3) < 0$$

Step 3 Find all boundary points by setting the factored polynomial equal to zero.

The boundary points are $x = -3$ and $x = \frac{1}{2}$ because these are the values that are solutions to $(2x - 1)(x + 3) = 0$.

Step 4 Plot the boundary points on a number line.

Note that we use open circles on our boundary points because these points are *not* part of the solution. (We are looking for values that make the expression strictly less than zero.)

Step 5 Now that the number line is divided into three intervals, pick a test value from each interval.

Interval 1: Test value $x = -4$

Interval 2: Test value $x = 0$

Interval 3: Test value $x = 1$

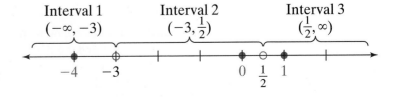

Step 6 **Substitute the test value into the polynomial, and determine whether the expression is positive or negative on the interval.**

$$x = -4: (2(-4) - 1)((-4) + 3) \Rightarrow (-)(-) = +,$$

expression is positive on $(-\infty, -3)$.

$$x = 0: (2(0) - 1)((0) + 3) \Rightarrow (-)(+) = -,$$

expression is negative on $\left(-3, \dfrac{1}{2}\right)$.

$$x = 1: (2(1) - 1)((1) + 3) \Rightarrow (+)(+) = +,$$

expression is positive on $\left(\dfrac{1}{2}, \infty\right)$.

$(2x - 1)(x + 3)$ is negative on this interval

$(2x - 1)(x + 3)$ is positive on these intervals

Step 7 **Determine the intervals that satisfy the inequality.**

Because we are looking for values of x that are less than zero (negative values), the solution must be the interval $\left(-3, \frac{1}{2}\right)$.

Work exercises 1–10 in this textbook or in the *MyMathLab*® Study Plan.

OBJECTIVE 2:   SOLVING RATIONAL INEQUALITIES

My video summary

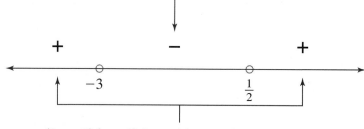 A **rational inequality** can be solved using a technique similar to the one used in Examples 1 and 2, except that the boundary points are found by setting both the polynomial in the numerator and the denominator equal to zero.

Example 3 Solve a Rational Inequality

Solve $\dfrac{x - 4}{x + 1} \geq 0$

Solution Because the inequality is already in completely factored form, we can skip steps 1 and 2 and go right to step 3.

Step 3 **Find all boundary points by setting the factored polynomial in the numerator and the denominator equal to zero.**

Numerator: $x - 4 = 0$, so $x = 4$ is a boundary point.
Denominator: $x + 1 = 0$, so $x = -1$ is a boundary point.

Step 4 **Plot the boundary points on a number line.**

The inequality is a greater than *or equal to* inequality; thus, the boundary point $x = 4$ is represented by a closed circle. Because division by zero is *never* permitted, the boundary points that are found by setting the polynomial in the denominator equal to zero must always be represented by an open circle on the number line.

Step 5 **Now that the number line is divided into three intervals, pick a test value from each interval.**

Interval 1: Test value $x = -2$

Interval 2: Test value $x = 0$

Interval 3: Test value $x = 5$

Step 6 **Substitute each test value into the rational expression and determine whether the expression is positive or negative on the interval.**

$$x = -2: \frac{(-2 - 4)}{(-2 + 1)} \Rightarrow \frac{(-)}{(-)} = +,$$

expression is positive on the interval $(-\infty, -1)$.

$$x = 0: \frac{(0 - 4)}{(0 + 1)} \Rightarrow \frac{(-)}{(+)} = -,$$

expression is negative on the interval $(-1, 4]$.

$$x = 5: \frac{(5 - 4)}{(5 + 1)} \Rightarrow \frac{(+)}{(+)} = +,$$

expression is positive on the interval $[4, \infty)$.

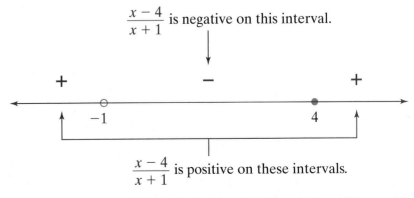

1.9 Polynomial and Rational Inequalities **53**

Step 7 Determine the intervals that satisfy the inequality.

Finally, because we are looking for values of x that are greater than or equal to zero, looking at the number line above we see that the solution to the inequality is $(-\infty, -1) \cup [4, \infty)$.

Note that we include 4 as a solution but not -1 because $x = -1$ makes the denominator equal to 0.

My video summary

📹 **Example 4 Solve a Rational Inequality with Nonzero Factors on Both Sides of the Inequality**

Solve $x > \dfrac{3}{x-2}$

Solution

Step 1 Move all terms to one side of the inequality, leaving zero on the other side.

$$x - \frac{3}{x-2} > 0$$

⬧**CAUTION** You cannot multiply both sides of the inequality by $x - 2$ to clear the fraction. This is because we do not know whether $x - 2$ is negative or positive; therefore, we do not know whether we would need to reverse the direction of the inequality.

Now that all nonzero terms are on the left-hand side, we must combine the terms by getting a common denominator:

$$x - \frac{3}{x-2} > 0 \qquad \textit{Rewrite the inequality.}$$

$$\frac{x(x-2)}{x-2} - \frac{3}{x-2} > 0 \qquad \textit{Rewrite using a common denominator of } x - 2.$$

$$\frac{x^2 - 2x - 3}{x-2} > 0 \qquad \textit{Combine terms.}$$

Step 2 Factor the nonzero side of the inequality.

$$\frac{(x-3)(x+1)}{x-2} > 0$$

Finish steps 3–7 on your own, and see if you can come up with the correct solution of $(-1, 2) \cup (3, \infty)$. Watch this video to see the entire solution.

Work exercises 11–20 in this textbook or in the *MyMathLab*® Study Plan.

1.9 **Exercises**

In Exercises 1–10, solve each polynomial inequality. Express each solution using interval notation.

1. $(x-1)(x+3) \geq 0$ **2.** $x^2 + 3x - 21 < 0$ **3.** $x(3x+2) \leq 0$ **4.** $2x^2 - 4x > 0$

5. $x^2 \leq 1$ **6.** $(x-1)(x+4)(x-3) \geq 0$ **7.** $3x^2 + x < 3x - 1$ **8.** $2x^3 > 24x - 2x^2$

9. $x^3 + x^2 - x \leq 1$ **10.** $x^3 \geq -2x^2 - x$

In Exercises 11–20, solve each rational inequality. Express each solution using interval notation.

11. $\dfrac{x + 3}{x - 1} \leq 0$

12. $\dfrac{2 - x}{3x + 9} \geq 0$

13. $\dfrac{x}{x - 1} > 0$

14. $\dfrac{x^2 - 9}{x + 2} \leq 0$

15. $\dfrac{4}{x + 1} \geq 2$

16. $\dfrac{x + 5}{2x - 3} > 1$

17. $\dfrac{x^2 - 8}{x + 4} \geq x$

18. $\dfrac{x}{2 - x} \leq \dfrac{1}{x}$

19. $\dfrac{x - 1}{x - 2} \geq \dfrac{x + 2}{x + 3}$

20. $\dfrac{x - 1}{x + 1} + \dfrac{x + 1}{x - 1} \leq \dfrac{x + 5}{x^2 - 1}$

CHAPTER TWO

The Rectangular Coordinate System, Lines, and Circles

CHAPTER TWO CONTENTS

2.1 The Rectangular Coordinate System

THINGS TO KNOW

Before working through this section, be sure that you are familiar with the following concept:

VIDEO   ANIMATION   INTERACTIVE

You Try It

1. Simplifying Radicals (Section R.3)

OBJECTIVES

1 Plotting Ordered Pairs

2 Graphing Equations by Plotting Points

3 Finding the Midpoint of a Line Segment Using the Midpoint Formula

4 Finding the Distance between Two Points Using the Distance Formula

Throughout chapter one, we solved several types of equations, including linear equations, quadratic equations, and rational equations. Each equation had something in common. They were all examples of **equations in one variable**. In this chapter, we study **equations involving two variables**. To motivate the discussion, consider the two-variable equation $x - 2y = -8$. What does it mean for this equation to have a solution? A solution to this equation consists of a *pair of numbers*, that is, an *x*-value and a *y*-value for which the equation is true.

For example, the pair of numbers $x = 2$ and $y = 5$ is a solution to this equation because substituting these two values into the equation produces a true statement:

$$x - 2y = -8 \qquad \textit{Write the two-variable equation.}$$
$$(2) - 2(5) \overset{?}{=} -8 \qquad \textit{Substitute } x = 2 \textit{ and } y = 5.$$
$$-8 = -8 \checkmark \qquad \textit{The statement is true.}$$

The pair of numbers $x = -2$ and $y = 3$ are also solutions to this equation:

$$x - 2y = -8 \qquad \textit{Write the two-variable equation.}$$
$$(-2) - 2(3) \overset{?}{=} -8 \qquad \textit{Substitute } x = -2 \textit{ and } y = 3.$$
$$-8 = -8 \checkmark \qquad \textit{The statement is true.}$$

In fact, there are infinitely many pairs of numbers that satisfy this equation. Table 1 illustrates a few solutions to this equation.

Table 1

x	y	$x - 2y = -8$
-2	3	$-2 - 2(3) = -8$
0	4	$0 - 2(4) = -8$
2	5	$2 - 2(5) = -8$
4	6	$4 - 2(6) = -8$

Each pair of values is called an **ordered pair** because you can see that the order does matter. We use the notation (x, y) to represent an ordered pair. Notice that the *x*-coordinate or **abscissa** is first, followed by the *y*-coordinate or **ordinate** listed second. To represent an ordered pair graphically, we use the rectangular coordinate system, also called the Cartesian coordinate system, named after the French mathematician **René Descartes**. The plane used in this system is called the **coordinate plane** or **Cartesian plane**. The horizontal axis (*x*-axis) and the vertical axis (*y*-axis) intersect at the **origin** and naturally divide the plane into four quadrants, labeled quadrants I, II, III, and IV. See Figure 1.

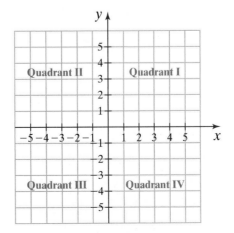

Figure 1 Rectangular coordinate system

2.1 The Rectangular Coordinate System   57

OBJECTIVE 1:   PLOTTING ORDERED PAIRS

My video summary

▣ Example 1 Plot Ordered Pairs

Plot the ordered pairs $(-2, 3)$, $(0, 4)$, $(2, 5)$, and $(4, 6)$, and state in which quadrant or on which axis each pair lies.

Solution To plot the point $(-2, 3)$, go two units to the left of the origin on the *x*-axis, and then move three units up, parallel to the *y*-axis. The point corresponding to the ordered pair $(-2, 3)$ is labeled A in Figure 2 and is located in quadrant II. To plot the point $(0, 4)$, start at the origin and move up four units. The point corresponding to the ordered pair $(0, 4)$ is labeled B in Figure 2 and lies on the *y*-axis. The points $(2, 5)$ and $(4, 6)$ are labeled C and D and are both located in quadrant I.

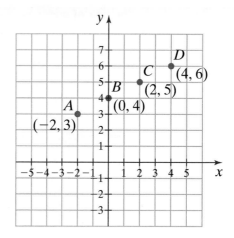

Figure 2

The four ordered pairs plotted in Figure 2 were all solutions to the equation $x - 2y = -8$. Connecting these points forms the **graph of the equation** $x - 2y = -8$. See Figure 3. The graph of the equation $x - 2y = -8$ is a straight line. We study the equations of lines extensively in Section 2.3.

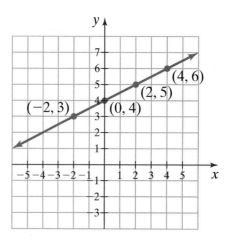

Figure 3 Graph of $x - 2y = -8$

Work exercises 1 and 2 in this textbook or in the *MyMathLab*® Study Plan.

OBJECTIVE 2:   GRAPHING EQUATIONS BY PLOTTING POINTS

Not every equation in two variables has a graph that is a straight line. One way to sketch the graph of an equation is to find several ordered pairs that satisfy the equation, plot those ordered pairs, and then connect the points with a smooth curve. We choose arbitrary values for one of the coordinates and then solve the equation for the other coordinate.

Example 2 Graph an Equation by Plotting Points

Sketch the graph of $y = x^2$.

Solution Arbitrarily choose several values of x and then find the corresponding values for y:

Let $x = -2$, then $y = (-2)^2 = 4$, and thus the ordered pair $(-2, 4)$ satisfies the equation $y = x^2$.

Let $x = -1$, then $y = (-1)^2 = 1$, and thus the ordered pair $(-1, 1)$ satisfies the equation $y = x^2$.

Let $x = 0$, then $y = (0)^2 = 0$, and thus the ordered pair $(0, 0)$ satisfies the equation $y = x^2$.

Table 2 shows several more ordered pairs that are solutions to the equation $y = x^2$. The graph of $y = x^2$ is called a parabola and is shown in Figure 4.

Table 2

x	$y = x^2$	Ordered pair which lies on the graph of $y = x^2$
-2	4	$(-2, 4)$
-1	1	$(-1, 1)$
0	0	$(0, 0)$
1	1	$(1, 1)$
2	4	$(2, 4)$

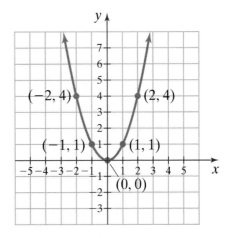

Figure 4 The graph of $y = x^2$

Work exercises 3–6 in this textbook or in the *MyMathLab*® Study Plan.

My video summary

▤ Example 3 Determine whether a Point Lies on the Graph of an Equation

Determine whether the following ordered pairs lie on the graph of the equation $x^2 + y^2 = 1$.

a. $(0, -1)$

b. $(1, 0)$

c. $\left(\dfrac{1}{3}, \dfrac{2}{3}\right)$

d. $\left(-\dfrac{\sqrt{2}}{2}, \dfrac{\sqrt{2}}{2}\right)$

Solution Watch the video to verify that the points $(0, -1)$, $(1, 0)$, and $\left(-\frac{\sqrt{2}}{2}, \frac{\sqrt{2}}{2}\right)$ lie on the graph of $x^2 + y^2 = 1$, and the point $\left(\frac{1}{3}, \frac{2}{3}\right)$ does not lie on the graph.

Work exercises 7–9 in this textbook or in the *MyMathLab*® Study Plan.

OBJECTIVE 3: FINDING THE MIDPOINT OF A LINE SEGMENT USING THE MIDPOINT FORMULA

Suppose we want to find the midpoint $M(x, y)$ of the line segment between the points $A(x_1, y_1)$ and $B(x_2, y_2)$. See Figure 5. To find this midpoint, we simply average the x- and y-coordinates, respectively. In other words, the x-coordinate of the midpoint is $\dfrac{x_1 + x_2}{2}$, whereas the y-coordinate of the midpoint is $\dfrac{y_1 + y_2}{2}$.

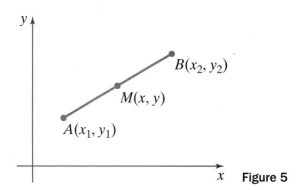

Figure 5

Midpoint Formula

The midpoint of the line segment from $A(x_1, y_1)$ to $B(x_2, y_2)$ is

$$\left(\frac{x_1 + x_2}{2}, \frac{y_1 + y_2}{2}\right).$$

Example 4 Find the Midpoint of a Line Segment

Find the midpoint of the segment whose endpoints are $(-3, 2)$ and $(4, 6)$.

Solution The midpoint of this line segment is $\left(\frac{-3 + 4}{2}, \frac{2 + 6}{2}\right) = \left(\frac{1}{2}, 4\right)$. See Figure 6.

Work exercises 10–13 in this textbook or in the *MyMathLab*® Study Plan.

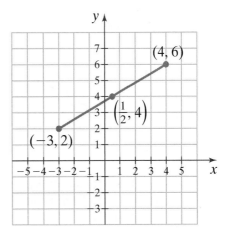

Figure 6

Example 5 Application of the Midpoint Formula

In geometry, it can be shown that four points in a plane form a **parallelogram** if the two diagonals of the quadrilateral formed by the four points **bisect** each other. Do the points $A(0, 4)$, $B(3, 0)$, $C(9, 1)$, and $D(6, 5)$ form a parallelogram?

Solution To prove that the two diagonals bisect each other, we can show that they share the same midpoint. We can see in Figure 7 that the diagonals are the segments AC and DB.

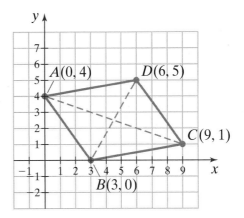

Figure 7

The midpoint of diagonal AC is $\left(\frac{0 + 9}{2}, \frac{4 + 1}{2}\right) = \left(\frac{9}{2}, \frac{5}{2}\right)$.

The midpoint of diagonal DB is $\left(\frac{6 + 3}{2}, \frac{5 + 0}{2}\right) = \left(\frac{9}{2}, \frac{5}{2}\right)$.

Because the diagonals share the same midpoint, it is implied that they bisect each other. Therefore, the quadrilateral is a parallelogram.

Work exercises 14–16 in this textbook or in the _MyMathLab_® Study Plan.

OBJECTIVE 4:   FINDING THE DISTANCE BETWEEN TWO POINTS USING THE DISTANCE FORMULA

My video summary

Recall that the Pythagorean theorem states that the sum of the squares of the two sides of a right triangle is equal to the square of the hypotenuse or $a^2 + b^2 = c^2$. See Figure 8.

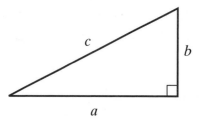

Figure 8 Pythagorean theorem:
$$a^2 + b^2 = c^2$$

We can use the Pythagorean theorem to find the distance between any two points in a plane. For example, suppose we want to find the distance, $d(A, B)$, between the points $A(x_1, y_1)$ and $B(x_2, y_2)$. To find the length $d(A, B)$ of the line segment AB, consider the point $C(x_1, y_2)$ in Figure 9, which is on the same vertical line segment as point A and the same horizontal line segment as point B. The triangle formed by points A, B, and C is a right triangle whose hypotenuse has length $d(A, B)$. The horizontal leg of the triangle has length $a = |x_2 - x_1|$, whereas the vertical leg of the triangle has length $b = |y_2 - y_1|$.

Therefore, by the Pythagorean theorem, $\underbrace{|x_2 - x_1|^2}_{a^2} + \underbrace{|y_2 - y_1|^2}_{b^2} = \underbrace{[d(A, B)]^2}_{c^2}$.

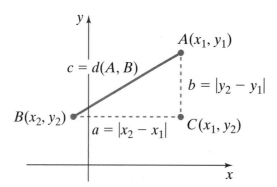

Figure 9 $c = d(A, B)$

$$[d(A, B)]^2 = |x_2 - x_1|^2 + |y_2 - y_1|^2 \qquad \textit{Use the Pythagorean theorem.}$$

$$d(A, B) = \pm\sqrt{|x_2 - x_1|^2 + |y_2 - y_1|^2} \qquad \textit{Use the square root property.}$$

However, because distance cannot be negative, exclude $-\sqrt{|x_2 - x_1|^2 + |y_2 - y_1|^2}$.

$$d(A, B) = \sqrt{|x_2 - x_1|^2 + |y_2 - y_1|^2} \qquad \textit{Use the positive square root only.}$$

$$d(A, B) = \sqrt{(x_2 - x_1)^2 + (y_2 - y_1)^2} \qquad \textit{For any real number } a,\ |a|^2 = a^2.$$

This formula is known as the distance formula.

Distance Formula

The distance between any two points $A(x_1, y_1)$ and $B(x_2, y_2)$ is given by the formula

$$d(A, B) = \sqrt{(x_2 - x_1)^2 + (y_2 - y_1)^2}.$$

Example 6 Use the Distance Formula to Find the Distance between Two Points

Find the distance, $d(A, B)$, between the points A and B.

$$A(-1, 5);\ B(4, -5)$$

Solution $d(A, B) = \sqrt{(x_2 - x_1)^2 + (y_2 - y_1)^2}$ *Use the distance formula.*

$\qquad\qquad = \sqrt{(4 - (-1))^2 + (-5 - 5)^2}$ *Substitute in the distance formula.*

$\qquad\qquad = \sqrt{5^2 + (-10)^2}$ *Combine terms.*

$\qquad\qquad = \sqrt{25 + 100}$ *Simplify.*

$\qquad\qquad = \sqrt{125}$ *Add.*

$\qquad\qquad = 5\sqrt{5}$ *Simplify the radical.*

The distance between the two given points is $d(A, B) = 5\sqrt{5}$ units. See Figure 10.

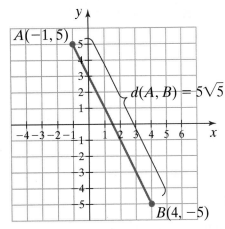

Figure 10 Distance from A to B is $d(A, B) = 5\sqrt{5}$.

Work exercises 17–20 in this textbook or in the *MyMathLab*® Study Plan.

My video summary

Example 7 Application of the Distance Formula

Verify that the points $A(3, -5)$, $B(0, 6)$, and $C(5, 5)$ form a right triangle.

Solution The three points do form a right triangle. Watch the video to see the solution to this example.

Work exercises 21–23 in this textbook or in the *MyMathLab*® Study Plan.

2.1 Exercises

In Exercises 1 and 2, plot each ordered pair in the Cartesian plane, and state in which quadrant or on which axis it lies.

1. a. $A(-1, 4)$ **b.** $B(-2, -2)$ **c.** $C(0, -3)$ **d.** $D\left(1, \dfrac{1}{2}\right)$ **e.** $E(2, 0)$

2. a. $A(2, -4)$ **b.** $B\left(-\dfrac{5}{2}, 1\right)$ **c.** $C(\sqrt{2}, \pi)$ **d.** $D\left(0, -\dfrac{9}{2}\right)$ **e.** $E(0, 0)$

In Exercises 3–6, sketch the graph for each equation by plotting the following points.

3. $y = 2x + 1$ **4.** $2y = x^3$ **5.** $y^2 + 3x = 0$ **6.** $y = 3x^2$

In Exercises 7–9, determine whether the indicated ordered pairs lie on the graph of the given equation.

7. $y = 2x^2 + 1$
  a. $(0, 1)$
  b. $(-1, 1)$
  c. $(1, 3)$

8. $y = \sqrt{x} - 1$
  a. $(4, 1)$
  b. $(1, 4)$
  c. $(0, 1)$

9. $y = |x|$
  a. $(-2, 2)$
  b. $(-9, 9)$
  c. $(3, 3)$

In Exercises 10–13, find the midpoint of the line segment joining points A and B.

10. $A(-1, 4)$; $B(-2, -2)$
  11. $A(2, -5)$; $B(4, 1)$
  12. $A(0, 1)$; $B\left(-3, \dfrac{1}{2}\right)$
  13. $A(a, b)$; $B(c, d)$

In Exercises 14–16, determine whether the points A, B, C, and D form a parallelogram.

14. $A(1, 4)$; $B(-2, -1)$; $C(4, 2)$; $D(7, 7)$
  15. $A(-2, 4)$; $B(-1, 2)$; $C(1, 1)$; $D(3, 3)$
16. $A(-2, -1)$; $B(2, 0)$; $C(3, 3)$; $D(-1, 2)$

In Exercises 17–20, find the distance, $d(A, B)$ between points A and B.

17. $A(1, 5)$; $B(-2, 1)$
  18. $A(3, 5)$; $B(-2, -2)$
  19. $A(2, -4)$; $B(-3, 6)$
  20. $A\left(-\dfrac{\sqrt{3}}{2}, \dfrac{\sqrt{17}}{2}\right)$; $B(0, 0)$

In Exercises 21–23, determine whether the points A, B, and C form a right triangle.

21. $A(-3, -3)$; $B(3, -1)$; $C(2, 2)$
  22. $A(1, 4)$; $B(-2, -1)$; $C(4, 2)$
  23. $A(1, 2)$; $B(2, 6)$; $C(9, 0)$

2.2 Circles

THINGS TO KNOW

Before working through this section, be sure that you are familiar with the following concepts:

VIDEO   ANIMATION   INTERACTIVE

You Try It
1. Solving Quadratic Equations by Completing the Square (Section 1.4)

You Try It
2. Solving Quadratic Equations Using the Square Root Property (Section 1.4)

You Try It
3. Finding the Midpoint of a Line Segment Using the Midpoint Formula (Section 2.1)

You Try It
4. Finding the Distance between Two Points Using the Distance Formula (Section 2.1)

OBJECTIVES

1 Writing the Standard Form of an Equation of a Circle

2 Sketching the Graph of a Circle

3 Converting the General Form of a Circle into Standard Form

A **circle** is the set of all points (x, y) in the Cartesian plane that are a fixed distance, r, from a fixed point, (h, k). The fixed distance, r, is called the **radius** of the circle, and the fixed point (h, k), is called the **center** of the circle. See Figure 11. To derive the equation of a circle, we use the distance formula that was discussed in Section 2.1. Watch the animation to see how the standard equation of a circle is derived.

My animation summary

The derivation of the standard form of the equation of a circle.

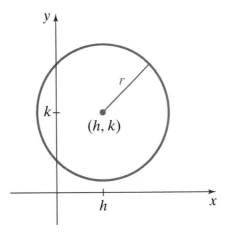

Figure 11 Circle with center, (h, k) and radius, r

The **standard form of an equation of a circle** with center, (h, k), and radius, r, is

$$(x - h)^2 + (y - k)^2 = r^2.$$

The standard form of an equation of a circle centered at the origin with radius, r, is

$$x^2 + y^2 = r^2.$$

OBJECTIVE 1:   WRITING THE STANDARD FORM OF AN EQUATION OF A CIRCLE

My video summary

Example 1 Find the Standard Form of an Equation of a Circle

Find the standard form of the equation of the circle whose center is $(-2, 3)$ and with radius 6.

Solution We know that the standard form of the equation of a circle is $(x - h)^2 + (y - k)^2 = r^2$. We are given $h = -2$, $k = 3$, and $r = 6$. Substituting these values into the equation, we get

$$(x - (-2))^2 + (y - 3)^2 = 6^2 \text{ or } (x + 2)^2 + (y - 3)^2 = 36$$

Work exercises 1–4 in this textbook or in the *MyMathLab*® Study Plan.

Example 2 Find the Equation of a Circle Given the Center and a Point on the Circle

Find the standard form of the equation of the circle whose center is $(0, 6)$ and that passes through the point $(4, 2)$.

Solution The graph of this circle is sketched in Figure 12. The center is labeled $A(0, 6)$, and the point on the circle is labeled $B(4, 2)$. To find the equation of this circle, we must calculate the radius. The radius can be found using the distance formula.

$$r = d(A, B) = \sqrt{(x_2 - x_1)^2 + (y_2 - y_1)^2} \qquad \textit{Write the distance formula.}$$

$$r = d(A, B) = \sqrt{(4 - 0)^2 + (2 - 6)^2} \qquad \textit{Substitute into the distance formula.}$$

$$r = d(A, B) = \sqrt{(4)^2 + (-4)^2} \qquad \textit{Combine terms.}$$

$$r = d(A, B) = \sqrt{32} \qquad \textit{Add.}$$

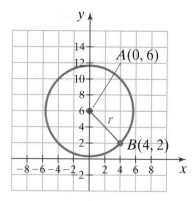

Figure 12

Now that we know the center and the radius, we can determine the equation of this circle.

$$(x - h)^2 + (y - k)^2 = r^2 \qquad \textit{Write the standard equation of a circle.}$$
$$(x - 0)^2 + (y - 6)^2 = \sqrt{32}\,^2 \qquad \textit{Substitute values for the center and the radius.}$$
$$x^2 + (y - 6)^2 = 32 \qquad \textit{Simplify.}$$

The equation of this circle in standard form is $x^2 + (y - 6)^2 = 32$.

My video summary

▦ Example 3 Find the Equation of a Circle Given the Endpoints of a Diameter

Find the standard form of the equation of the circle that contains endpoints of a diameter at $(-4, -3)$ and $(2, -1)$.

Solution We can use the **midpoint formula** to determine the center of this circle. The radius can be found using the **distance formula**. Watch this video to verify that the equation of this circle is $(x + 1)^2 + (y + 2)^2 = 10$.

Work exercises 5–10 in this textbook or in the *MyMathLab*® Study Plan.

OBJECTIVE 2:   SKETCHING THE GRAPH OF A CIRCLE

My video summary

▦ Once you know the center and the radius of a circle, it is fairly straightforward to graph the circle.

Example 4 Sketch the Graph of a Circle

Find the center and the radius, and sketch the graph of the circle $(x - 1)^2 + (y + 2)^2 = 9$. Also find any **intercepts**.

Solution The circle $(x - 1)^2 + (y + 2)^2 = 9$ has center $(h, k) = (1, -2)$ and $r = 3$.

◇ **CAUTION** Note that the y-coordinate of the center ($k = -2$) is negative because $(y + 2)^2 = (y - (-2))^2$.

To sketch the graph of this circle, plot the center, and then locate a few points on the circle. The points that are easiest to locate are the points that are three units left and right of the center and three units up and down from the center. Complete the graph by drawing the circle through these four points, as in Figure 13.

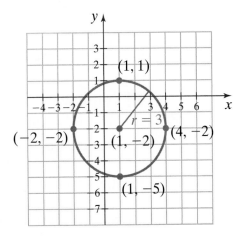

Figure 13 Graph of $(x - 1)^2 + (y + 2)^2 = 9$

To find the *x*-intercepts, set $y = 0$ and solve for *x*.

$$(x - 1)^2 + (0 + 2)^2 = 9 \qquad \text{\textit{Set y = 0.}}$$
$$(x - 1)^2 + 4 = 9 \qquad \text{\textit{Simplify.}}$$
$$(x - 1)^2 = 5 \qquad \text{\textit{Subtract 4 from both sides.}}$$
$$x - 1 = \pm\sqrt{5} \qquad \text{\textit{Use the square root property.}}$$
$$x = 1 \pm \sqrt{5} \qquad \text{\textit{Simplify.}}$$

The two *x*-intercepts are $x \approx 3.24$ and $x \approx -1.24$.

To find any *y*-intercepts, set $x = 0$ and solve for *y*. You should verify that the *y*-intercepts are $y = -2 + 2\sqrt{2} \approx .83$ and $y = -2 - 2\sqrt{2} \approx -4.83$. Watch the video to see this example worked out in its entirety.

Work exercises 11–17 in this textbook or in the MyMathLab® Study Plan.

OBJECTIVE 3:   CONVERTING THE GENERAL FORM OF A CIRCLE INTO STANDARD FORM

As you can see from Example 4, if the equation of a circle is in standard form, it is not too difficult to determine the center and the radius. For example, the center of the circle $(x + 3)^2 + (y - 1)^2 = 49$ is at $(-3, 1)$, and the radius is 7. Let's now take this same equation, eliminate both parentheses, and simplify:

$$(x + 3)^2 + (y - 1)^2 = 49 \qquad \text{\textit{Write the original equation in standard form.}}$$
$$x^2 + 6x + 9 + y^2 - 2y + 1 = 49 \qquad \text{\textit{Remove parentheses.}}$$
$$x^2 + 6x + y^2 - 2y + 10 = 49 \qquad \text{\textit{Combine like terms.}}$$
$$x^2 + y^2 + 6x - 2y - 39 = 0 \qquad \text{\textit{Subtract 49 from both sides and rearrange terms.}}$$

We see that the equation $(x + 3)^2 + (y - 1)^2 = 49$ is equivalent to the equation $x^2 + y^2 + 6x - 2y - 39 = 0$; however, determining the center and the radius of this equation is less obvious. The equation $x^2 + y^2 + 6x - 2y - 39 = 0$ is said to be in **general form**.

> The **general form of the equation of a circle*** is $Ax^2 + By^2 + Cx + Dy + E = 0$, where $A, B, C, D,$ and E are real numbers and $A = B$.

Before working through Example 5, it may be beneficial to review how to **complete the square**, which was discussed in Section 1.4.

My video summary

▦ Example 5 Convert the General Form of a Circle into Standard Form

Write the equation $x^2 + y^2 - 8x + 6y + 16 = 0$ in standard form; find the center, radius, and intercepts, and sketch the graph.

Solution Because we are going to complete the square on both x and y, we first rearrange the terms, leaving some room to complete the square, and move any constants to the right-hand side.

$$x^2 - 8x \qquad + y^2 + 6y \qquad = -16 \qquad \text{\textit{Rearrange the terms.}}$$

$$x^2 - 8x + 16 + y^2 + 6y + 9 = -16 + 16 + 9 \qquad \text{\textit{Complete the square on } x}$$
and y. Remember to add 16 and 9 to both sides.

$$\left(\tfrac{1}{2} \cdot (-8)\right)^2 = 16 \qquad \left(\tfrac{1}{2} \cdot 6\right)^2 = 9$$

$$(x - 4)^2 + (y + 3)^2 = 9 \qquad \text{\textit{Factor the left side.}}$$

We have now converted the general form of the circle into standard form. The center is at $(4, -3)$, and the radius is $r = 3$. Watch the video to verify that there are no y-intercepts. The only x-intercept is $x = 4$. The graph of the circle is shown in Figure 14.

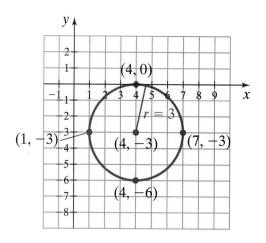

Figure 14 Graph of $x^2 + y^2 - 8x + 6y + 16 = 0$

My animation summary

◪ Example 6 Convert the General Form of a Circle into Standard Form, Where $A \neq 1$ and $B \neq 1$

Write the equation $4x^2 + 4y^2 + 4x - 8y + 1 = 0$ in standard form; find the center, radius, and intercepts, and sketch the graph.

---

*Every equation of a circle can be written in general form. However, not every equation of the form $Ax^2 + By^2 + Cx + Dy + E = 0$ has a graph that is a circle. For example, the equation $2x^2 + 2y^2 + 1 = 0$ has no graph. The graph of the equation $x^2 + y^2 + 2x - 4y + 5 = 0$ is the single point $(-1, 2)$.

Solution Work through the animation to see that this equation is equivalent to the following equation in standard form:

$$\left(x + \frac{1}{2}\right)^2 + (y - 1)^2 = 1; \text{ center} = \left(-\frac{1}{2}, 1\right) \text{ and } r = 1$$

Work through the animation to verify that the one x-intercept is found at $\left(-\frac{1}{2}, 0\right)$, whereas the y-intercepts are found at $\left(0, 1 + \frac{\sqrt{3}}{2}\right)$ and $\left(0, 1 - \frac{\sqrt{3}}{2}\right)$. See the graph in Figure 15.

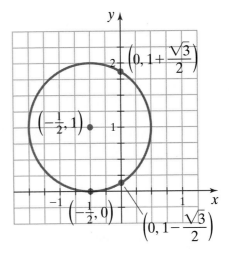

Figure 15 Graph of $4x^2 + 4y^2 + 4x - 8y + 1 = 0$

Work exercises 18–28 in this textbook or in the *MyMathLab*® Study Plan.

2.2 Exercises

In Exercises 1–10, write the standard form of the equation of each circle described.

1. Center $(0, 0), r = 1$

2. Center $(-2, 3), r = 4$

3. Center $(1, -4), r = 3$

4. Center $\left(-\frac{1}{4}, -\frac{1}{3}\right), r = 2$

5. Center $(0, 2)$, passes through the point $(4, -1)$

6. Center $(-4, 7)$, passes through the point $(2, 1)$

7. The endpoints of a diameter are $(0, 1)$ and $(6, -3)$.

8. The endpoints of a diameter are $(2, -6)$ and $(6, 1)$.

9. Center $(2, -3)$ and tangent to the x-axis.

10. Center $(-4, 1)$ and tangent to the y-axis.

In Exercises 11–17, find the center, radius, and intercepts of each circle and then sketch the graph.

11. $x^2 + y^2 = 1$

12. $x^2 + (y - 2)^2 = 4$

13. $(x - 1)^2 + (y + 5)^2 = 16$

14. $(x + 2)^2 + (y + 4)^2 = 36$

15. $(x - 4)^2 + (y + 7)^2 = 12$

16. $(x + 1)^2 + (y - 3)^2 = 20$

17. $\left(x - \frac{1}{4}\right)^2 + \left(y + \frac{1}{2}\right)^2 = 4$

In Exercises 18–28, find the center, radius, and intercepts of each circle and then sketch the graph.

18. $x^2 + y^2 + 2x - 4y + 1 = 0$

19. $x^2 + y^2 - 10x + 6y + 18 = 0$

20. $x^2 + y^2 - 4x - 8y + 19 = 0$

21. $x^2 + y^2 + 2y - 8 = 0$

22. $x^2 + y^2 - 6x - 12y - 5 = 0$

23. $x^2 + y^2 - 3x - y - \dfrac{1}{2} = 0$

24. $x^2 + y^2 + \dfrac{2}{3}x - \dfrac{1}{2}y - \dfrac{7}{18} = 0$

25. $2x^2 + 2y^2 - 4x + 8y + 2 = 0$

26. $16x^2 + 16y^2 - 16x + 8y - 11 = 0$

27. $144x^2 + 144y^2 - 72x - 96y - 551 = 0$

28. $36x^2 + 36y^2 + 12x + 72y - 35 = 0$

2.3 Lines

THINGS TO KNOW

Before working through this section, be sure that you are familiar with the following concepts:

VIDEO   ANIMATION   INTERACTIVE

You Try It

1. Solving Linear Equations (Section 1.1)

You Try It

2. Plotting Ordered Pairs (Section 2.1)

OBJECTIVES

1 Determining the Slope of a Line

2 Sketching a Line Given a Point and the Slope

3 Finding the Equation of a Line Using the Point–Slope Form

4 Finding the Equation of a Line Using the Slope–Intercept Form

5 Writing the Equation of a Line in Standard Form

6 Finding the Slope and the y-Intercept of a Line in Standard Form

7 Sketching Lines by Plotting Intercepts

8 Finding the Equation of a Horizontal Line and a Vertical Line

OBJECTIVE 1:   DETERMINING THE SLOPE OF A LINE

My video summary

In this section, we study equations of lines that lie in the Cartesian plane. Before we learn about a line's equation, we must first establish a way to measure the steepness of a line. In mathematics, the steepness of a line can be measured by computing the line's **slope**. Every nonvertical line has slope. (Vertical lines are said to have no slope and are discussed in detail at the end of this section.) A line going up from left to right has **positive slope**, a line going down from left to right has **negative slope**, horizontal lines have **zero slope**, and vertical lines have **no slope** or **undefined slope**. See Figure 16.

(a) Positive slope

(b) Negative slope

(c) Zero slope

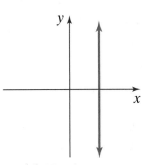

(d) No slope or
undefined slope

Figure 16

The slope can be computed by comparing the vertical change (the **rise**) to the horizontal change (the **run**). Given any two points on the line, the slope m, can be computed by taking the quotient of the rise over the run.

DEFINITION Slope

If $x_1 \neq x_2$, the **slope** of a nonvertical line passing through distinct points (x_1, y_1) and (x_2, y_2) is given by

$$m = \frac{rise}{run} = \frac{\text{change in } y}{\text{change in } x} = \frac{y_2 - y_1}{x_2 - x_1}.$$

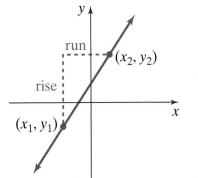

Example 1 Find the Slope of a Line

Find the slope of the line that passes through the indicated ordered pairs:

a. $(-2, 3)$ and $(2, -5)$ **b.** $(6, -4)$ and $(-5, 1)$

Solution a. $m = \dfrac{y_2 - y_1}{x_2 - x_1} = \dfrac{-5 - 3}{2 - (-2)} = \dfrac{-8}{2 + 2} = -\dfrac{8}{4} = -2$

b. $m = \dfrac{y_2 - y_1}{x_2 - x_1} = \dfrac{1 - (-4)}{-5 - 6} = \dfrac{1 + 4}{-11} = \dfrac{5}{-11} = -\dfrac{5}{11}$

Work exercises 1–4 in this textbook or in the *MyMathLab*® Study Plan.

OBJECTIVE 2:   SKETCHING A LINE GIVEN A POINT AND THE SLOPE

If we know a point on a line and the slope, we can quickly sketch the line. Watch this video to see how to sketch the line described in Example 2.

My video summary

🎬 **Example 2 Sketch a Line Given a Point and the Slope**

Sketch the line with slope $m = \frac{2}{3}$ that passes through the point $(-1, -4)$. Also, find three more points located on the line.

Solution To sketch the line, plot the ordered pair $(-1, -4)$. We can now use the slope to plot another point. Starting at $(-1, -4)$, we rise two units and then run three units. In other words, we add 2 to the y-coordinate and 3 to the x-coordinate to get $(2, -2)$. We can continue to find more points on the line by adding 2 to each successive y-coordinate and 3 to each successive x-coordinate to get the points listed in Figure 17.

$(-1, -4)$

run 3   rise 2

$(2, -2)$

run 3   rise 2

$(5, \ \ 0)$

run 3   rise 2

$(8, \ \ 2)$

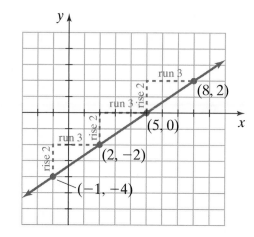

Figure 17 Line with slope passing through the point $(-1, -4)$.

Work exercises 5–8 in this textbook or in the *MyMathLab*® Study Plan.

OBJECTIVE 3:   FINDING THE EQUATION OF A LINE USING THE POINT–SLOPE FORM

Given the slope m of a line and a point on the line, (x_1, y_1), we can use what is known as the **point–slope form of the equation** to determine the equation of the line. The point–slope form of a line is given in the following box. Watch the video to see exactly how this equation is derived.

My video summary

▦ Derivation of the Point–Slope Form of the Equation of a Line

---

Point–Slope Form of the Equation of a Line

Given the slope of a line m and a point on the line (x_1, y_1), the point–slope form of the equation of a line is given by $y - y_1 = m(x - x_1)$.

Example 3 Find the Equation of a Line Given a Point and the Slope

Find an equation in point–slope form of the line with slope $m = \frac{2}{3}$ that passes through the point $(-1, -4)$.

Solution This is the same line that was discussed in Example 2. Because we are given the slope of the line and a point on that line, we can use the point–slope form to produce an equation for this specific line. Substitute $m = \frac{2}{3}$ and $(x_1, y_1) = (-1, -4)$ into the point–slope form $y - y_1 = m(x - x_1)$ to get the equation

$$y - (-4) = \frac{2}{3}(x - (-1)) \text{ or } y + 4 = \frac{2}{3}(x + 1)$$

Work exercises 9–12 in this textbook or in the *MyMathLab*® Study Plan.

The line used in Examples 2 and 3 is sketched in Figure 18. We see from Example 3 that the equation of this line in point–slope form is $y + 4 = \frac{2}{3}(x + 1)$. Is this the only equation of this line? Because the point $(8, 2)$ is also a point on this line, we can use the point–slope form with $m = \frac{2}{3}$ and $(x_1, y_1) = (8, 2)$ to obtain the equation $y - 2 = \frac{2}{3}(x - 8)$. We have now found two equations that describe the same line! Are these equations equivalent? Solving each equation for y, we see, in fact, that these equations are equivalent.

Equations $y + 4 = \frac{2}{3}(x + 1)$ and $y - 2 = \frac{2}{3}(x - 8)$ are equivalent

$y + 4 = \dfrac{2}{3}(x + 1)$	*Write the original equations.*	$y - 2 = \dfrac{2}{3}(x - 8)$
$y + 4 = \dfrac{2}{3}x + \dfrac{2}{3}$	*Use the distributive property.*	$y - 2 = \dfrac{2}{3}x - \dfrac{16}{3}$
$y = \dfrac{2}{3}x + \dfrac{2}{3} - 4$	*Move constant term to the right.*	$y = \dfrac{2}{3}x - \dfrac{16}{3} + 2$
$y = \dfrac{2}{3}x + \dfrac{2}{3} - \dfrac{12}{3}$	*Get a common denominator.*	$y = \dfrac{2}{3}x - \dfrac{16}{3} + \dfrac{6}{3}$
$y = \dfrac{2}{3}x - \dfrac{10}{3}$	← *The two equations are equivalent.* →	$y = \dfrac{2}{3}x - \dfrac{10}{3}$

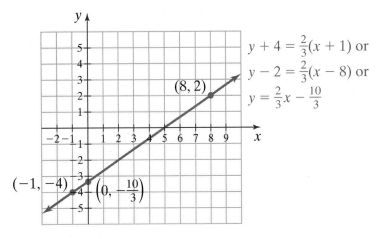

$y + 4 = \frac{2}{3}(x + 1)$ or

$y - 2 = \frac{2}{3}(x - 8)$ or

$y = \frac{2}{3}x - \frac{10}{3}$

Figure 18

It does not matter which point on the line we choose to determine the equation, as long as we use the correct slope.

OBJECTIVE 4:   FINDING THE EQUATION OF A LINE USING THE
SLOPE–INTERCEPT FORM

The equation $y = \frac{2}{3}x - \frac{10}{3}$ sketched in Figure 18 is said to be in *slope–intercept form* because $\frac{2}{3}$ describes the slope of the line, whereas the constant $-\frac{10}{3}$ is the *y*-intercept. The slope–intercept form of a line is extremely important because every nonvertical line has exactly one slope–intercept equation.

Slope–Intercept Form of the Equation of a Line

Given the slope of a line m and the *y*-intercept b, the slope–intercept form of the equation of a line is given by $y = mx + b$.

Click here to see how we can derive the slope–intercept form from the point–slope form.

My video summary

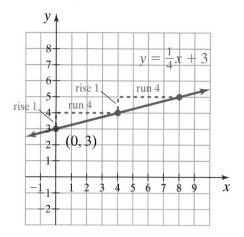 **Derivation of the Slope–Intercept Form of the Equation of a Line**

Example 4 Find the Equation of a Line Given the Slope and *y*-Intercept

Find the equation of the line with slope $\frac{1}{4}$ and *y*-intercept 3, and write the equation in slope–intercept form.

Solution To find the equation of this line, we substitute $m = \frac{1}{4}$ and $b = 3$ into the slope–intercept form.

$$y = mx + b \qquad \textit{Write the slope–intercept form.}$$

$$y = \frac{1}{4}x + 3 \qquad \textit{Substitute } m = \frac{1}{4} \textit{ and } b = 3.$$

We can sketch the graph of the equation $y = \frac{1}{4}x + 3$ by plotting the *y*-intercept at the point $(0, 3)$ and then using the slope of $m = \frac{1}{4}$ to plot one more point. We then draw a straight line through the two points to complete the sketch. See Figure 19.

Figure 19 Graph of $y = \frac{1}{4}x + 3$

Work exercises 13–16 in this textbook or in the *MyMathLab* Study Plan.

OBJECTIVE 5: **WRITING THE EQUATION OF A LINE IN STANDARD FORM**

Consider the equation $y = \frac{1}{4}x + 3$ from Example 4. Suppose we want to eliminate the fractions so all **coefficients** were integers. We can do so by multiplying both sides of the equation by the **LCD** of 4.

$$4 \cdot y = 4 \cdot \left(\frac{1}{4}x + 3\right) \qquad \textit{Multiply both sides of the equation by 4.}$$

$$4y = x + 12 \qquad \textit{Use the distributive property.}$$

$$x - 4y = -12 \qquad \textit{Subtract } 4y \textit{ from both sides,}$$
$$\textit{and subtract 12 from both sides.}$$

Although this equation $x - 4y = -12$ describes the exact same line as the equation $y = \frac{1}{4}x + 3$, it may look a little nicer because all coefficients are integers. The equation $x - 4y = -12$ is said to be in **standard form**.

Standard Form Equation of a Line

The standard form of an equation of a line is given by $Ax + By = C$, where A, B, and C are real numbers such that A and B are both not zero.

Note: Every equation of a line in two variables can be written in standard form. Furthermore, fractions can always be eliminated by multiplying the equation by the LCD. Therefore, the standard form of a line seen in this text always includes nonfractional coefficients, and A is always greater than or equal to zero.

My video summary

▦ **Example 5 Find the Equation of a Line Given Two Points on the Line**

Find the equation of the line passing through the points $(-1, 3)$ and $(2, -4)$. Write the equation in point–slope form, slope–intercept form, and standard form.

Solution First, we always find the slope.

$$m = \frac{-4 - 3}{2 - (-1)} = \frac{-7}{2 + 1} = -\frac{7}{3}$$

Next, select either of the given points. Let's choose the first point $(-1, 3)$. Now that we have a point and the slope, we can use the point–slope form.

$$y - y_1 = m(x - x_1) \qquad \textit{Write the point–slope equation.}$$

$$y - 3 = -\frac{7}{3}(x + 1) \qquad \textit{Substitute } m = -\frac{7}{3} \textit{ and } (x_1, y_1) = (-1, 3).$$

To find the slope–intercept form of the line, we solve the equation $y - 3 = -\frac{7}{3}(x + 1)$ for y to get $y = -\frac{7}{3}x + \frac{2}{3}$. Click here to see how this is done.

Finally, write the equation in standard form by eliminating the fractions of the slope–intercept form.

$$3 \cdot y = 3 \cdot \left(-\frac{7}{3}x + \frac{2}{3}\right) \qquad \textit{Multiply by 3.}$$

$$3y = -7x + 2 \qquad \textit{Use the distributive property.}$$

$$7x + 3y = 2 \qquad \textit{Add } 7x \textit{ to both sides.}$$

Three equations of this line are

1. Point–slope form: $y - 3 = -\frac{7}{3}(x + 1)$ (or $y + 4 = -\frac{7}{3}(x - 2)$ or infinitely many others)

2. Slope–intercept form: $y = -\frac{7}{3}x + \frac{2}{3}$

3. Standard form: $7x + 3y = 2$

This line is sketched and labeled in Figure 20.

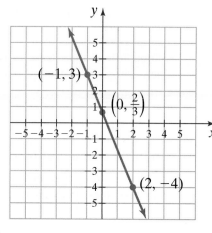

$y - 3 = -\frac{7}{3}(x + 1)$ or

$y + 4 = -\frac{7}{3}(x - 2)$ or

$y = -\frac{7}{3}x + \frac{2}{3}$ or

$7x + 3y = 2$

Figure 20 Line through the points $(-1, 3)$ and $(2, -4)$

Work exercises 17–20 in this textbook or in the *MyMathLab*® Study Plan.

OBJECTIVE 6:   FINDING THE SLOPE AND THE *y*-INTERCEPT OF A LINE IN STANDARD FORM

My video summary

■ Suppose we are given the standard form of a line $Ax + By = C$ with $B \neq 0$ and want to solve for y. To solve for y, we subtract Ax from both sides and divide by B.

$$Ax + By = C \qquad \text{Write the standard form of a line.}$$
$$By = -Ax + C \qquad \text{Subtract } Ax \text{ from both sides.}$$
$$\frac{\cancel{B}y}{\cancel{B}} = -\frac{A}{B}x + \frac{C}{B} \qquad \text{Divide both sides by B.}$$

The standard form of a line $Ax + By = C$ with $B \neq 0$ is equivalent to the equation $y = \overset{m}{-\frac{A}{B}}x + \overset{b}{\frac{C}{B}}$, which is the equation of a line in slope–intercept form. Thus, given the standard form of a line $Ax + By = C$ with $B \neq 0$, the slope of the line is $m = -\frac{A}{B}$, and the *y*-intercept is $b = \frac{C}{B}$. Watch the video to see how to find the slope and *y*-intercept of a line given in standard form and to see how to work Example 6.

Given the standard form of a line $Ax + By = C$ with $B \neq 0$, the slope is given by $m = -\frac{A}{B}$, and the *y*-intercept is $\frac{C}{B}$.

Example 6 Find the Slope and *y*-Intercept of a Line in Standard Form

Find the slope and *y*-intercept, and sketch the line $3x - 2y = 6$.

76   **Chapter 2**   The Rectangular Coordinate System, Lines, and Circles

Solution Because $A = 3$ and $B = -2$, the slope is $m = -\frac{A}{B} = -\frac{3}{-2} = \frac{3}{2}$. The y-intercept is $\frac{C}{B} = \frac{6}{-2} = -3$. To sketch this line, plot the y-intercept, and then rise 3 and run 2 to plot the point $(2, 0)$. Now connect the points with a straight line, as in Figure 21. Watch the video to see this solution worked out in its entirety.

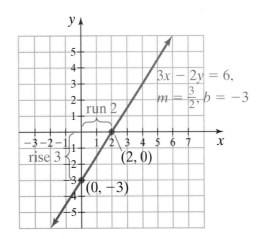

Figure 21 Line $3x - 2y = 6$

Work exercises 21–24 in this textbook or in the **MyMathLab®** Study Plan.

OBJECTIVE 7:   SKETCHING LINES BY PLOTTING INTERCEPTS

In Example 6, we were able to sketch the graph of the line $3x - 2y = 6$ by first plotting two points, and then drawing a straight line through them. These two points happened to be the x- and y-intercepts. The method of sketching a line by plotting intercepts is very useful and is illustrated again in Example 7.

My video summary

▦ **Example 7 Sketch a Line by Plotting Intercepts**

Sketch the line $2x - 5y = 8$ by plotting intercepts.

Solution

Find the x-intercept.

The x-intercept is found by setting $y = 0$ and solving for x:

$$2x - 5(0) = 8$$
$$2x = 8$$
$$x = 4$$

The x-intercept is 4.

Find the y-intercept.

The y-intercept is found by setting $x = 0$ and solving for y:

$$2(0) - 5y = 8$$
$$-5y = 8$$
$$y = -\frac{8}{5}$$

The y-intercept is $-\frac{8}{5}$.

To sketch the line, plot the two intercepts by plotting the points $(4, 0)$ and $\left(0, -\frac{8}{5}\right)$, and then draw a straight line between them, as in Figure 22.

2.3 Lines   77

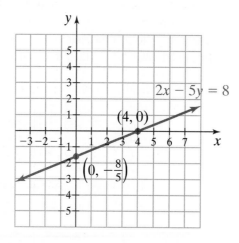

Figure 22 Line $2x - 5y = 8$

Work exercises 25–28 in this textbook or in the *MyMathLab*® Study Plan.

OBJECTIVE 8:   FINDING THE EQUATIONS OF HORIZONTAL AND VERTICAL LINES

Horizontal Lines

My video summary

▦ Suppose we want to determine the equation of the horizontal line that contains the point (a, b). To find this equation, we must first determine the slope. Because the line must also pass through the point $(0, b)$, we see that the slope of this line is $m = \frac{b-b}{a-b} = \frac{0}{a} = 0$. Using the slope–intercept form of a line with $m = 0$ and y-intercept b, we see that the equation is $y = 0x + b$ or $y = b$. See Figure 23.

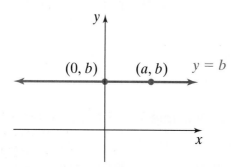

Figure 23 Equation of a horizontal line is $y = b$.

Therefore, we know that for any horizontal line that contains the point (a, b), the equation of that line is $y = b$, and the slope is $m = 0$.

Vertical Lines

▦ Vertical lines have either **no slope** or an **undefined slope**. We can see this by looking at the vertical line that passes through the point (a, b). Because this line also passes through the x-intercept at the point $(a, 0)$, we see that the slope of this line is $m = \frac{b-0}{a-a} = \frac{b}{0}$, which is not a real number as division by zero is not defined. Because the x-coordinate of this vertical line is always equal to a regardless of the y-coordinate, we say that the equation of a vertical line is $x = a$. See Figure 24.

Therefore, we know that for any vertical line that contains the point (a, b), the equation of that line is $x = a$, and the slope is undefined.

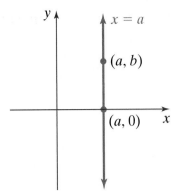

Figure 24 Equation of a vertical line is $x = a$.

Example 8 Find the Equations of Horizontal and Vertical Lines

a. Find the equation of the horizontal line passing through the point $(-1, 3)$.
b. Find the equation of the vertical line passing through the point $(-1, 3)$.

Solution

a. A horizontal line has the equation $y = b$, where b is the y-coordinate of any point on the line. Therefore, the equation is $y = 3$.
b. A vertical line has the equation $x = a$, where a is the x-coordinate of any point on the line. Therefore, the equation is $x = -1$. Both lines are sketched in Figure 25.

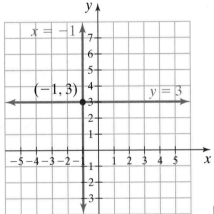

Figure 25 Graph of $y = 3$ and $x = -1$

Work exercises 29–32 in this textbook or in the **MyMathLab**® Study Plan.

Summary of Forms of Equations of Lines

1. $y - y_1 = m(x - x_1)$ **Point–Slope Form**
   Slope is m, and (x_1, y_1) is a point on the line.

2. $y = mx + b$ **Slope–Intercept Form**
   Slope is m, and y-intercept is at $(0, b)$.

3. $Ax + By = C$ **Standard Form**
   A, B, and C are real numbers, with A and B both not zero and $A \geq 0$.

4. $y = b$ **Horizontal Line**
   Slope is zero, and y-intercept is at $(0, b)$.

5. $x = a$ **Vertical Line**
   Slope is undefined, and x-intercept is at $(a, 0)$.

2.3 Exercises

In Exercises 1–4, find the slope of the line passing through the given points.

1. $(-5, 2)$ and $(2, -6)$

2. $(0, -7)$ and $(-4, -9)$

3.

4.
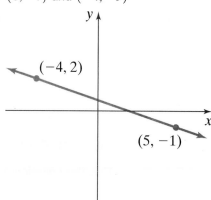

In Exercises 5–8, sketch the line with the given slope that passes through the indicated point.

5. Slope $= \dfrac{1}{2}$; line passes through the point $(-2, -2)$

6. Slope $= -\dfrac{4}{3}$; line passes through the point $(3, 8)$

7. Slope $= 0$; line passes through the point $(-5, -1)$

8. Slope is undefined; line passes through the point $(8, 0)$.

In Exercises 9–12, find the point–slope form of the line with the given slope that passes through the indicated point.

9. Slope $= \dfrac{1}{2}$; line passes through the point $(-2, -2)$

10. Slope $= -\dfrac{4}{3}$; line passes through the point $(3, 8)$

11. Slope $= 3$; line passes through the point $(-5, 3)$

12. Slope $= \dfrac{5}{11}$; line passes through the point $(-99, 88)$

In Exercises 13–16, find the slope–intercept form of the line with the given slope and y-intercept.

13. Slope $= 1$; y-intercept $= -2$

14. Slope $= -\dfrac{1}{6}$; y-intercept $= \dfrac{1}{2}$

15. Slope $= \dfrac{7}{9}$; y-intercept $= -\dfrac{8}{7}$

16. Slope $= 0$; y-intercept $= 5$

In Exercises 17–20, find the equation of the line passing through the indicated two points. Write the equation in point–slope form, slope–intercept form, and standard form.

17. $(1, 2)$ and $(-2, 5)$ **18.** $(-5, 7)$ and $(3, -5)$ **19.** $\left(-\dfrac{1}{2}, 1\right)$ and $\left(-3, \dfrac{2}{3}\right)$ **20.** $(-3, 4)$ and $(2, 4)$

In Exercises 21–24, given the equation of a line in standard form, determine the slope and y-intercept, and sketch the line.

21. $4x - y = 12$ **22.** $x - 3y = -9$ **23.** $5x + 7y = 12$ **24.** $8x - 11y = -23$

In Exercises 25–28, sketch the given line by plotting intercepts.

25. $4x - y = 12$ **26.** $x - 3y = -9$ **27.** $4x + 3y = 6$ **28.** $5x + 4y = -10$

29. Find the equation of the horizontal line passing through the point $(5, -2)$.

30. Find the equation of the vertical line passing through the point $(5, -2)$.

In Exercises 31 and 32, find the equation of the given line.

31.

32.

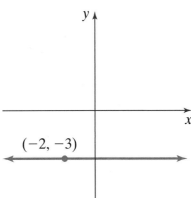

2.4 **Parallel and Perpendicular Lines**

THINGS TO KNOW

Before working through this section, be sure that you are familiar with the following concepts:

VIDEO ANIMATION INTERACTIVE

 1. Determining the Slope of a Line (Section 2.3)

You Try It

 2. Finding the Equation of a Line Using the Point–Slope Form (Section 2.3)

You Try It

 3. Writing the Equation of a Line in Standard Form (Section 2.3)

You Try It

 4. Finding the Slope and the y-Intercept of a Line in Standard Form (Section 2.3)

You Try It

OBJECTIVES

1 Understanding the Definition of Parallel Lines

2 Understanding the Definition of Perpendicular Lines

3 Determining Whether Two Lines Are Parallel, Perpendicular, or Neither

4 Finding the Equations of Parallel and Perpendicular Lines

5 Solving a Geometric Application of Parallel and Perpendicular Lines

....

Given any two *distinct* lines in the Cartesian plane, the two lines will either intersect or not. In this section, we investigate the nature of two lines that do not intersect (**parallel lines**) and then discuss the special case of two lines that intersect at a right angle (**perpendicular lines**). These two cases are interesting because we need only know the slope of the two lines to determine whether the lines are parallel, perpendicular, or neither. Let's first consider two lines in the plane that do not intersect.

OBJECTIVE 1: UNDERSTANDING THE DEFINITION OF PARALLEL LINES

As stated, two lines are parallel if they do not intersect, or in other words, the lines do not share any common points. In Section 2.3, we learned about the slope of a line. It is imperative that you can compute the slope of a line given its equation. Click here to review this concept. Because parallel lines do not intersect, the ratio of the vertical change (rise) to the horizontal change (run) of each line must be equivalent (Figure 26). In other words, parallel lines have the same slope!

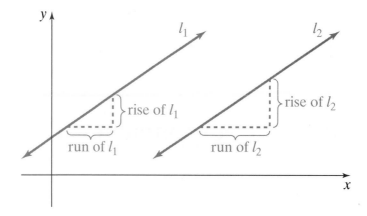

The ratio of the vertical rise to the horizontal run

of two parallel lines are equal: $\dfrac{\text{rise of } l_1}{\text{run of } l_1} = \dfrac{\text{rise of } l_2}{\text{run of } l_2}$

Figure 26

Theorem

Two distinct nonvertical lines in the Cartesian plane are parallel if and only if they have the same slope.

We can prove this theorem using a little geometry. We can show that if l_1 is parallel to l_2, then triangles $\triangle ABC$ and $\triangle DEF$ are similar. See Figure 27. Because the ratio of two corresponding sides of similar triangles are proportional, it follows that the slope of $l_1 = \frac{BC}{AB} = \frac{EF}{DE} =$ slope of l_2. Conversely, if the slope of l_1 is the same as the slope of l_2, then the two triangles are similar. Therefore, $\angle CAB \cong \angle FDE$ and thus the lines must be parallel.

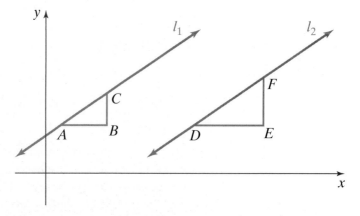

Figure 27 $\triangle ABC$ and $\triangle DEF$ are similar, so $\frac{BC}{AB} = \frac{EF}{DE}$. This implies that the slopes are equal.

Example 1 Show That Two Lines Are Parallel

Show that the lines $y = -\frac{2}{3}x - 1$ and $4x + 6y = 12$ are parallel.

Solution The line $y = -\frac{2}{3}x - 1$ is in **slope-intercept form** with slope $m = -\frac{2}{3}$. The line $4x + 6y = 12$ is in **standard form** with slope

$$m = -\frac{A}{B} = -\frac{4}{6} = -\frac{2}{3}.$$

Because the lines have the same slope, they must be parallel. See Figure 28.

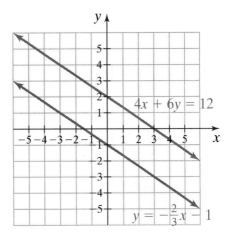

Figure 28 Parallel lines $y = -\frac{2}{3}x - 1$ and $4x + 6y = 12$

OBJECTIVE 2:   UNDERSTANDING THE DEFINITION OF PERPENDICULAR LINES

If two *distinct* lines are not parallel, then they must intersect at a single point. If the two lines intersect at a right angle (90 degrees), then the lines are said to be **perpendicular**. The slopes of perpendicular lines have a special relationship that is stated in the following theorem.

Theorem

Two nonvertical lines in the Cartesian plane are perpendicular if and only if the product of their slopes is -1.

This theorem states that if line 1 with slope m_1 is perpendicular to line 2 with slope m_2, then $m_1 m_2 = -1$. Because the product of the slopes of nonvertical perpendicular lines is -1, it follows that $m_1 = -\frac{1}{m_2}$. This means that if $m_1 = \frac{a}{b}$, then $m_2 = -\frac{b}{a}$. In other words, the slopes of nonvertical perpendicular lines are **negative reciprocals** of each other, as in Figure 29.

Example 2 Show That Two Lines Are Perpendicular

Show that the lines $3x - 6y = -12$ and $2x + y = 4$ are perpendicular.

Solution Both lines are in **standard form**. The first line, $3x - 6y = -12$, has slope $m_1 = -\frac{A}{B} = -\frac{3}{-6} = \frac{1}{2}$. The slope of the second line, $2x + y = 4$, has slope $m_2 = -\frac{A}{B} = -\frac{2}{1}$. Because the slopes of these lines are negative reciprocals of each other, this implies that the lines are perpendicular. See Figure 30.

2.4 Parallel and Perpendicular Lines   83

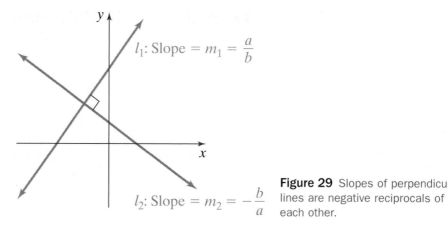

l_1: Slope $= m_1 = \dfrac{a}{b}$

l_2: Slope $= m_2 = -\dfrac{b}{a}$

Figure 29 Slopes of perpendicular lines are negative reciprocals of each other.

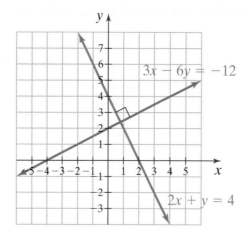

$3x - 6y = -12$

$2x + y = 4$

Figure 30 Perpendicular lines $3x - 6y = -12$ and $2x + y = 4$

Summary of Parallel and Perpendicular Lines

Given two nonvertical lines l_1 and l_2 such that the slope of line l_1 is $m_1 = \frac{a}{b}$,

1. l_2 is parallel to l_1 if and only if $m_2 = \frac{a}{b}$.
2. l_2 is perpendicular to l_1 if and only if $m_2 = -\frac{b}{a}$.

Note: Any two distinct vertical lines are parallel to each other, whereas any horizontal line is perpendicular to any vertical line.

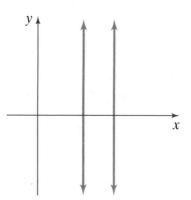

Figure 31 Any two distinct vertical lines are parallel.

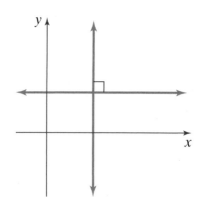

Figure 32 Horizontal lines are perpendicular to vertical lines.

OBJECTIVE 3:   DETERMINING WHETHER TWO LINES ARE PARALLEL,
PERPENDICULAR, OR NEITHER

Given any two distinct lines, we are now able to quickly determine whether the lines are parallel to each other, perpendicular to each other, or neither by simply evaluating their slopes.

My video summary

Example 3 Determine Whether Two Lines Are Parallel, Perpendicular, or Neither

For each of the following pairs of lines, determine whether the lines are parallel, perpendicular, or neither.

a. $3x - y = 4$ **b.** $y = \dfrac{1}{2}x + 3$ **c.** $x = -1$

$x + 3y = 7$ $x + 2y = 1$ $x = 3$

Solution Watch the video to verify that

a. The lines are perpendicular.
b. The lines are neither parallel nor perpendicular.
c. The lines are parallel.

Work exercises 1–6 in this textbook or in the *MyMathLab*® Study Plan.

OBJECTIVE 4:   FINDING THE EQUATIONS OF PARALLEL AND PERPENDICULAR LINES

My video summary

Example 4 Find the Equation of a Parallel Line

Find the equation of the line parallel to the line $2x + 4y = 1$ that passes through the point $(3, -5)$. Write the answer in **point–slope form, slope–intercept form,** and **standard form.**

Solution Watch the video to see that the equations of the line parallel to $2x + 4y = 1$ that passes through the point $(3, -5)$ in all three forms are

$$y + 5 = -\frac{1}{2}(x - 3) \quad \text{\textit{Point–slope form.}}$$

$$y = -\frac{1}{2}x - \frac{7}{2} \quad \text{\textit{Slope–intercept form.}}$$

$$x + 2y = -7 \quad \text{\textit{Standard form.}}$$

Work exercises 7–11 in this textbook or in the *MyMathLab*® Study Plan.

My video summary

Example 5 Find the Equation of a Perpendicular Line

Find the equation of the line perpendicular to the line $y = -5x + 2$ that passes through the point $(3, -1)$. Write the answer in slope–intercept form.

Solution The line $y = -5x + 2$ is in slope–intercept form with slope -5. We are looking for the line perpendicular to this line. Therefore, the slope of the new line must be the negative reciprocal of -5. Because we can rewrite -5 as $-\frac{5}{1}$, the negative reciprocal is $\frac{1}{5}$. Using a slope of $\frac{1}{5}$ and the point $(3, -1)$, we can obtain the point–slope equation $y + 1 = \frac{1}{5}(x - 3)$. Solving this equation for y, we obtain the intercept form of slope: $y = \frac{1}{5}x - \frac{8}{5}$. You can see in Figure 33 how these two lines intersect at a right angle.

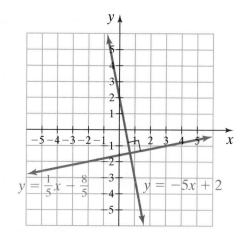

Figure 33 Graph of perpendicular lines
$y = -5x + 2$ and $y = \frac{1}{5}x - \frac{8}{5}$

Work exercises 12–16 in this textbook or in the *MyMathLab*® Study Plan.

OBJECTIVE 5:   SOLVING A GEOMETRIC APPLICATION OF PARALLEL AND PERPENDICULAR LINES

In Section 2.1, we used the midpoint formula to determine whether four points in the plane form a parallelogram. (See Section 2.1, Example 5.) Another way to determine whether four points in the plane form a parallelogram is to determine the slopes of the sides opposite each other. If the slopes of the segments opposite one another are equal, then the four points must form a parallelogram. A special parallelogram in which all four sides have the same length is called a *rhombus*. In geometry, it can be shown that if the diagonals of a parallelogram are perpendicular, then the parallelogram is a rhombus. See Figures 34 and 35.

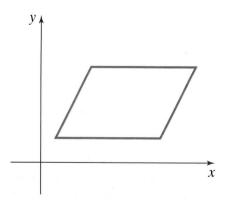

Opposite sides of a parallelogram are parallel.   **Figure 34** Parallelogram

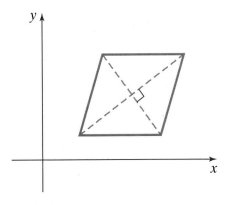

If the diagonals of a parallelogram are perpendicular, then it is also a rhombus.   **Figure 35** Rhombus

Example 6 Determine whether Four Points in the Plane Form a Parallelogram or a Rhombus

My video summary

Do the points $A(0, 4)$, $B(3, 0)$, $C(9, 1)$, and $D(6, 5)$ form a parallelogram? If the points form a parallelogram, then is the parallelogram a rhombus?

Solution These four points do form a parallelogram because the slopes of the segments opposite one another are equal. Verify this for yourself, or watch the video to see the worked out solution. The parallelogram formed by these four points is not a rhombus because diagonal $\overline{AC}$ has slope $-\frac{1}{3}$, whereas diagonal $\overline{BD}$ has slope $\frac{5}{3}$ and $\left(-\frac{1}{3}\right)\left(\frac{5}{3}\right) = -\frac{5}{9} \neq -1$. Therefore, the diagonals are *not* perpendicular. Watch the video to see the entire solution.

Work exercises 17–20 in this textbook or in the *MyMathLab* Study Plan.

2.4 Exercises

In Exercises 1–6, decide whether the pair of lines are parallel, perpendicular, or neither.

1. $-4x - 2y = -9$
$5x - 10y = 7$

2. $y = 3x - 2$
$9x - 3y = 1$

3. $y = -2$
$3y = 12$

4. $x = -1$
$y = -1$

5. $4x - 3y = 7$
$3x - 4y = -6$

6. $2x + 3y = 3$
$y = -\frac{2}{3}x + 1$

In Exercises 7–9, find the equation of a line described as follows, and express your answer in point–slope form, slope–intercept form, and standard form.

7. Find the equation of the line parallel to the line $y = \frac{1}{4}x - 2$ that passes through the point $(-2, 3)$.

8. Find the equation of the line parallel to the line $3x - 5y = 1$ that passes through the point $(1, -4)$.

9. Find the equation of the line parallel to the line $y + 1 = 3(x - 5)$ that passes through the point $(-2, -5)$.

In Exercises 10 and 11, find the equation of a line described as follows, and express your answer in standard form.

10. Find the equation of the line parallel to the line $x = 7$ that passes through the point $(11, -3)$.

11. Find the equation of the line parallel to the line $y = 8$ that passes through the point $(11, -3)$.

In Exercises 12–14, find the equation of a line described as follows, and express your answer in slope–intercept form.

12. Find the equation of the line perpendicular to the line $y = \frac{1}{4}x - 2$ that passes through the point $(-2, 3)$.

13. Find the equation of the line perpendicular to the line $3x - 5y = 1$ that passes through the point $(1, -4)$.

14. Find the equation of the line perpendicular to the line $y + 1 = 3(x - 5)$ that passes through the point $(-2, -5)$.

In Exercises 15 and 16, find the equation of a line described as follows, and express your answer in standard form.

15. Find the equation of the line perpendicular to the line $x = 7$ that passes through the point $(11, -3)$.

16. Find the equation of the line perpendicular to the line $y = 8$ that passes through the point $(11, -3)$.

In Exercises 17–20, use slope to determine whether the quadrilateral with the given vertices forms a parallelogram. If it is a parallelogram, determine whether it is also a rhombus.

17. $A(-1, 2), B(3, 3), C(4, 7), D(0, 6)$

18. $A(2, 2), B(3, -1), C(5, 2), D(4, 5)$

19. $A(1, 1), B(5, 2), C(8, 5), D(3, 3)$

20. $A(1, 1), B(4, -3), C(7, 1), D(4, 5)$

CHAPTER THREE
Functions

CHAPTER THREE CONTENTS

3.1 Relations and Functions

THINGS TO KNOW

Before working through this section, be sure that you are familiar with the following concepts:

VIDEO ANIMATION INTERACTIVE

You Try It

1. Describing Intervals of Real Numbers
   (Section R.1)

You Try It

2. Solving Linear Inequalities
   (Section 1.7)

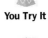
You Try It

3. Solving Polynomial Inequalities
   (Section 1.9)

You Try It

4. Solving Rational Inequalities
   (Section 1.9)

You Try It

5. Graphing Equations by Plotting Points
   (Section 2.1)

OBJECTIVES

1 Understanding the Definitions of Relations and Functions

2 Determining Whether Equations Represent Functions

3 Using Function Notation; Evaluating Functions

4 Using the Vertical Line Test

5 Determining the Domain of a Function Given the Equation

Functions appear all around us and are a fundamental part of mathematics. Functions occur when one quantity depends on another. Here are just a few examples:

The revenue generated by selling a certain product depends on the number of items sold.

The volume of a cube depends on the length of an edge of the cube.

For a person who works by the hour, the gross pay depends on the number of hours worked.

If you think about it, you can probably come up with countless situations where one thing, event, or occurrence depends on another. Before we officially define a function, we must first define a **relation**.

OBJECTIVE 1:   UNDERSTANDING THE DEFINITIONS OF RELATIONS
AND FUNCTIONS

My video summary

> **DEFINITION Relation**
>
> A **relation** is a correspondence between two sets A and B such that each element of set A corresponds to one or more elements of set B. Set A is called the **domain** of the relation, and set B is called the **range** of the relation.

Suppose that Adam, Patrice, and Scott are college roommates and their weights are 180 lb, 190 lb, and 180 lb, respectively. The pairing of the roommates' names with their weights is a relation. In this example, we are relating a set of names with a set of weights. The set of names {Adam, Patrice, Scott} is the domain of this relation, whereas their weights {180 lb, 190 lb} is the range. Notice that 180 lb is listed only once in the range. There is no need to list an element of either set more than once. See Figure 1.

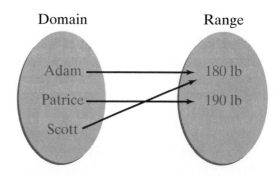

Figure 1

A relation can be represented as a set of ordered pairs with the elements of the domain listed first. The previous relation can be written as the set of ordered pairs {(Adam, 180 lb), (Patrice, 190 lb), (Scott, 180 lb)}.

Let's now switch the domain and range of this relation. The relation in Figure 2 can be written as the set of ordered pairs {(180 lb, Adam), (180 lb, Scott), (190 lb, Patrice)}.

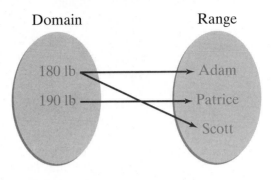

Figure 2

Suppose that a delivery man was to deliver a package to the roommate who weighs 180 lb. Because there are two people who weigh 180 lb, there is no way to know for sure who this package belongs to. This is an example of a relation that is not well defined. A well-defined relation has exactly one output (range) value for any input (domain) value. Well-defined relations are called **functions**.

DEFINITION Function

A **function** is a relation such that for each element in the domain, there is *exactly one* corresponding element in the range. In other words, a function is a well-defined relation.

The relation in Figure 1 is a function because for every name in the domain, there is exactly one corresponding weight in the range. The relation in Figure 2 is not a function because there is a domain value (180 lb) that corresponds to two range values (Adam and Scott).

My video summary

🎞 **Example 1 Determine whether a Relation Is a Function**

Determine whether each relation is a function, and then find the domain and range.

a.

b.

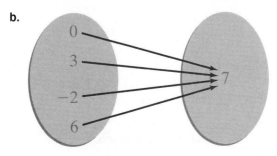

   c. $\{(3, 7), (-3, 2), (-4, 5), (1, 4), (3, -4)\}$
   d. $\{(-1, 2), (0, 2), (1, 5), (2, 3), (3, 2)\}$

Solution

a. This relation is a function. For each element in the domain, there is *exactly one* corresponding element in the range. The domain is {Phoebe, Amy, Catie}, and the range is $\{5'5'', 3'1''\}$.

b. This relation is a function. For each element in the domain, there is *exactly one* corresponding element in the range. The domain is $\{-2, 0, 3, 6\}$, and the range is $\{7\}$.

Note: The elements of the domain and range are typically listed in ascending order when using set notation.

c. This relation is *not* a function because the element 3 in the domain corresponds to 7 and -4 in the range. The domain of this relation is $\{-4, -3, 1, 3\}$. The range is $\{-4, 2, 4, 5, 7\}$.

d. This relation is a function. For each element in the domain, there is *exactly one* corresponding element in the range. The domain is $\{-1, 0, 1, 2, 3\}$. The range is $\{2, 3, 5\}$.

If the domain and range of a relation are sets of real numbers, then the relation can be represented by plotting the ordered pairs in the Cartesian plane. The set of all x-coordinates is the domain of the relation, and the set of all y-coordinates is the range of the relation.

Example 2 Determine whether a Relation Is a Function

Use the domain and range of the following relation to determine whether the relation is a function.

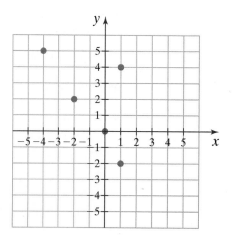

Solution The set of ordered pairs of this graph are $\{(-4, 5), (-2, -2), (0, 0), (1, -2), (1, 4)\}$. The domain is the set of all x-coordinates: $\{-4, -2, 0, 1\}$. The range is the set of all y-coordinates $\{-2, 0, 4, 5\}$. This relation is *not* a function because the element 1 in the domain corresponds to two different elements (-2 and 4) in the range.

Work exercises 1–6 in this textbook or in the *MyMathLab*® **Study Plan.**

OBJECTIVE 2:   DETERMINING WHETHER EQUATIONS REPRESENT FUNCTIONS

In Example 2, we see that a *finite* number of ordered pairs in the Cartesian plane can represent a relation. Relations can also be described by *infinitely* many ordered pairs in the plane. Recall from Section 2.1 that the solution set of an equation in two variables is the set of all ordered pairs (x, y) for which the equation is true. For example, the ordered pairs $(1, 1)$, $(\sqrt{3}, 0)$, and $(1, -1)$ all satisfy the equation $x^2 + 2y^2 = 3$. (Click here to verify.) Although this equation is a relation, it is not a function because $x = 1$ corresponds to both $y = 1$ and $y = -1$. To determine whether an equation represents a function, we must show that for any value in the domain, there is exactly one corresponding value in the range.

My interactive video summary

Example 3 Determine whether an Equation Represents a Function

Determine whether the following equations represent y as a function of x:

a. $3x - 2y = -12$ **b.** $y = 3x^2 - x + 2$ **c.** $(x + 3)^2 + y^2 = 16$

Solution The equations in parts a and b both represent y as a function of x. Watch the interactive video to see parts a and b worked out in detail.

To determine whether the equation $(x + 3)^2 + y^2 = 16$ from part c represents y as a function of x, we can solve for y.

Solving the equation for y, we obtain $y = \pm\sqrt{16 - (x + 3)^2}$. Click here to see all steps. Notice the symbol "$\pm$," which precedes the radical in the previous equation. This symbol indicates that there are two possible y-values for a given x-value. For example, if $x = -3$, then the corresponding y-values are

$$y = \pm\sqrt{16 - (-3 + 3)^2} = \pm\sqrt{16 - (0)^2} = \pm\sqrt{16} = \pm 4.$$

Therefore, the value of $x = -3$ in the domain corresponds to $y = 4$ and $y = -4$ in the range. Because there are two y-values that correspond to a single x-value, the original equation $(x + 3)^2 + y^2 = 16$ does not represent a function. Recall from Section 2.2 that the equation $(x + 3)^2 + y^2 = 16$ is the equation of a circle in **standard form** with center $(-3, 0)$ and $r = 4$. We can sketch this circle using the techniques discussed in Section 2.2, or we can use a graphing utility to graph the two equations $y = \sqrt{16 - (x + 3)^2}$ and $y = -\sqrt{16 - (x + 3)^2}$. For all x-values on the interval $(-7, 1)$, there are two corresponding values of y. Watch the interactive video to see all three solutions worked out in detail.

Using Technology

$y_1 = \sqrt{16 - (x + 3)^2}$

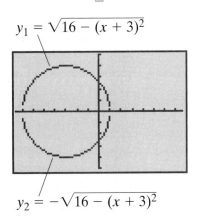

$y_2 = -\sqrt{16 - (x + 3)^2}$

The graph of $(x + 3)^2 + y^2 = 16$ was created using a graphing utility. The top half of the circle is $y = \sqrt{16 - (x + 3)^2}$, whereas the bottom half of the circle represents the equation $y = -\sqrt{16 - (x + 3)^2}$.

Work exercises 7–12 in this textbook or in the *MyMathLab*® Study Plan.

OBJECTIVE 3:   USING FUNCTION NOTATION; EVALUATING FUNCTIONS

My video summary

In Example 3, we see that the equation $3x - 2y = -12$ could be explicitly solved for y to obtain the function $y = \frac{3}{2}x + 6$. (See the interactive video from Example 3a.) When an equation is explicitly solved for y, we say that "y is a function of x" or that the variable y depends on the variable x. Thus, x is the **independent variable**, and y is the **dependent variable**.

Instead of using the variable y, letters such as f, g, or h (and others) are commonly used for functions. For example, suppose we want to name a function f. Then, for any x-value in the domain, we call the y-value (or function value) $f(x)$. The symbol $f(x)$ is read as "the value of the function f at x" or simply "f of x." In our example, the function $y = \frac{3}{2}x + 6$ can be written as $f(x) = \frac{3}{2}x + 6$. The notation $f(x)$ is called the **function notation**.

In our example, if $f(x) = \frac{3}{2}x + 6$, then the expression $f(-4)$ represents the range value (y-coordinate) when the domain value (x-coordinate) is -4. To find this value, replace x with -4 in the equation $f(x) = \frac{3}{2}x + 6$ to get $f(-4) = \frac{3}{2}(-4) + 6 = -6 + 6 = 0$. Thus, the ordered pair $(-4, 0)$ must lie on the graph of f. Table 1 shows this function evaluated at several values of x. The function $f(x) = \frac{3}{2}x + 6$ is a linear function whose graph is sketched in Figure 3.

CAUTION   The symbol $f(x)$ does not mean f times x. The notation $f(x)$ refers to the value of the function at x.

Table 1

Domain Value x	Range Value $f(x) = \dfrac{3}{2}x + 6$	Ordered Pair $(x, f(x))$
-4	$f(-4) = \dfrac{3}{2}(-4) + 6 = 0$	$(-4, 0)$
0	$f(0) = \dfrac{3}{2}(0) + 6 = 6$	$(0, 6)$
1	$f(1) = \dfrac{3}{2}(1) + 6 = \dfrac{15}{2}$	$\left(1, \dfrac{15}{2}\right)$
-3	$f(-3) = \dfrac{3}{2}(-3) + 6 = \dfrac{3}{2}$	$\left(-3, \dfrac{3}{2}\right)$

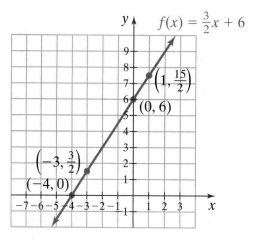

Figure 3 The graph of the function $f(x) = \frac{3}{2}x + 6$

94   Chapter 3   Functions

My interactive video summary

▶ Example 4 Use Function Notation

Rewrite these equations using function notation where y is a function of x. Then answer the question following each equation.

a. $3x - y = 5$
What is the value of $f(4)$?

b. $x^2 - 2y + 1 = 0$
Does the point $(-2, 1)$ lie on the graph of this function?

c. $y + 7 = 0$
What is $f(x)$ when $x = 3$?

Solution You should watch the interactive video to verify the following:

a. This equation can be rewritten in function notation as
$f(x) = 3x - 5$ and $f(4) = 7$.

b. This equation can be rewritten in function notation as $f(x) = \frac{1}{2}x^2 + \frac{1}{2}$. The point $(-2, 1)$ is *not* on the graph of this function because $f(-2) \neq 1$.

c. The equation $y + 7 = 0$ can be rewritten using function notation as $f(x) = -7$. When $x = 3$, the value of $f(x)$ is -7.

Example 5 Evaluate a Function

Given that $f(x) = x^2 + x - 1$, evaluate the following:

a. $f(0)$ **b.** $f(-1)$ **c.** $f(x + h)$ **d.** $\dfrac{f(x + h) - f(x)}{h}$

Solution

a. Substitute $x = 0$ into the function to get

$$f(0) = (0)^2 + (0) - 1 = -1$$

b. Substitute $x = -1$ into the function to get

$$f(-1) = (-1)^2 + (-1) - 1$$
$$= 1 - 1 - 1$$
$$= -1$$

⬥ **CAUTION** The expression $(-1)^2$ does *not* equal -1^2.

c. Substitute $x = x + h$ into the function to get

$$f(x + h) = (x + h)^2 + (x + h) - 1$$
$$= x^2 + 2xh + h^2 + x + h - 1$$

My video summary

 d. The expression $\dfrac{f(x + h) - f(x)}{h}$ is called the **difference quotient** and is very important in calculus. First, from part c, we see that $f(x + h) = x^2 + 2xh + h^2 + x + h - 1$. So,

$$\frac{f(x + h) - f(x)}{h} = \frac{\overbrace{x^2 + 2xh + h^2 + x + h - 1}^{f(x+h)} - \overbrace{[x^2 + x - 1]}^{f(x)}}{h}$$
$$= \frac{x^2 + 2xh + h^2 + x + h - 1 - x^2 - x + 1}{h}$$

3.1 Relations and Functions   **95**

$$= \frac{2xh + h^2 + h}{h}$$

$$= \frac{h(2x + h + 1)}{h}$$

$$= 2x + h + 1$$

Therefore, $\dfrac{f(x + h) - f(x)}{h} = 2x + h + 1.$

Notice the use of the letter h in part d of Example 5. Function notation is not restricted to the use of x for the independent variable or to the use of f for the name of the function, which is the dependent variable. Other letters commonly used in place of x are h, t, r, a, and b. Letters such as A for area, V for volume, and d for distance are often used to represent the names of functions. In fact, any letter can be used to represent values of either the domain or the range.

Work exercises 13–18 in this textbook or in the *MyMathLab*® Study Plan.

OBJECTIVE 4:   USING THE VERTICAL LINE TEST

My video summary

We see in Figure 3 (page 94) that the graph of the function $f(x) = \frac{3}{2}x + 6$ is created by plotting several ordered pairs in the Cartesian plane. Although every function has a graph, it is *not* the case that every graph in the plane represents a function. A graph does not represent a function if two or more points on the graph with the same first coordinate have different second coordinates. The **vertical line test** can be used to quickly determine whether a graph represents a function. See Figure 4.

> **Vertical Line Test**
>
> A graph in the Cartesian plane is the graph of a function if and only if no vertical line intersects the graph more than once.

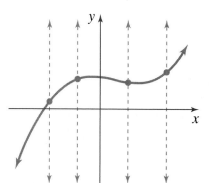

This graph is a function.
(No vertical line intersects the graph more than once.)

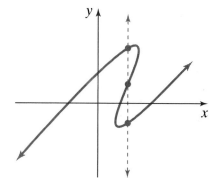

This graph is not a function.
(The graph does not pass the vertical line test.)

Figure 4

Example 6 Use the Vertical Line Test

Use the vertical line test to determine which of the following graphs represents the graph of a function.

a.

b.

c.

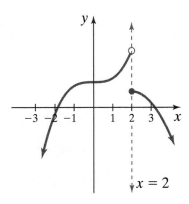

Solution

a. This graph represents a function because no vertical lines intersect the graph more than once.

b. At first glance, it may appear that this graph does not represent a function. However, we see that at the point where $x = 2$, there is only one corresponding y-value. The open dot indicates that the function is not defined at that point. The vertical line $x = 2$ only intersects the graph once; therefore, the graph represents the graph of a function.

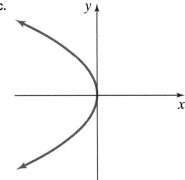

c. This graph does not represent a function because we can find several vertical lines that intersect the graph more than once.

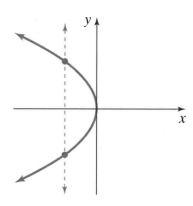

Work exercises 19–27 in this textbook or in the *MyMathLab*® Study Plan.

OBJECTIVE 5:   DETERMINING THE DOMAIN OF A FUNCTION
GIVEN THE EQUATION

My video summary

The domain of a function $y = f(x)$ is the set of all values of x for which the function is defined. In other words, a number $x = a$ is in the domain of a function f if $f(a)$ is a real number. For example, the domain of $f(x) = x^2$ is all real numbers because for any real number $x = a$, the value of $f(a) = a^2$ is also a real number.

It is very helpful to classify a function to determine its domain. For example, the function $f(x) = x^2$ belongs to a class of functions called **polynomial functions**. The domain of every polynomial function is all real numbers. Polynomial functions are discussed in great detail in Chapter Four.

DEFINITION Polynomial Function

The function $f(x) = a_n x^n + a_{n-1} x^{n-1} + a_{n-2} x^{n-2} + \cdots + a_1 x + a_0$ is a **polynomial function** of degree n, where n is a nonnegative integer and $a_0, a_1, a_2, \ldots, a_n$ are real numbers. The domain of every polynomial function is $(-\infty, \infty)$.

Many functions can have restricted domains. For example, the quotient of two polynomial functions is called a **rational function**. The rational function $f(x) = \frac{x+5}{x-4}$ is defined everywhere except when $x = 4$ because the value $f(4) = \frac{9}{0}$ is undefined. Therefore, the domain of a rational function consists of all real numbers for which the denominator does not equal zero. We investigate rational functions further in Chapter Four.

DEFINITION Rational Function

A **rational function** is a function of the form $f(x) = \frac{g(x)}{h(x)}$, where g and h are polynomial functions such that $h(x) \neq 0$.

The domain of a rational function is the set of all real numbers such that $h(x) \neq 0$.

Note: If $h(x) = c$, where c is a real number, then we will consider the function $f(x) = g(x)/h(x) = g(x)/c$ to be a polynomial function.

Root functions can also have restricted domains. Consider the root function $f(x) = \sqrt{x+1}$. The number $x = 3$ is in the domain because $f(3) = \sqrt{3+1} = \sqrt{4} = 2$ is a real number. However, -5 is not in the domain because $f(-5) = \sqrt{-5+1} = \sqrt{-4} = 2i$ is not a real number. Therefore, the domain of $f(x) = \sqrt{x+1}$ consists of all values of x for which the **radicand** is greater than or equal to zero. The domain of f is the solution to the inequality $x + 1 \geq 0$.

$$x + 1 \geq 0$$
$$x \geq -1 \quad \textit{Subtract 1 from both sides.}$$

Therefore, the domain of f is $[-1, \infty)$. Root functions with roots that are odd numbers such as 3 or 5 can have negative radicands. Therefore, the domain of a root function of the form $f(x) = \sqrt[n]{g(x)}$, where n is an odd positive integer, consists of all real numbers for which $g(x)$ is defined.

DEFINITION Root Function

The function $f(x) = \sqrt[n]{g(x)}$ is a **root function**, where n is an integer such that $n \geq 2$.

1. If n is *even*, the domain is the solution to the inequality $g(x) \geq 0$.
2. If n is *odd*, the domain is the set of all real numbers for which $g(x)$ is defined.

Following is a quick guide for finding the domain of three specific types of functions:

Class of Function	Form	Domain
Polynomial functions	$f(x) = a_n x^n + a_{n-1} x^{n-1} + \cdots + a_1 x + a_0$	Domain is $(-\infty, \infty)$.
Rational functions	$f(x) = \dfrac{g(x)}{h(x)}$ where g and h are polynomial functions such that $h(x) \neq 0$	Domain is all real numbers such that $h(x) \neq 0$.
Root functions	$f(x) = \sqrt[n]{g(x)}$, where $g(x)$ is a function and n is an integer such that $n \geq 2$.	1. If n is even, the domain is the solution to the inequality $g(x) \geq 0$. 2. If n is odd, the domain is the set of all real numbers for which g is defined.

We study other classes of functions later in this text. For now, we limit our discussion to polynomial functions, rational functions, and root functions.

My interactive video summary

⬚ **Example 7 Find the Domain of a Function Given the Equation**

Classify each function as a polynomial function, rational function, or root function, and then find the domain. Write the domain in interval notation.

a. $f(x) = 2x^2 - 5x$ **b.** $f(x) = \dfrac{x}{x^2 - x - 6}$ **c.** $h(x) = \sqrt{x^2 - 2x - 8}$

d. $f(x) = \sqrt[3]{5x - 9}$ **e.** $k(x) = \sqrt[4]{\dfrac{x-2}{x^2 + x}}$

Solution Work through the interactive video to verify the following:

a. The function $f(x) = 2x^2 - 5x$ is a polynomial function. The domain of f is all real numbers or $(-\infty, \infty)$.

b. The function $f(x) = \frac{x}{x^2 - x - 6}$ is a rational function. The domain of f is $\{x \mid x \neq -2, x \neq 3\}$. The domain in interval notation is $(-\infty, -2) \cup (-2, 3) \cup (3, \infty)$.

c. The function $h(x) = \sqrt{x^2 - 2x - 8}$ is a root function. The domain of h in interval notation is $(-\infty, -2] \cup [4, \infty)$.

d. The function $f(x) = \sqrt[3]{5x - 9}$ is a root function. The domain of f is all real numbers or $(-\infty, \infty)$.

e. The function $k(x) = \sqrt[4]{\frac{x - 2}{x^2 + x}}$ is a root function. The domain of k in interval notation is $(-1, 0) \cup [2, \infty)$.

Work exercises 28–46 in this textbook or in the _MyMathLab_® Study Plan.

3.1 Exercises

In Exercises 1–6, find the domain and range of each relation, and then determine whether each relation represents a function.

1.

2.
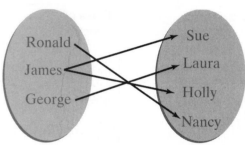

3. $\{(-1, 5), (0, 2), (1, 5), (2, 3), (1, 7)\}$

4. $\{(-1, 5), (0, 2), (1, 5), (2, 3), (1, 7)\}$

5.

6.
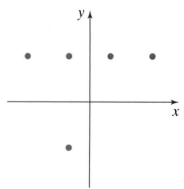

In Exercises 7–12, determine whether the equation represents y as a function of x.

7. $x^2 + y = 1$

8. $x^2 + y^2 = 1$

9. $x^3 + y^3 = 1$

10. $3xy = 6$

11. $3x - 4y = 10$

12. $6x + 1 = y^4$

In Exercises 13–18, evaluate the function at the indicated values.

13. $f(x) = x^2 + 1$

a. $f(2)$ b. $f(-2)$ c. $f(x - 1)$

14. $f(t) = \dfrac{t}{t + 1}$

    a. $f(-2)$ **b.** $f(7)$ **c.** $f(a)$

15. $V(r) = \dfrac{4}{3}\pi r^3$

    a. $V(3)$ **b.** $V(6)$ **c.** $V(r + 1)$

16. $f(x) = x^2 - 2x$

    a. $f(x + 1)$ **b.** $f(x + h)$ **c.** $\dfrac{f(x + h) - f(x)}{h}$

17. $g(t) = t^3 - t$

    a. $g(t + 1)$ **b.** $g(t + h)$ **c.** $\dfrac{g(t + h) - g(t)}{h}$

18. $g(x) = \sqrt{x} - 2x$

    a. $g(4)$ **b.** $g(x + h)$ **c.** $\dfrac{g(x + h) - g(x)}{h}$

In Exercises 19–27, use the vertical line test to determine whether the graph represents a function.

19.

20.

21.

22.

23.

24.

25.

26.

27.
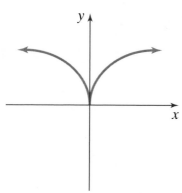

In Exercises 28–46, classify each function as a polynomial function, rational function, or root function, and then find the domain. Write the domain in interval notation.

28. $f(t) = 6t^4 + t^3 - 2t^2 + t - 5$

29. $h(t) = \dfrac{8 - t}{t^2}$

30. $f(x) = 8x^2 - x$

31. $g(x) = \sqrt{3 - x}$

32. $f(t) = \sqrt[3]{t - 5}$

33. $f(x) = \dfrac{2x - 1}{2x + 1}$

34. $f(x) = \sqrt[5]{\dfrac{x + 1}{x - 5}}$

35. $g(x) = 2x - 1$

36. $h(x) = \dfrac{x^2 + 4}{x^2 + x - 42}$

37. $f(x) = 4$

38. $h(t) = \sqrt[5]{t^4 - 1}$

39. $f(t) = \sqrt[4]{t^4 - 1}$

40. $g(t) = \dfrac{t + 5}{t^3 - 3t^2 - 4t}$

41. $f(x) = \sqrt{x^2 + 2x - 15}$

42. $h(x) = x^3 - 1$

43. $g(x) = \sqrt{\dfrac{x^2 - 9}{5}}$

44. $f(x) = \sqrt{\dfrac{5}{x^2 - 9}}$

45. $h(x) = \dfrac{1}{x^2 + 1}$

46. $f(x) = \sqrt[6]{\dfrac{2x - 6}{x^2 - x - 20}}$

3.2 Properties of a Function's Graph

THINGS TO KNOW

Before working through this section, be sure that you are familiar with the following concepts:

VIDEO   ANIMATION   INTERACTIVE

  You Try It
  1. Solving Higher-Order Polynomial Equations
     (Section 1.6)

  You Try It
  2. Understanding the Definitions of Relations and
     Functions (Section 3.1)

  You Try It
  3. Using the Vertical Line Test
     (Section 3.1)

  You Try It
  4. Determining the Domain of a Function Given
     the Equation (Section 3.1)

OBJECTIVES

1 Determining the Intercepts of a Function

2 Determining the Domain and Range of a Function from Its Graph

3 Determining whether a Function Is Increasing, Decreasing, or Constant

4 Determining Relative Maximum and Relative Minimum Values of a Function

5 Determining whether a Function Is Even, Odd, or Neither

6 Determining Information about a Function from a Graph

OBJECTIVE 1:   DETERMINING THE INTERCEPTS OF A FUNCTION

My video summary

In Section 3.1, we discussed that every function has a graph in the Cartesian plane. In this section, we take a closer look at a function's graph by discussing some of its characteristics. A good place to start is to talk about a function's intercepts. An **intercept** of a function is a point on the graph of a function where the graph either crosses or touches a coordinate axis. There are two types of intercepts:

1. The y-intercept, which is the y-coordinate of the point where the graph crosses or touches the y-axis

2. The x-intercepts, which are the x-coordinates of the points where the graph crosses or touches the x-axis

The y-intercept

A function can have, *at most,* one y-intercept. The y-intercept exists if $x = 0$ is in the domain of the function. The y-intercept can be found by evaluating $f(0)$.

Example 1 Determine the y-Intercept of a Function

Find the y-intercept of the function $f(x) = -3x + 2$.

Solution Because $f(x) = -3x + 2$ is a polynomial function, we know that the domain of f is composed of all real numbers. Thus, $x = 0$ is in the domain of f. The y-intercept is $f(0) = -3(0) + 2 = 2$. We say that the y-intercept of f is found at the point $(0, 2)$. The graph of f is the line sketched in Figure 5. You can see that the graph intersects the y-axis at $(0, 2)$.

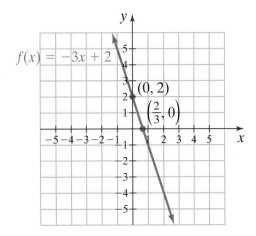

Figure 5

THE x-INTERCEPT(S)

A function may have several (even infinitely many) x-intercepts. The x-intercepts, also called **real zeros**, can be found by finding all *real solutions* to the equation $f(x) = 0$. Although a function may have several zeros, only the real zeros are x-intercepts. The only real zero of $f(x) = -3x + 2$ is $x = \frac{2}{3}$ because the solution to the equation $-3x + 2 = 0$ is $x = \frac{2}{3}$. Notice from the graph in Figure 5 that the graph crosses the x-axis at the point $\left(\frac{2}{3}, 0\right)$.

Example 2 Find the Intercepts of a Function

Find all intercepts of the function $f(x) = x^3 - 2x^2 + x - 2$.

Solution Because $f(0) = -2$, the y-intercept is -2. To find the x-intercepts of f, we must find the real solutions to the equation $f(x) = 0$ or $x^3 - 2x^2 + x - 2 = 0$. To solve this equation, we can factor by grouping and solve to get the solution set $\{2, i, -i\}$. Click here to see the worked out solution. Because $x = 2$ is the only real solution, 2 is the only x-intercept.

Note: The solutions $x = i$ and $x = -i$ are **zeros** of f because $f(i) = 0$ and $f(-i) = 0$; however, these values do not represent x-intercepts because they are not real numbers.

Using Technology

The graph of $f(x) = x^3 - 2x^2 + x - 2$ was created using a graphing utility. Notice that the graph crosses the x-axis at the point $(2, 0)$ and crosses the y-axis at $(0, -2)$. Therefore, the x-intercept is 2, and the y-intercept is -2.

Work exercises 1–6 in this textbook or in the *MyMathLab*® Study Plan.

OBJECTIVE 2: DETERMINING THE DOMAIN AND RANGE OF A FUNCTION FROM ITS GRAPH

My video summary

In Section 3.1, we see that every function has a graph. Figure 6 illustrates how we can use a function's graph to determine the domain and range. The domain is the set of all input values (x-values), and the range is the set of all output values

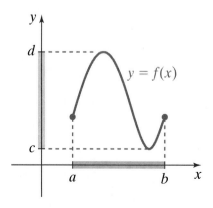

Figure 6 Domain and range of $y = f(x)$

(*y*-values). In Figure 6, the domain is the interval $[a, b]$ while the range is the interval $[c, d]$.

My video summary

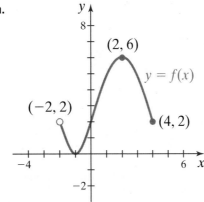

■ **Example 3** Find the Domain and Range of a Function from Its Graph

Use the graph of the following functions to determine the domain and range.

a.

b.

c.

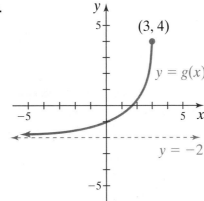

Solution

a. The open dot representing the ordered pair $(-2, 2)$ indicates that this point is not included on the graph, whereas the closed dot representing the ordered pair $(4, 2)$ indicates that this point is included on the graph. Therefore, the domain includes all *x*-values between -2, not including -2, and 4 inclusive. Thus, the domain in set-builder notation is $\{x \mid -2 < x \leq 4\}$ or the interval $(-2, 4]$. The range includes all *y*-values between 0 and 6 inclusive because the points $(-1, 0)$ and $(2, 6)$ are both included on the graph. The range is $\{y \mid 0 \le y \le 6\}$ or the interval $[0, 6]$.

b. The vertical dashed line at $x = 5$ is called a **vertical asymptote** and indicates that the function will get arbitrarily close to the line $x = 5$ but will never intersect the line. Therefore, the domain of f in set-builder notation is $\{x \mid -4 < x < 5\}$ or the interval $(-4, 5)$. The range includes all y-values strictly greater than -3. Thus, the range is $\{y \mid y > -3\}$ or the interval $(-3, \infty)$.

c. To determine the domain of g, notice that the graph extends indefinitely to the left and includes all values of x less than or equal to 3. Thus, the domain is $\{x \mid x \le 3\}$ or the interval $(-\infty, 3]$. The horizontal dashed line $y = -2$ is called a **horizontal asymptote** and indicates that the function will get arbitrarily close to the line $y = -2$. Thus, the range includes all y-values greater than -2 but less than or equal to 4. So, the range of g is $\{y \mid -2 < y \le 4\}$ or the interval $(-2, 4]$.

Work exercises 7–12 in this textbook or in the MyMathLab® Study Plan.

OBJECTIVE 3:   DETERMINING WHETHER A FUNCTION IS INCREASING, DECREASING, OR CONSTANT

My video summary

A function f is said to be **increasing** on an **open interval** if the value of $f(x)$ gets larger as x gets larger on the interval. The graph of f rises from left to right on the interval in which f is increasing. Likewise, a function f is said to be **decreasing** on an open interval if the value of $f(x)$ gets smaller as x gets larger on the interval. The graph of f falls from left to right on the interval in which f is decreasing. A graph is **constant** on an open interval if the values of $f(x)$ do not change as x gets larger on the interval. In this case, the graph is a horizontal line on the interval. Figure 7 illustrates all three cases.

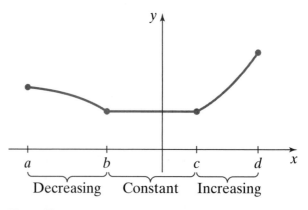

Figure 7 The function is increasing on the interval (c, d). The function is decreasing on the interval (a, b). The function is constant on the interval (b, c).

My video summary

Example 4 Determine whether a Function Is Increasing, Decreasing, or Constant from Its Graph

Given the graph of $y = f(x)$, determine whether the function is increasing, decreasing, or constant.

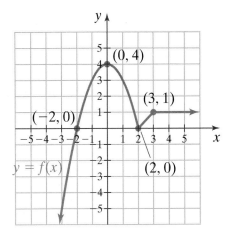

Solution The function is increasing on the interval $(-\infty, 0)$ and on the interval $(2, 3)$. The function is decreasing on the interval $(0, 2)$. The function is constant on the interval $(3, \infty)$. See the video to see the solution worked out in detail.

Work exercises 13–16 in this textbook or in the *MyMathLab*® Study Plan.

OBJECTIVE 4: DETERMINING RELATIVE MAXIMUM AND RELATIVE MINIMUM VALUES OF A FUNCTION

My video summary

When a function changes from increasing to decreasing at a point $(c, f(c))$, then f is said to have a relative maximum at $x = c$. The relative maximum value is $f(c)$. See Figure 8(a). Similarly, when a function changes from decreasing to increasing at a point $(c, f(c))$, then f is said to have a relative minimum at $x = c$. The relative minimum value is $f(c)$. See Figure 8(b).

(a)

(b)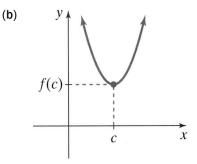

The relative maximum occurs at $x = c$, and the relative maximum value is $f(c)$.

The relative minimum occurs at $x = c$, and the relative minimum value is $f(c)$.

Figure 8

Note: The word "relative" indicates that the function obtains a maximum or minimum value relative to some **open interval**. It is not necessarily the maximum (or minimum) value of the function on the entire domain.

Example 5 Determine Information from the Graph of a Function

Use the graph of $y = f(x)$ to answer each question.

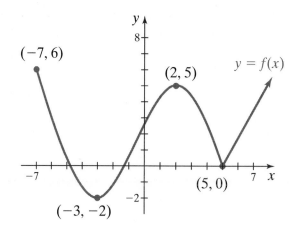

a. On what interval(s) is f increasing?
b. On what interval(s) is f decreasing?
c. At what number(s) does f have a relative minimum?
d. At what number(s) does f have a relative maximum?
e. What are the relative minima?
f. What are the relative maxima?

Solution

a. The function f is increasing on the interval $(-3, 2)$. The function is also increasing for values of x greater than 5; that is, f is also increasing on the interval $(5, \infty)$.

b. The function f is decreasing on the intervals $(-7, -3)$ and $(2, 5)$.

c. The function f obtains a relative minimum at $x = -3$ and $x = 5$.

d. The function f obtains a relative maximum at $x = 2$.

e. The relative minima are $f(-3) = -1$ and $f(5) = 0$.

f. The relative maximum is $f(2) = 4$.

CAUTION In Example 5, the value $f(-7) = 6$ does not represent a relative maximum. This is because the function does not change from increasing to decreasing at the point $(-7, 6)$. Also, there is no open interval that includes $x = -7$ because the function is not defined for values of x less than -7.

Work exercises 17–21 in this textbook or in the *MyMathLab*® Study Plan.

OBJECTIVE 5: DETERMINING WHETHER A FUNCTION IS EVEN, ODD, OR NEITHER

My video summary

The graphs of some functions are symmetric about the y-axis. For example, the graphs of the three functions sketched in Figure 9 are all symmetric about the y-axis.

Imagine folding each graph of Figure 9 along the y-axis. If each graph was folded along the y-axis, the left and right halves of each graph would match. Functions whose graphs are symmetric about the y-axis are called *even functions*. For any point (x, y) on each graph, the point $(-x, y)$ also lies on the graph. Therefore, for any x value in the domain, $f(x) = f(-x)$.

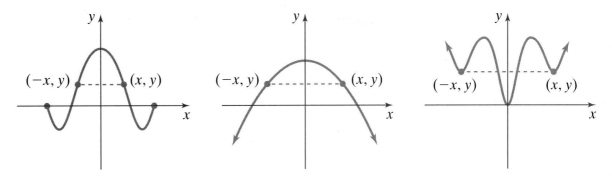

Figure 9 Functions whose graphs are symmetric about the y-axis are called even functions.

DEFINITION Even Functions

A function f is **even** if for every x in the domain, $f(x) = f(-x)$. Even functions are symmetric about the y-axis. For each point (x, y) on the graph, the point $(-x, y)$ is also on the graph.

Functions can also be symmetric about the origin. The functions sketched in Figure 10 illustrate this type of symmetry. Functions whose graphs are symmetric about the origin are called *odd functions*. If a function is an odd function, then for any point (x_1, y_1) on the graph of f in quadrant I, there is a corresponding point $(-x_1, -y_1)$ on the graph of f in quadrant III. Similarly, for any point (x_2, y_2) on the graph of f in quadrant II, there is a corresponding point $(-x_2, -y_2)$ on the graph of f in quadrant IV.

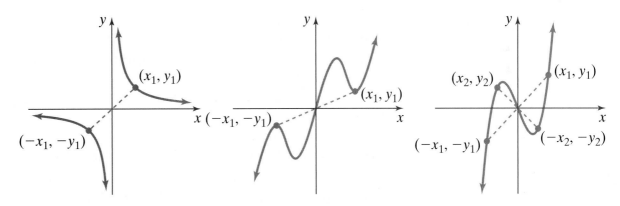

Figure 10 Functions whose graphs are symmetric about the origin are called odd functions.

For any point (x, y) on each graph in Figure 10, the point $(-x, -y)$ also lies on the graph. Therefore, for any x value in the domain, $f(x) = -f(-x)$ or, equivalently, $-f(x) = f(-x)$.

DEFINITION Odd Functions

A function f is **odd** if for every x in the domain, $-f(x) = f(-x)$. Odd functions are symmetric about the origin. For each point (x, y) on the graph, the point $(-x, -y)$ is also on the graph.

3.2 Properties of a Function's Graph   109

Example 6 Determine Even and Odd Functions from the Graph

Determine whether each function is even, odd, or neither.

a.

b.

c.
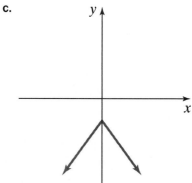

Solution

a. The graph is neither even nor odd because the graph has no symmetry.

b. The graph is symmetric about the origin; therefore, this is the graph of an odd function.

c. This graph is symmetric about the y-axis; thus, this represents the graph of an even function.

Work exercises 22–25 in this textbook or in the *MyMathLab*® Study Plan.

My video summary

Example 8 Determine Even and Odd Functions Algebraically

Determine whether each function is even, odd, or neither.

a. $f(x) = x^3 + x$

b. $g(x) = \dfrac{1}{x^2} + 7|x|$

c. $h(x) = 2x^5 - \dfrac{1}{x}$

d. $G(x) = x^2 + 4x$

Solution Watch the video to verify that the functions in parts a and c are odd functions. The function in part b is even, whereas the function in part d is neither even nor odd.

Work exercises 26–30 in this textbook or in the *MyMathLab*® Study Plan.

OBJECTIVE 6:   DETERMINING INFORMATION ABOUT A FUNCTION FROM A GRAPH

 My animation summary

We are now ready to take the concepts learned in this section to answer questions about a function's graph. Carefully work through the animation of Example 9.

Example 9 **Determine Information about a Function from a Graph**

Use the graph of $y = f(x)$ to answer each question.

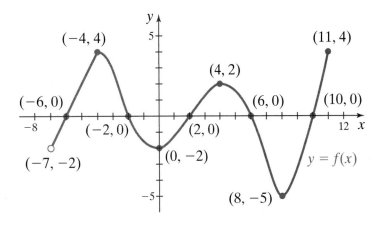

a. What is the y-intercept?
b. What are the real zeros of f?
c. Determine the domain and range of f.
d. Determine the interval(s) on which f is increasing, decreasing, and constant.
e. For what value(s) of x does f obtain a relative maximum? What are the relative maxima?
f. For what value(s) of x does f obtain a relative minimum? What are the relative minima?
g. Is f even, odd, or neither?
h. For what values of x is $f(x) = 4$?
i. For what values of x is $f(x) < 0$?

Solution Carefully work through the animation to see the solution worked out in detail.

Work exercises 31 and 32 in this textbook or in the _MyMathLab_® Study Plan

3.2 Exercises

In Exercises 1–6, find the x-intercept(s) and the y-intercept of each function.

1. $f(x) = 2x - 1$

2. $g(x) = 3$

3. $f(t) = t^2 - t - 6$

4. $F(x) = 2x^2 - 7x - 3$

5. $h(x) = x^3 + 2x^2 - x - 2$

6. $f(x) = 2x^3 + 11x^2 - 6x$

In Exercises 7–12, use the graph to determine the domain and range of each function.

7.

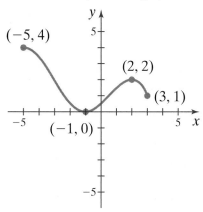

$(-5, 4)$
$(2, 2)$
$(3, 1)$
$(-1, 0)$

8.

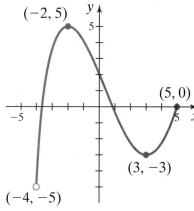

$(-2, 5)$
$(5, 0)$
$(3, -3)$
$(-4, -5)$

9.

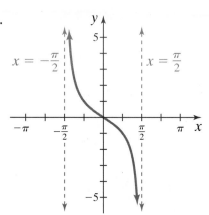

$x = -\dfrac{\pi}{2}$
$x = \dfrac{\pi}{2}$

10.

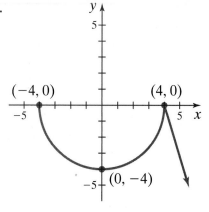

$(-4, 0)$
$(4, 0)$
$(0, -4)$

11.

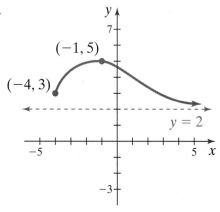

$(-1, 5)$
$(-4, 3)$
$y = 2$

12.

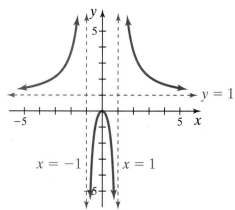

$y = 1$
$x = -1$
$x = 1$

In Exercises 13–16, determine the interval(s) for which each function is (a) increasing, (b) decreasing, and (c) constant.

13.

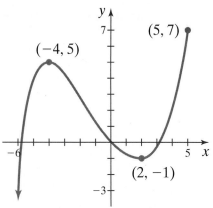

$(5, 7)$
$(-4, 5)$
$(2, -1)$

14.

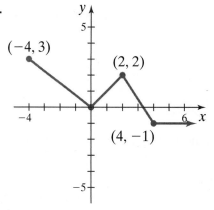

$(-4, 3)$
$(2, 2)$
$(4, -1)$

15.

16.

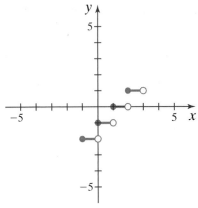

In Exercises 17–21, (a) find the number(s) for which the function obtains a relative minimum, (b) find the relative minimum value, (c) find the number(s) for which the function obtains a relative maximum, and (d) find the relative maximum value.

17.

18.

19.

20.

21.

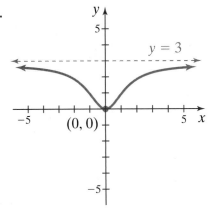

In Exercises 22–30 determine whether each function is even, odd, or neither.

22.

23.

24.

25.

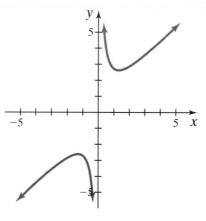

26. $f(x) = 2x^4 - x^2 + 1$ **27.** $g(x) = x^5 - x^2$ **28.** $h(x) = x^5 - x^3$

29. $F(x) = \dfrac{2x^2 - 4}{x}$ **30.** $G(x) = \sqrt{9 - x^2}$

31. Use the graph of $y = f(x)$ to answer each of the following questions:

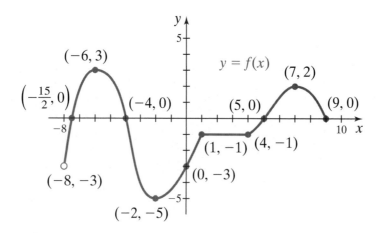

a. What is the domain of f?

b. What is the range of f?

c. On what interval(s) is f increasing, decreasing, or constant?

d. For what value(s) of x does f obtain a relative minimum? What are the relative minima?

e. For what value(s) of x does f obtain a relative maximum? What are the relative maxima?

f. What are the real zeros of f?

g. What is the y-intercept?

h. For what values of x is $f(x) \leq 0$?

i. For how many values of x is $f(x) = -3$?

j. What is the value of $f(2)$?

32. Use the graph of $y = f(x)$ to answer each of the following questions:

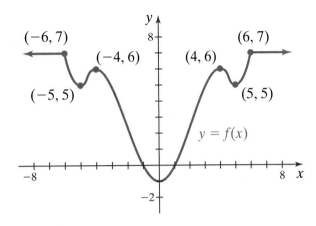

a. What is the domain of f?

b. What is the range of f?

c. On what interval(s) is f increasing, decreasing, or constant?

d. For what value(s) of x does f obtain a relative minimum? What are the relative minima?

e. For what value(s) of x does f obtain a relative maximum? What are the relative maxima?

f. What are the x-intercept(s) of f?

g. Is f even, odd, or neither?

h. For what values of x is $f(x) > 0$?

i. True or false: $f(-3) > f(2)$

j. For what value(s) of x is $f(x) = 7$?

3.3 Graphs of Basic Functions; Piecewise Functions

THINGS TO KNOW

Before working through this section, be sure that you are familiar with the following concepts:

VIDEO   ANIMATION   INTERACTIVE

You Try It

1. Determining the Domain of a Function Given the Equation (Section 3.1)

You Try It

2. Determining the Domain and Range of a Function from Its Graph (Section 3.2)

OBJECTIVES

1 Sketching the Graphs of the Basic Functions

2 Analyzing Piecewise-Defined Functions

3 Solving Applications of Piecewise-Defined Functions

OBJECTIVE 1:   SKETCHING THE GRAPHS OF THE BASIC FUNCTIONS

There are several functions whose graphs are worth memorizing. In this section, you learn about the graphs of nine basic functions. We begin by discussing the graphs of two specific linear functions. Recall that a linear function has the form $f(x) = mx + b$, where m is the slope of the line and b represents the y-coordinate of the y-intercept.

We start our discussion of the basic functions by looking at the constant linear function, that is, the linear function with $m = 0$, the graph of which is a horizontal line.

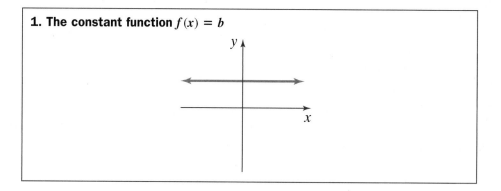

1. The constant function $f(x) = b$

The **identity function** defined by $f(x) = x$ is a linear function with $m = 1$ and $b = 0$. This function assigns to each number in the domain the exact same number in the range.

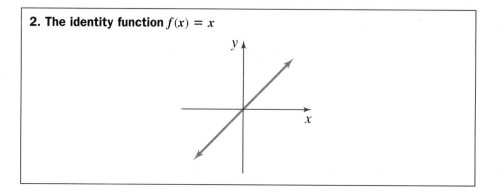

2. The identity function $f(x) = x$

Of course, you should be able to sketch the graph of any linear function of the form $f(x) = mx + b$. Click here to review how to sketch a line that is given in **slope–intercept form**.

The **square function**, $f(x) = x^2$, assigns to each real number in the domain the square of that number in the range.

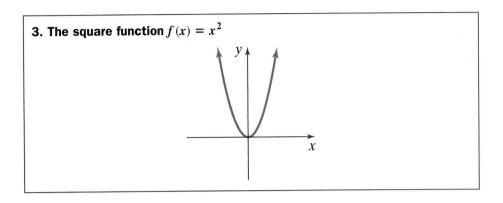

3. The square function $f(x) = x^2$

The **cube function**, $f(x) = x^3$, assigns to each real number in the domain the cube of that number in the range.

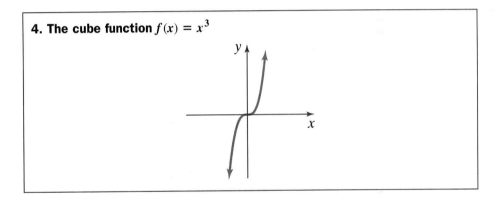

4. The cube function $f(x) = x^3$

The **absolute value function**, $f(x) = |x|$, assigns to each real number in the domain the absolute value of that number in the range.

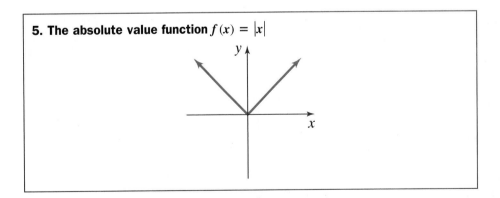

5. The absolute value function $f(x) = |x|$

The **square root function**, $f(x) = \sqrt{x}$, is only defined for values of x that are greater than or equal to zero. It assigns to each real number in the domain the square root of that number in the range.

6. The square root function $f(x) = \sqrt{x}$

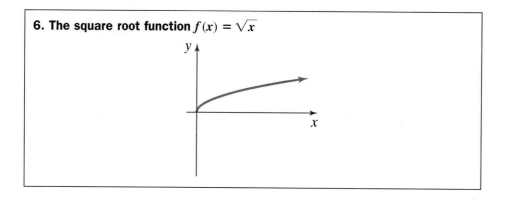

Unlike the square root function, which is only defined for values of x greater than or equal to zero, the **cube root function**, $f(x) = \sqrt[3]{x}$, is defined for all real numbers and assigns to each number in the domain the cube root of that number in the range.

7. The cube root function $f(x) = \sqrt[3]{x}$

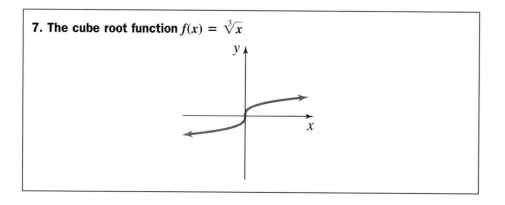

The reciprocal function, $f(x) = \frac{1}{x}$, is a **rational function** whose domain is $\{x \mid x \neq 0\}$. It assigns to each number a in the domain its reciprocal, $\frac{1}{a}$, in the range. The reciprocal function has two asymptotes. The y-axis (the line $x = 0$) is a **vertical asymptote**, and the x-axis (the line $y = 0$) is a **horizontal asymptote**.

8. The reciprocal function $f(x) = \frac{1}{x}$

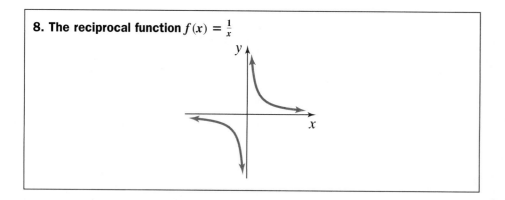

The greatest integer function, $f(x) = [\![x]\!]$, (sometimes denoted $f(x) = \text{int}(x)$) assigns to each number in the domain the greatest integer less than or equal to that number in the range. For example, $[\![3.001]\!] = 3$, whereas $[\![2.999]\!] = 2$.

9. The greatest integer function $f(x) = [\![x]\!]$

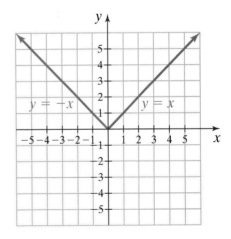

Work exercises 1–9 in this textbook or in the **MyMathLab** ® Study Plan.

OBJECTIVE 2:   ANALYZING PIECEWISE-DEFINED FUNCTIONS

The absolute value function, $f(x) = |x|$, can also be defined by a rule that has two different pieces.

$$f(x) = |x| = \begin{cases} x \text{ if } x \geq 0 \\ -x \text{ if } x < 0 \end{cases}$$

You can see by the following graph that the left-hand piece is actually part of the line $y = -x$, whereas the right-hand piece is part of the line $y = x$.

Functions defined by a rule that has more than one piece are called piecewise-defined functions.

My animation summary

Example 1 Sketch a Piecewise-Defined Function

Sketch the function $f(x) = \begin{cases} x^2 & \text{if } x < 1 \\ 1 - x & \text{if } x \geq 1 \end{cases}$.

Solution The graph of this piecewise function is sketched as follows. Watch the animation to see how to sketch this function.

3.3 Graphs of Basic Functions; Piecewise Functions   119

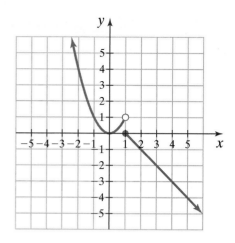

My video summary

📼 **Example 2** Analyze a Piecewise-Defined Function

$$\text{Let } f(x) = \begin{cases} 1 & \text{if } x < -1 \\ \sqrt[3]{x} & \text{if } -1 \le x < 0 \\ \dfrac{1}{x} & \text{if } x > 0 \end{cases}.$$

a. Evaluate $f(-3)$, $f(-1)$, and $f(2)$.
b. Sketch the graph of f.
c. Determine the domain of f.
d. Determine the range of f.

Solution Watch the video to view the solution.

Work exercises 10–15 in this textbook or in the *MyMathLab*® Study Plan.

My video summary

📼 **Example 3** Find a Rule That Defines a Piecewise Function

Find a rule that describes the following piecewise function:

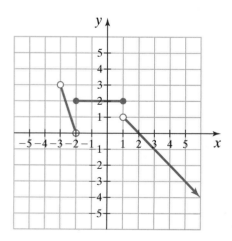

Solution Watch this video to see that we can define this piecewise function as

$$f(x) = \begin{cases} -3x - 6 & \text{if } -3 < x < -2 \\ 2 & \text{if } -2 \le x \le 1 \\ -x + 2 & \text{if } x > 1 \end{cases}$$

Work exercises 16–20 in this textbook or in the *MyMathLab*® Study Plan.

OBJECTIVE 3:   SOLVING APPLICATIONS OF PIECEWISE-DEFINED FUNCTIONS

Example 4 A Flat Tax Plan

Steve Forbes, a presidential candidate in 1996, proposed a flat tax to replace the existing U.S. income tax system. In his tax proposal, every adult would pay $0.00 in taxes on the first $13,000 earned. They would then pay a flat tax of 17% on everything over $13,000. Forbes' tax plan is actually a piecewise-defined function.

a. According to Forbes' plan, how much in taxes are owed for someone earning $50,000?

b. Find the piecewise function, $T(x)$, that describes the amount of taxes paid, T, as a function of the dollars earned, x, for Forbes' tax plan.

c. Sketch the piecewise function, $T(x)$.

Solution

a. On $50,000, a person would pay 0% on the first $13,000 and 17% on the remaining $37,000 $(50,000 - 13,000)$. Thus, the taxes owed would be $0.17(37,000) = \$6,290$.

b. For values of x less than 13,000, $T(x) = 0$. When x is greater than 13,000, the interest rate is 17% or $T(x) = 0.17(x - 13,000)$ for $x > 13,000$. Thus, the rule for computing the tax owed is

$$T(x) = \begin{cases} 0 & \text{if } x \le 13{,}000 \\ 0.17(x - 13{,}000) & \text{if } x > 13{,}000 \end{cases}.$$

c. The graph of $T(x)$ is sketched as follows.

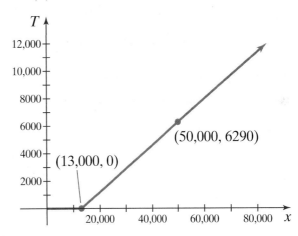

My video summary

▦ Example 5 Rent a Car

Cheapo Rental Car Co. charges a flat rate of $80 to rent a car for up to 2 days. The company charges an additional $20 for each additional day (or part of a day).

a. How much does it cost to rent a car for $2\frac{1}{2}$ days? 3 days?

b. Write and sketch the piecewise function, $C(x)$, that describes the cost of renting a car as a function of the time, x, in days rented for values of x less than or equal to 5.

Solution The rule for this piecewise function is $C(x) = \begin{cases} 85 & \text{if } x \le 2 \\ 105 & \text{if } 2 < x \le 3 \\ 125 & \text{if } 3 < x \le 4 \\ 145 & \text{if } 4 < x \le 5 \end{cases}$

and is sketched as follows. Watch the video for a detailed explanation.

3.3 Graphs of Basic Functions; Piecewise Functions   121

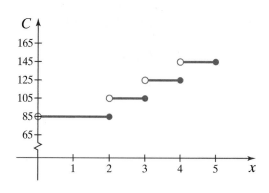

Work exercises 21–24 in this textbook or in the *MyMathLab*® Study Plan.

3.3 Exercises

In Exercises 1–9, sketch the graph of each function, and identify as many properties on the right that apply to each:

1. $f(x) = x^2$

2. $f(x) = \sqrt[3]{x}$

3. $f(x) = 2x - 1$

4. $f(x) = |x|$

5. $f(x) = -4$

6. $f(x) = x$

7. $f(x) = x^3$

8. $f(x) = \sqrt{x}$

9. $f(x) = \dfrac{1}{x}$

Properties

a. The function f is an even function.

b. The function f is an odd function.

c. The function f is increasing on the interval $(-\infty, \infty)$.

d. The function f is a linear function.

e. The domain of f is $(-\infty, \infty)$.

f. The range of f is $(-\infty, \infty)$.

In Exercises 10–15, (a) find $f(-2)$, $f(0)$, and $f(2)$; (b) sketch the graph of each piecewise-defined function; (c) determine the domain of f; and (d) determine the range of f.

10. $f(x) = \begin{cases} 1 & \text{if } x < 2 \\ -3 & \text{if } x \geq 2 \end{cases}$

11. $f(x) = \begin{cases} 2x - 1 & \text{if } x \leq 1 \\ x + 2 & \text{if } x > 1 \end{cases}$

12. $f(x) = \begin{cases} x^2 & \text{if } x \leq 0 \\ \dfrac{1}{x} & \text{if } x > 0 \end{cases}$

13. $f(x) = \begin{cases} 3 & \text{if } x < -1 \\ |x| & \text{if } -1 \geq x < 2 \\ 1 & \text{if } x \geq 2 \end{cases}$

14. $f(x) = \begin{cases} x^2 & \text{if } x < 0 \\ 2 & \text{if } x = 0 \\ \sqrt{x} & \text{if } x > 0 \end{cases}$

15. $f(x) = \begin{cases} [\![x]\!] & \text{if } -3 \leq x \leq -1 \\ x + 2 & \text{if } -1 < x < 1 \\ x^3 & \text{if } x \geq 1 \end{cases}$

In Exercises 16–20, find the rule that describes each piecewise-defined function.

16.

17.

18.

19.

20.

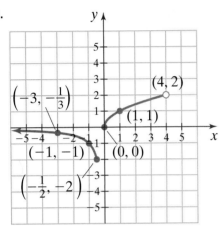

21. The currency in Norway is called the krone (plural: kroner) and is abbreviated NOK. Following is a table describing the 1999 class 1 individual tax rate on personal income in Norway.

INCOME	TAX RATE
NOK 0–26,300	0%
Over NOK 26,300	28%

a. How many Kroner in taxes are owed for an individual earning NOK 30,000?

b. Find the piecewise function, $T(x)$, that describes the amount of taxes paid, T, as a function of Kroner earned, x, for Norwegians paying class 1 personal income tax in Norway in 1999.

c. Sketch the piecewise function, $T(x)$.

22. An electric company charges $9.50 per month plus $0.07 per kilowatt-hour (kWhr) for the first 350 kWhr used per month and $0.05 per kWhr for anything over 350 kWhr used.

a. How much is owed for using 280 kWhr per month?

b. How much is owed for using 400 kWhr per month?

c. Write a piecewise function, $C(x)$, where C is the cost charged per month as a function of the monthly kilowatt-hours used, x.

d. Sketch the piecewise function.

23. In 2007, the postage charged for a first-class mail parcel was $1.13 for the first ounce (or fraction of an ounce) and $0.17 for each additional ounce (or fraction of an ounce).

a. How much does it cost to send a 2.4-ounce first-class letter?

b. Write a piecewise function, $P(x)$, where P is the postage charged as a function of its weight in ounces, x, for values of x less than or equal to 5 ounces.

c. Sketch the piecewise function.

Source: United States Postal Service.

24. A certain cell phone company charges $25.00 per month for the first 500 minutes of use and $0.05 for each additional minute (or part of a minute).

a. How much is owed to the cell phone company if a person used 502.5 minutes?

b. Write a piecewise function, $C(x)$, where C is the cost charged per month as a function of the minutes, x, used for values of x less than or equal to 505 minutes.

c. Sketch the piecewise function.

3.4 **Transformations of Functions**

THINGS TO KNOW

Before working through this section, be sure that you are familiar with the following concepts:

VIDEO   ANIMATION   INTERACTIVE

You Try It
1. Determining the Domain of a Function Given the Equation (Section 3.1)

You Try It
2. Determining the Domain and Range of a Function from Its Graph (Section 3.2)

You Try It
3. Determining Whether a Function Is Even, Odd, or Neither (Section 3.2)

You Try It
4. Sketching the Graphs of the Basic Functions (Section 3.3)

OBJECTIVES

1 Using Vertical Shifts to Graph Functions

2 Using Horizontal Shifts to Graph Functions

3 Using Reflections to Graph Functions

4 Using Vertical Stretches and Compressions to Graph Functions

5 Using Horizontal Stretches and Compressions to Graph Functions

6 Using Combinations of Transformations to Graph Functions

In this section, we learn how to sketch the graphs of new functions using the graphs of known functions. Starting with the graph of a known function, we "transform" it into a new function by applying various transformations. Before we begin our discussion about transformations, it is critical that you know the graphs of the basic functions that are discussed in Section 3.3. Take a moment to review the basic functions.

REVIEW OF THE BASIC FUNCTIONS

Click on any function to review its graph.

The **identity function**	$f(x) = x$
The **cube function**	$f(x) = x^3$
The **square root function**	$f(x) = \sqrt{x}$
The **reciprocal function**	$f(x) = \dfrac{1}{x}$
The **square function**	$f(x) = x^2$
The **absolute value function**	$f(x) = |x|$
The **cube root function**	$f(x) = \sqrt[3]{x}$

OBJECTIVE 1: USING VERTICAL SHIFTS TO GRAPH FUNCTIONS

My video summary ✎

▦ Example 1 Vertically Shift a Function

Sketch the graphs of $f(x) = |x|$ and $g(x) = |x| + 2$.

Solution Table 2 shows that for every value of x, the y-coordinate of the function g is always 2 greater than the y-coordinate of the function f. The two functions are sketched in Figure 11. The graph of $g(x) = |x| + 2$ is exactly the same as the graph of $f(x) = |x|$, except the graph of g is shifted *up* two units.

Table 2

x	$f(x) = |x|$	$g(x) = |x| + 2$
−3	3	5
−2	2	4
−1	1	3
0	0	2
1	1	3
2	2	4
3	3	5

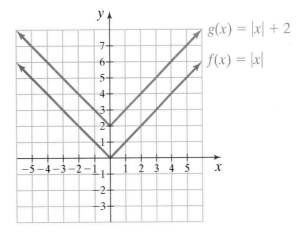

Figure 11

We see from Example 1 that if c is a positive number, then $y = f(x) + c$ is the graph of f shifted *up* c units. It follows that for $c > 0$, the graph of $y = f(x) - c$ is the graph of f shifted *down* c units.

Vertical Shifts of Functions

If c is a positive real number,

> The graph of $y = f(x) + c$ is obtained by shifting the graph of $y = f(x)$ vertically upward c units.

> The graph of $y = f(x) - c$ is obtained by shifting the graph of $y = f(x)$ vertically downward c units.

My animation summary

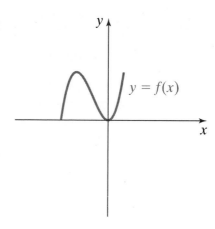

Sketch $y = f(x) + c$.   Sketch $y = f(x) - c$.

Work exercises 1–8 in this textbook or in the *MyMathLab*® Study Plan.

OBJECTIVE 2:   USING HORIZONTAL SHIFTS TO GRAPH FUNCTIONS

My video summary

To illustrate a horizontal shift, let $f(x) = x^2$ and $g(x) = (x + 2)^2$. Tables 3 and 4 show tables of values for f and g, respectively. The graphs of f and g are sketched in Figure 12. The graph of g is the graph of f shifted to the *left* two units.

Table 3

x	$f(x) = x^2$
-2	4
-1	1
0	0
1	1
2	4

Table 4

x	$g(x) = (x + 2)^2$
-4	4
-3	1
-2	0
-1	1
0	4

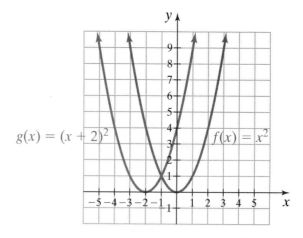

Figure 12

It follows that if c is a positive number, then $y = f(x + c)$ is the graph of f shifted to the *left* c units. For $c > 0$, the graph of $y = f(x - c)$ is the graph of f shifted to the *right* c units. At first glance, it appears that the rule for horizontal shifts is the opposite of what seems natural. Substituting $x + c$ for x causes the graph of $y = f(x)$ to be shifted to the left, whereas substituting $x - c$ for x causes the graph to shift to the right c units.

My animation summary

Horizontal Shifts of Functions

If c is a positive real number,

The graph of $y = f(x + c)$ is obtained by shifting the graph of $y = f(x)$ horizontally to the left c units.

The graph of $y = f(x - c)$ is obtained by shifting the graph of $y = f(x)$ horizontally to the right c units.

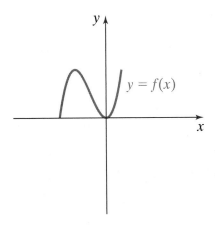

Sketch $y = f(x + c)$.   Sketch $y = f(x - c)$.

Work exercises 9–16 in this textbook or in the *MyMathLab*® Study Plan.

My animation summary

Example 2 Combine Horizontal and Vertical Shifts

Use the graph of $y = x^3$ to sketch the graph of $g(x) = (x - 1)^3 + 2$.

Solution The graph of g is obtained by shifting the graph of the basic function $y = x^3$ first horizontally to the right one unit and then vertically upward two units. When doing a problem with multiple transformations, it is good practice to always perform the vertical transformation last. Click on the animate button to see how to sketch the graph of $g(x) = (x - 1)^3 + 2$.

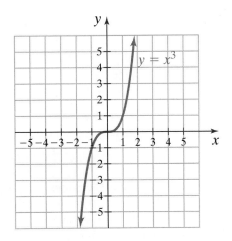

Work exercises 17–24 in this textbook or in the *MyMathLab*® Study Plan.

OBJECTIVE 3:   USING REFLECTIONS TO GRAPH FUNCTIONS

My video summary

Given the graph of $y = f(x)$, what does the graph of $y = -f(x)$ look like?

Using a graphing utility with $y_1 = x^2$ and $y_2 = -x^2$, we can see that the graph of $y_2 = -x^2$ is the graph of $y_1 = x^2$ reflected about the **x-axis**.

Using Technology

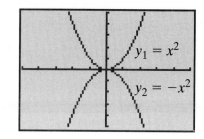

My animation summary

Reflections of Functions about the x-Axis

The graph of $y = -f(x)$ is obtained by reflecting the graph of $y = f(x)$ about the x-axis.

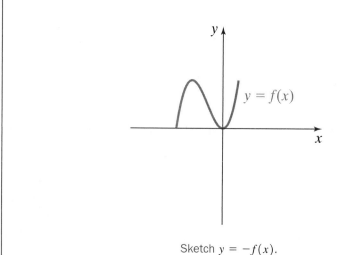

Sketch $y = -f(x)$.

Functions can also be reflected about the y-axis. Given the graph of $y = f(x)$, the graph of $y = f(-x)$ will be the graph of $y = f(x)$ reflected about the y-axis. Using a graphing utility, we illustrate a y-axis reflection by letting $y_1 = \sqrt{x}$ and $y_2 = \sqrt{-x}$. You can see that the functions are mirror images of each other about the y-axis.

Using Technology

My animation summary

Reflections of Functions about the y-Axis

The graph of $y = f(-x)$ is obtained by reflecting the graph of $y = f(x)$ about the y-axis.

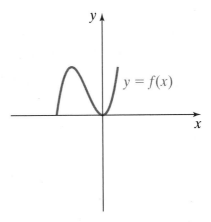

Sketch $y = f(-x)$.

Example 3 Sketch Functions Using Reflections and Shifts

Use the graph of the basic function $y = \sqrt[3]{x}$ to sketch each graph.

a. $g(x) = -\sqrt[3]{x} - 2$ **b.** $h(x) = \sqrt[3]{1 - x}$.

Solution

a. Starting with the graph of $y = \sqrt[3]{x}$, we can obtain the graph of $g(x) = -\sqrt[3]{x} - 2$ by performing two transformations:

1. Reflect about the x-axis.

2. Vertically shift down two units.

See Figure 13.

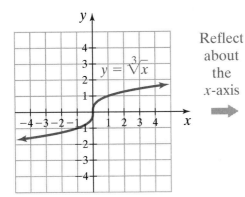

Start with the graph of the basic function $y = \sqrt[3]{x}$.

Reflect about the x-axis

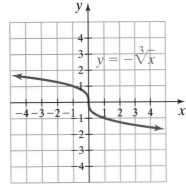

Sketch the graph of $y = -\sqrt[3]{x}$.

Vertical shift down two units

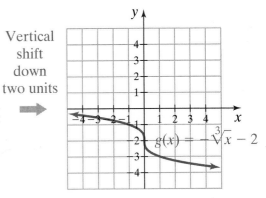

Sketch the graph of $g(x) = -\sqrt[3]{x} - 2$.

Figure 13

b. Starting with the graph of $y = \sqrt[3]{x}$, we can obtain the graph of $h(x) = \sqrt[3]{1 - x}$ by performing two transformations:

1. Horizontal shift left one unit.

2. Reflect about the y-axis.

See Figure 14.

3.4 Transformations of Functions   **129**

Start with the graph of the
basic function $y = \sqrt[3]{x}$.

Sketch the graph of
$y = \sqrt[3]{x + 1}$.

Replace x with $-x$, and
sketch the graph of
$h(x) = \sqrt[3]{-x + 1} = \sqrt[3]{1 - x}.$

Figure 14

Work exercises 25–38 in this textbook or in the **MyMathLab** Study Plan.

OBJECTIVE 4:   USING VERTICAL STRETCHES AND COMPRESSIONS TO GRAPH FUNCTIONS

My video summary

📼 Example 4 Vertical Stretch and Compression

Use the graph of $f(x) = x^2$ to sketch the graph of $g(x) = 2x^2$.

Solution Notice in Table 5 that for each value of x, the y-coordinate of g is two times as large as the corresponding y-coordinate of f. We can see in Figure 15 that the graph of $f(x) = x^2$ is vertically stretched by a factor of 2 to obtain the graph of $g(x) = 2x^2$. In other words, for each point (a, b) on the graph of f, the graph of g contains the point $(a, 2b)$.

TABLE 5

x	$f(x) = x^2$	$g(x) = 2x^2$
-2	4	8
-1	1	2
0	0	0
1	1	2
2	4	8

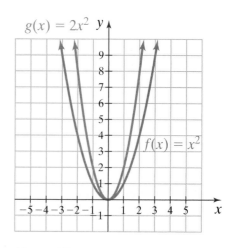

Figure 15

It follows from Example 4 that if $a > 1$, the graph of $y = af(x)$ is a **vertical stretch** of the graph of $y = f(x)$ and is obtained by multiplying each y-coordinate on the graph of f by a factor of a. If $0 < a < 1$, then the graph of $y = af(x)$ is a **vertical compression** of the graph of $y = f(x)$. Table 6 and Figure 16 show the relationship between the graphs of the functions $f(x) = x^2$ and $h(x) = \frac{1}{2}x^2$.

TABLE 6

x	$f(x) = x^2$	$h(x) = \frac{1}{2}x^2$
-2	4	2
-1	1	$\frac{1}{2}$
0	0	0
1	1	$\frac{1}{2}$
2	4	2

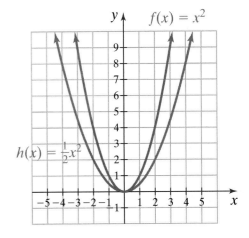

Figure 16

Vertical Stretches and Compressions of Functions

Suppose a is a positive real number:

The graph of $y = af(x)$ is obtained by multiplying each y-coordinate of $y = f(x)$ by a. If $a > 1$, the graph of $y = af(x)$ is a vertical stretch of the graph of $y = f(x)$. If $0 < a < 1$, the graph of $y = af(x)$ is a vertical compression of the graph of $y = f(x)$.

My animation summary

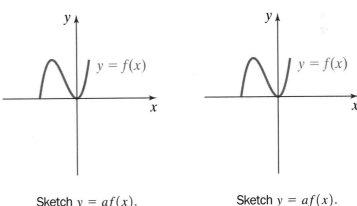

Sketch $y = af(x)$.
$(a > 1)$

Sketch $y = af(x)$.
$(0 < a < 1)$

Work exercises 39–46 in this textbook or in the _MyMathLab_® Study Plan.

OBJECTIVE 5: **USING HORIZONTAL STRETCHES AND COMPRESSIONS TO GRAPH FUNCTIONS**

My video summary

The final transformation to discuss is a horizontal stretch or compression. A function, $y = f(x)$, will be horizontally stretched or compressed when x is multiplied by a positive number, $a \neq 1$, to obtain the new function, $y = f(ax)$.

Horizontal Stretches and Compressions of Functions

If a is a positive real number,

For $a > 1$, the graph of $y = f(ax)$ is obtained by dividing each x-coordinate of $y = f(x)$ by a. The resultant graph is a horizontal compression.

For $0 < a < 1$, the graph of $y = f(ax)$ is obtained by dividing each x-coordinate of $y = f(x)$ by a. The resultant graph is a horizontal stretch.

My animation summary

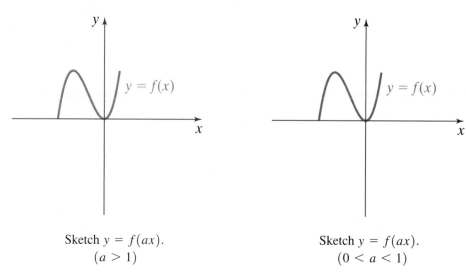

Sketch $y = f(ax)$.   Sketch $y = f(ax)$.
$(a > 1)$ $(0 < a < 1)$

Example 5 Horizontal Stretch and Compression

Use the graph of $f(x) = \sqrt{x}$ to sketch the graphs of $g(x) = \sqrt{4x}$ and $h(x) = \sqrt{\frac{1}{4}x}$.

Solution The graph of $f(x) = \sqrt{x}$ contains the ordered pairs $(0, 0)$, $(1, 1)$, $(4, 2)$. To sketch the graph of $g(x) = \sqrt{4x}$, we must divide each previous x-coordinate by 4. Therefore, the ordered pairs $(0, 0)$, $(\frac{1}{4}, 1)$, $(1, 2)$ must lie on the graph of g. You can see that the graph of g is a horizontal compression of the graph of f. See Figure 17.

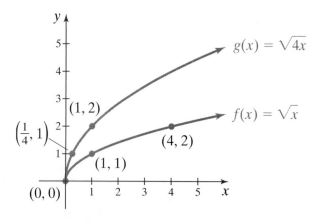

Figure 17 Graphs of $f(x) = \sqrt{x}$ and $g(x) = \sqrt{4x}$

Similarly, to sketch the graph of $h(x) = \sqrt{\frac{1}{4}x}$, we divide the x-coordinates of the ordered pairs of f by $\frac{1}{4}$ to get the ordered pairs $(0, 0)$, $(4, 1)$, $(16, 2)$. You can see that the graph of h is a horizontal stretch of the graph of f. See Figure 18.

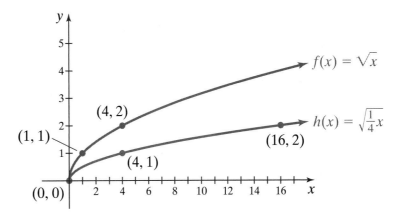

Figure 18 Graphs of $f(x) = \sqrt{x}$ and $g(x) = \sqrt{\frac{1}{4}x}$

Work exercises 47–51 in this textbook or in the *MyMathLab*® Study Plan.

OBJECTIVE 6: USING COMBINATIONS OF TRANSFORMATIONS TO GRAPH FUNCTIONS

You may encounter functions that combine many (if not all) of the transformations discussed in this section. When sketching a function that involves multiple transformations, it is important to follow a certain "order of operations." Following is the order in which each transformation is performed in this text:

1. Horizontal shifts
2. Horizontal stretches/compressions
3. Reflection about y-axis
4. Vertical stretches/compressions
5. Reflection about x-axis
6. Vertical shifts

Different ordering is possible for transformations 2 through 5, but you should always perform the horizontal shift first and the vertical shift last.

My animation summary

 Example 6 Combine Transformations

Use transformations to sketch the graph of $f(x) = -2(x + 3)^2 - 1$.

Solution Watch the animation to see how to sketch the function $f(x) = -2(x + 3)^2 - 1$ as seen in Figure 19.

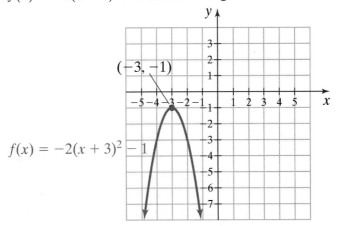

Figure 19 Graph of $f(x) = -2(x + 3)^2 - 1$

3.4 Transformations of Functions   **133**

My interactive video summary

▶ **Example 7** Combine Transformations

Use the graph of $y = f(x)$ to sketch each of the following functions.

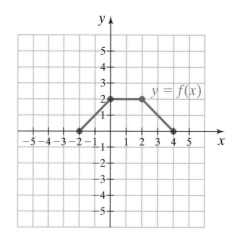

a. $y = -f(2x)$ **b.** $y = 2f(x - 3) - 1$ **c.** $y = -\dfrac{1}{2}f(2 - x) + 3$

Solution Watch the interactive video to see any one of the solutions worked out in detail.

a. The graph of $y = -f(2x)$ can be obtained from the graph of $y = f(x)$ using two transformations: (1) a horizontal compression and (2) a reflection about the x-axis. The resultant graph is sketched in Figure 20.

b. The graph of $y = 2f(x - 3) - 1$ can be obtained from the graph of $y = f(x)$ using three transformations: (1) a horizontal shift to the right three units, (2) a vertical stretch by a factor of 2, and (3) a vertical shift down one unit. The resultant graph is sketched in Figure 21.

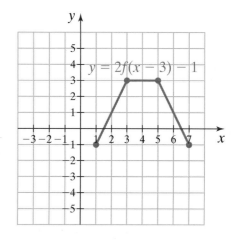

Figure 20 Graph of $y = -f(2x)$

Figure 21 Graph of $y = 2f(x - 3) - 1$

c. The graph of $y = -\frac{1}{2}f(2 - x) + 3$ can be obtained from the graph of $y = f(x)$ using five transformations: (1) a horizontal shift to the left two units, (2) a reflection about the y-axis, (3) a vertical compression by a factor of $\frac{1}{2}$, (4) a reflection about the x-axis, and (5) a vertical shift up three units. The resultant graph is sketched in Figure 22.

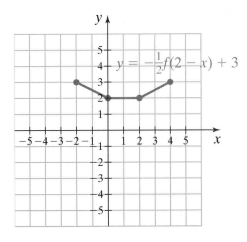

Figure 22 Graph of $y = -\dfrac{1}{2}f(2 - x) + 3$

Work exercises 52–62 in this textbook or in the *MyMathLab*® Study Plan.

3.4 Exercises

In Exercises 1–6, use the graph of a known basic function and vertical shifting to sketch each function.

1. $f(x) = x^2 - 1$ **2.** $y = \sqrt{x} + 2$ **3.** $g(x) = \dfrac{1}{x} + 3$

4. $h(x) = \sqrt[3]{x} - 2$ **5.** $y = |x| - 3$ **6.** $g(x) = x^3 + 1$

7. Use the graph of $y = f(x)$ to sketch the graph of $y = f(x) - 1$. Label at least three points on the new graph.

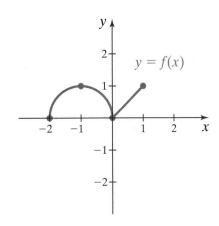

8. Use the graph of $y = f(x)$ to sketch the graph of $y = f(x) + 2$. Label at least three points on the new graph.

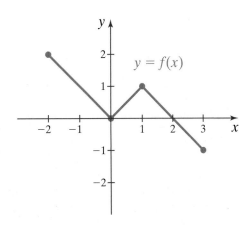

In Exercises 9–14, use the graph of a known basic function and horizontal shifts to sketch each function.

9. $f(x) = \sqrt[3]{x} - 2$ **10.** $g(x) = \dfrac{1}{x+3}$ **11.** $y = \sqrt{x-4}$

12. $h(x) = (x+1)^3$ **13.** $k(x) = |x-1|$ **14.** $y = (x-3)^2$

15. Use the graph of $y = f(x)$ to sketch the graph of $y = f(x-2)$. Label at least three points on the new graph.

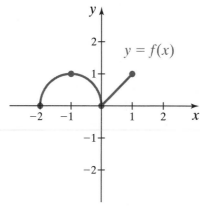

16. Use the graph of $y = f(x)$ to sketch the graph of $y = f(x+2)$. Label at least three points on the new graph.

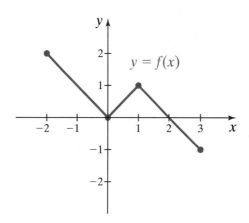

In Exercises 17–22, use the graph of a known basic function and a combination of horizontal and vertical shifts to sketch each function.

17. $y = (x+1)^2 - 2$ **18.** $f(x) = (x-3)^2 + 1$ **19.** $y = \sqrt{x+3} + 2$

20. $h(x) = \dfrac{1}{x-2} + 3$ **21.** $f(x) = |x+2| + 2$ **22.** $y = \sqrt[3]{x+1} - 1$

23. Use the graph of $y = f(x)$ to sketch the graph of $y = f(x-2) - 1$. Label at least three points on the new graph.

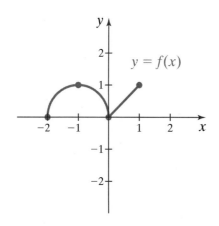

24. Use the graph of $y = f(x)$ to sketch the graph of $y = f(x + 1) + 2$. Label at least three points on the new graph.

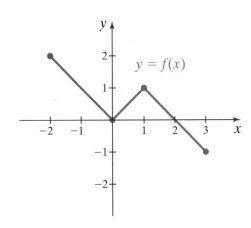

In Exercises 25–36, use the graph of a known basic function and a combination of horizontal shifts, reflections, and vertical shifts to sketch each function.

25. $g(x) = -x^2 - 2$　　**26.** $f(x) = -\dfrac{1}{x} + 1$　　**27.** $h(x) = \sqrt{2 - x}$　　**28.** $f(x) = |-1 - x|$

29. $h(x) = \sqrt[3]{-x - 2}$　　**30.** $g(x) = -x^3 + 1$　　**31.** $h(x) = -\sqrt[3]{x} + 2$　　**32.** $g(x) = (3 - x)^2$

33. $h(x) = -\sqrt{x} - 1$　　**34.** $f(x) = -|x| + 1$　　**35.** $g(x) = (1 - x)^3$　　**36.** $f(x) = \dfrac{1}{2 - x}$

37. Use the graph of $y = f(x)$ to sketch the graph of $y = -f(x)$. Label at least three points on the new graph.

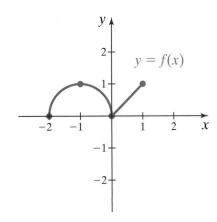

38. Use the graph of $y = f(x)$ to sketch the graph of $y = f(-x)$. Label at least three points on the new graph.

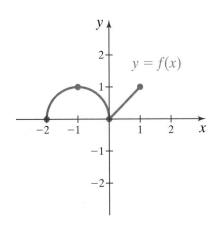

In Exercises 39–44, use the graph of a known basic function and a vertical stretch or vertical compression to sketch each function.

39. $f(x) = \frac{1}{4}|x|$ **40.** $g(x) = 2\sqrt{x}$ **41.** $f(x) = \frac{1}{3}x^3$

42. $f(x) = \frac{1}{2}\sqrt[3]{x}$ **43.** $g(x) = 3x^2$ **44.** $y = \frac{1}{2x}$

45. Use the graph of $y = f(x)$ to sketch the graph of $y = 3f(x)$. Label at least three points on the new graph.

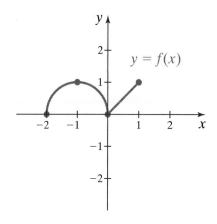

46. Use the graph of $y = f(x)$ to sketch the graph of $y = \frac{1}{2}f(x)$. Label at least three points on the new graph.

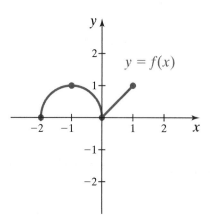

In Exercises 47–49, use the graph of a known basic function and a horizontal stretch or horizontal compression to sketch each function.

47. $y = \left|\frac{1}{4}x\right|$ **48.** $f(x) = \sqrt{2x}$ **49.** $g(x) = \sqrt[3]{3x}$

50. Use the graph of $y = f(x)$ to sketch the graph of $y = f(2x)$. Label at least three points on the new graph.

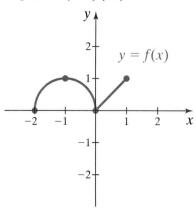

51. Use the graph of $y = f(x)$ to sketch the graph of $y = f(\frac{1}{2}x)$. Label at least three points on the new graph.

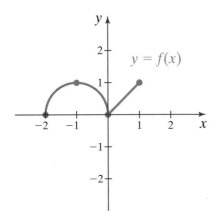

In Exercises 52–57, use the graph of a known basic function and a combination of transformations to sketch each function.

52. $f(x) = -(x - 2)^2 + 3$

53. $g(x) = \dfrac{1}{2}|x + 1| - 1$

54. $y = \dfrac{1}{x - 3} + 2$

55. $f(x) = 2\sqrt[3]{x} - 1$

56. $g(x) = -\dfrac{1}{2}(2 - x)^3 + 1$

57. $h(x) = 2\sqrt{4 - x} + 5$

In Exercises 58–62, use the graph of $y = f(x)$ to sketch each function. Label at least three points on each graph.

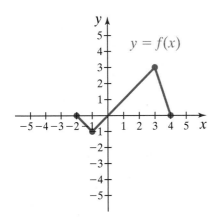

58. $y = -f(-x) - 1$

59. $y = \dfrac{1}{2}f(2 - x)$

60. $y = -2f(x + 1) + 2$

61. $y = 3 - 3f(x + 3)$

62. $y = -f(1 - x) - 2$

3.5 The Algebra of Functions; Composite Functions

THINGS TO KNOW

Before working through this section, be sure that you are familiar with the following concepts:

VIDEO   ANIMATION   INTERACTIVE

You Try It

1. Solving Compound Inequalities (Section 1.7)

You Try It

2. Solving Polynomial Inequalities (Section 1.9)

You Try It

3. Solving Rational Inequalities (Section 1.9)

You Try It

4. Determining the Domain of a Function Given the Equation (Section 3.1)

You Try It

5. Determining the Domain and Range of a Function from Its Graph (Section 3.2)

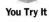
You Try It

6. Sketching the Graphs of the Basic Functions (Section 3.3)

OBJECTIVES

1 Evaluating a Combined Function

2 Finding the Intersection of Intervals

3 Finding Combined Functions and Their Domains

4 Forming and Evaluating Composite Functions

5 Determining the Domain of Composite Functions

My video summary

In this section, we learn how to create a new function by combining two or more existing functions. First, we combine functions by adding, subtracting, multiplying, or dividing two existing functions. The addition, subtraction, multiplication, and division of functions is also known as the *algebra of functions*.

Algebra of Functions

Let f and g be functions, then for all x such that both $f(x)$ and $g(x)$ are defined, the sum, difference, product, and quotient of f and g exist and are defined as follows:

1. The sum of f and g: $\qquad (f + g)(x) = f(x) + g(x)$

2. The difference of f and g: $\qquad (f - g)(x) = f(x) - g(x)$

3. The product of f and g: $\qquad (fg)(x) = f(x)g(x)$

4. The quotient of f and g: $\qquad \left(\dfrac{f}{g}\right)(x) = \dfrac{f(x)}{g(x)}$ for all $g(x) \neq 0$

OBJECTIVE 1:   EVALUATING A COMBINED FUNCTION

Example 1 Evaluate Combined Functions

Let $f(x) = \dfrac{12}{2x + 4}$ and $g(x) = \sqrt{x}$. Find each of the following.

a. $(f + g)(1)$ **b.** $(f - g)(1)$ **c.** $(fg)(4)$ **d.** $\left(\dfrac{f}{g}\right)(4)$

Solution

a. Because $f(1) = \dfrac{12}{2(1) + 4} = \dfrac{12}{6} = 2$ and $g(1) = \sqrt{1} = 1$, it follows

from the sum of f and g that

$$(f + g)(1) = f(1) + g(1)$$
$$= 2 + 1$$
$$= 3$$

b. $(f - g)(1) = f(1) - g(1) = 2 - 1 = 1$

c. $f(4) = \dfrac{12}{2(4) + 4} = \dfrac{12}{12} = 1$ and $g(4) = \sqrt{4} = 2$, so

$(fg)(4) = f(4)g(4) = (1)(2) = 2$

d. $\left(\dfrac{f}{g}\right)(4) = \dfrac{f(4)}{g(4)} = \dfrac{1}{2}$

Work exercises 1–10 in this textbook or in the *MyMathLab*® Study Plan.

My interactive video summary

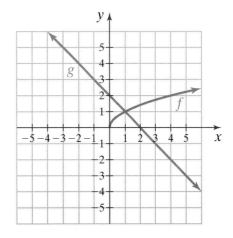

▶ Example 2 Evaluate Combined Functions Using a Graph

Use the graph to evaluate each expression or state that it is undefined.

a. $(f + g)(1)$ **b.** $(f - g)(0)$ **c.** $(fg)(4)$ **d.** $\left(\dfrac{f}{g}\right)(2)$

Solution Work through the interactive video to verify that

a. $(f + g)(1) = 2$ **b.** $(f - g)(0) = -2$

c. $(fg)(4) = -4$ **d.** $\left(\dfrac{f}{g}\right)(2)$ is undefined

Work exercises 11–14 in this textbook or in the *MyMathLab*® Study Plan.

OBJECTIVE 2: FINDING THE INTERSECTION OF INTERVALS

Before learning how to find the domain of $f + g, f - g, fg$, or $\frac{f}{g}$, you must be able to find the intersection of two or more intervals. The notation $A \cap B$ is used to represent the **intersection** of sets A and B. To find the intersection of two intervals, we must find the set of all numbers *common* to both intervals.

My video summary

▦ Example 3 Find the Intersection of Intervals

Find the intersection of the following intervals and graph the set on a number line.

a. $[0, \infty) \cap (-\infty, 5]$ **b.** $((-\infty, -2) \cup (-2, \infty)) \cap [-4, \infty)$

Solution Watch the video to verify the solutions.

a. $[0, \infty) \cap (-\infty, 5] = [0, 5]$

b. $((-\infty, -2) \cup (-2, \infty)) \cap [-4, \infty) = [-4, -2) \cup (-2, \infty)$

Work exercises 15–20 in this textbook or in the *MyMathLab*® Study Plan.

3.5 The Algebra of Functions; Composite Functions   141

OBJECTIVE 3:   FINDING COMBINED FUNCTIONS AND THEIR DOMAINS

My video summary

Consider the graph of f and g in Figure 23. Notice that the domain of f is the interval $[1, \infty)$ and the domain of g is the interval $[-1, 4)$. What is the domain of $f + g$? Because $(f + g)(x) = f(x) + g(x)$, for a number a to be in the domain of $f + g$ means that both $f(a)$ and $g(a)$ must be defined. For example, because $x = 0$ is not in the domain of f, $f(0)$ is undefined and $x = 0$ is *not* in the domain of $f + g$. Similarly, because $x = 4$ is not in the domain of g, $x = 4$ cannot be in the domain of $f + g$. Thus, for a number a to be in the domain of $f + g$, a must be in the domain of f, *and* a must be in the domain of g. In other words, the domain of $f + g$ must be the intersection of the two domains. The domain of $f + g$ in Figure 23 is $[1, 4)$.

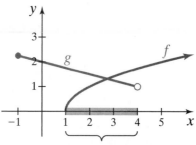

The intersection of the domain of f
and the domain of g is $[1, 4)$.   **Figure 23**

Domain of $f + g, f - g, fg,$ and $\frac{f}{g}$

Suppose f is a function with domain A and g is a function with domain B, then

1.   The domain of the sum, $f + g$, is the set of all x in $A \cap B$.

2.   The domain of the difference, $f - g$, is the set of all x in $A \cap B$.

3.   The domain of the product, fg, is the set of all x in $A \cap B$.

4.   The domain of the quotient, $\frac{f}{g}$, is the set of all x in $A \cap B$ such that $g(x) \neq 0$.

My video summary

Example 4 Find $f + g, f - g, fg,$ and $\frac{f}{g}$ and Their Domains

Let $f(x) = \dfrac{x + 2}{x - 3}$ and $g(x) = \sqrt{4 - x}$.

Find a) $f + g$, b) $f - g$, c) fg, d) $\frac{f}{g}$, and the domain of each.

Solution The domain of f is the interval $(-\infty, 3) \cup (3, \infty)$, and the domain of g is the interval $(-\infty, 4]$. The intersection of these two intervals is $(-\infty, 3) \cup (3, 4]$.

a. $(f + g)(x) = \dfrac{x + 2}{x - 3} + \sqrt{4 - x}$

The domain of $f + g$ is the set of all x in the intersection of the domain of f and the domain of g. Thus, the domain of $f + g$ is $(-\infty, 3) \cup (3, 4]$.

b. $(f - g)(x) = \dfrac{x + 2}{x - 3} - \sqrt{4 - x}$

The domain of $f - g$ is the set of all x in the intersection of the domain of f and the domain of g. Thus, the domain of $f - g$ is $(-\infty, 3) \cup (3, 4]$.

c. $(fg)(x) = \dfrac{(x + 2)\sqrt{4 - x}}{x - 3}$

The domain of fg is the set of all x in the intersection of the domain of f and the domain of g. Thus, the domain of fg is $(-\infty, 3) \cup (3, 4]$.

d. $\left(\dfrac{f}{g}\right)(x) = \dfrac{\dfrac{x + 2}{x - 3}}{\sqrt{4 - x}} = \dfrac{x + 2}{(x - 3)\sqrt{4 - x}}$

The domain of $\frac{f}{g}$ is the set of all x in the intersection of the domain of f and the domain of g such that $g(x) \neq 0$. Because $g(x) = 0$ when $x = 4$, we must exclude $x = 4$ from the domain of $\frac{f}{g}$. Thus, the domain of $\frac{f}{g}$ is $(-\infty, 3) \cup (3, 4)$.

Work exercises 21–32 in this textbook or in the *MyMathLab*® Study Plan.

OBJECTIVE 4:   FORMING AND EVALUATING COMPOSITE FUNCTIONS

My video summary

Consider the function $f(x) = x^2$ and $g(x) = 2x + 1$. How could we find $f(g(x))$? To find $f(g(x))$, we substitute $g(x)$ for x in the function f to get

$$f(g(x)) = f(2x + 1) = (2x + 1)^2$$

substitute $g(x)$ into f

The diagram in Figure 24 shows that given a number x, we first apply it to the function g to obtain $g(x)$. We then substitute $g(x)$ into f to get the result.

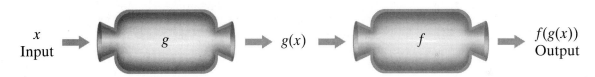

x
Input g $g(x)$ f $f(g(x))$
Output

Figure 24 Composition of f and g

The function $f(g(x))$ is called a *composite function* because one function is "composed" of another function.

DEFINITION Composite Function

Given functions f and g, the **composite function**, $f \circ g$, (also called the **composition of f and g**) is defined by

$$(f \circ g)(x) = f(g(x))$$

provided $g(x)$ is in the domain of f.

CAUTION   The composition of f and g does not equal the product of f and g: $(f \circ g)(x) \neq fg(x)$. Also, the composition of f and g does not necessarily equal the composition of g and f, although this equality does exist for certain pairs of functions.

My interactive video summary

▶ Example 5 Form and Evaluate Composite Functions

Let $f(x) = 4x + 1$, $g(x) = \dfrac{x}{x - 2}$ and $h(x) = \sqrt{x + 3}$.

a. Find the function $f \circ g$.
b. Find the function $g \circ h$.
c. Find the function $h \circ f \circ g$.
d. Evaluate $(f \circ g)(4)$, or state that it is undefined.
e. Evaluate $(g \circ h)(1)$, or state that it is undefined.
f. Evaluate $(h \circ f \circ g)(6)$, or state that it is undefined.

Solution Work through the interactive video to verify the following:

a. $(f \circ g)(x) = 4\left(\dfrac{x}{x - 2}\right) + 1 = \dfrac{5x - 2}{x - 2}$

b. $(g \circ h)(x) = \dfrac{\sqrt{x + 3}}{\sqrt{x + 3} - 2}$

c. $(h \circ f \circ g)(x) = \sqrt{\dfrac{5x - 2}{x - 2} + 3} = \sqrt{\dfrac{8x - 8}{x - 2}}$

d. $(f \circ g)(4) = 9$

e. $(g \circ h)(1)$ is undefined

f. $(h \circ f \circ g)(6) = \sqrt{10}$

Work exercises 33–56 in this textbook or in the _MyMathLab_® Study Plan.

My interactive video summary

▶ Example 6 Evaluate Composite Functions Using a Graph

Use the graph to evaluate each expression:

a. $(f \circ g)(4)$ **b.** $(g \circ f)(-3)$ **c.** $(f \circ f)(-1)$
d. $(g \circ g)(4)$ **e.** $(f \circ g \circ f)(1)$

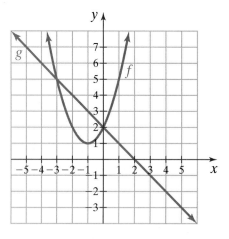

Solution Work through the interactive video to verify that

a. $(f \circ g)(4) = 2,$ **b.** $(g \circ f)(-3) = -3,$
c. $(f \circ f)(-1) = 5,$ **d.** $(g \circ g)(4) = 4,$ and
e. $(f \circ g \circ f)(1) = 5$

Work exercises 57 and 58 in this textbook or in the _MyMathLab_® Study Plan.

OBJECTIVE 5:   DETERMINING THE DOMAIN OF COMPOSITE FUNCTIONS

My video summary

 Suppose f and g are functions. For a number x to be in the domain of $f \circ g$, x must be in the domain of g *and* $g(x)$ must be in domain of f. Follow these two steps to find the domain of $f \circ g$:

1. Find the domain of g.
2. Exclude from the domain of g all values of x for which $g(x)$ is not in the domain of f.

My interactive video summary

Example 7 Find the Domain of a Composite Function

Let $f(x) = \dfrac{-10}{x-4}$ and $g(x) = \sqrt{5-x}$.

a. Find the domain of $f \circ g$.   **b.** Find the domain of $g \circ f$.

Solution

a. First, form the composite function $(f \circ g)(x) = \dfrac{-10}{\sqrt{5-x}-4}$. To find the domain of $f \circ g$, we follow these two steps:

Step 1   Find the domain of g.
The domain of $g(x) = \sqrt{5-x}$ is $(-\infty, 5]$. The domain of $f \circ g$ cannot contain any values of x that are not in this interval; in other words, the domain of $f \circ g$ is a **subset** of $(-\infty, 5]$.

Step 2   Exclude from the domain of g all values of x for which $g(x)$ is not in the domain of f.
All real numbers except 4 are in the domain of f. This implies that $g(x)$ cannot equal 4 because $g(x)$ equal to 4 would make the denominator of f, $x-4$, equal to 0. Thus, we must exclude all values of x such that $g(x) = 4$.

$$g(x) = 4$$
$$\sqrt{5-x} = 4 \qquad \textit{Substitute } \sqrt{5-x} \textit{ for } g(x).$$
$$5 - x = 16 \qquad \textit{Square both sides.}$$
$$x = -11 \qquad \textit{Solve for } x.$$

We must *exclude* $x = -11$ from the domain of $f \circ g$. Therefore, the domain of $f \circ g$ is all values of x less than 5 such that $x \neq -11$, or the interval $(-\infty, -11) \cup (-11, 5]$. You may want to view the interactive video to view the solution in more detail.

b. You should carefully work through the interactive video to verify that the domain of $(g \circ f)(x) = \sqrt{5 + \dfrac{10}{x-4}}$ is the interval $(-\infty, 2] \cup (4, \infty)$.

Using Technology

vertical asymptote $x = -11$

This screen shows the graph of $(f \circ g)(x) = \dfrac{-10}{\sqrt{5-x}-4}$. The **vertical asymptote**, $x = -11$, is shown in the graph. You can see from the graph that the domain is $(-\infty, -11) \cup (-11, 5]$.

3.5 The Algebra of Functions; Composite Functions   **145**

vertical asymptote $x = 4$

This screen represents the graph of $(g \circ f)(x) = \sqrt{5 + \dfrac{10}{x - 4}}$. The **vertical asymptote**, $x = 4$, is shown in the graph. You can see that the domain is $(-\infty, 2] \cup (4, \infty)$.

Work exercises 59–66 in this textbook or in the MyMathLab® Study Plan.

3.5 Exercises

In Exercises 1–10, evaluate the following given that $f(x) = \sqrt{x + 1}$, $g(x) = x^2 - 2$, and $h(x) = \dfrac{1}{2x + 3}$.

1. $(f + g)(3)$ **2.** $(f - g)(8)$ **3.** $(g + h)(-2)$ **4.** $\left(\dfrac{f}{g}\right)(0)$ **5.** $(fh)(15)$

6. $(h - g)(4)$ **7.** $\left(\dfrac{g}{f}\right)(-1)$ **8.** $\left(\dfrac{f}{h}\right)(2)$ **9.** $(gf)(8)$ **10.** $(hh)\left(\dfrac{3}{2}\right)$

In Exercises 11–14, use the graph to evaluate the given expression, or state that it is undefined.

11.

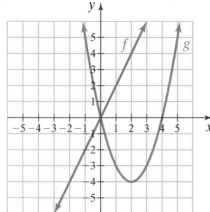

a. $(f + g)(0)$ b. $(f - g)(2)$

c. $(fg)(-1)$ d. $\left(\dfrac{f}{g}\right)(1)$

12.

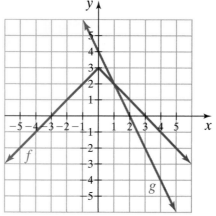

a. $(f + g)(4)$ b. $(g - f)(1)$

c. $(fg)(0)$ d. $\left(\dfrac{f}{g}\right)(2)$

13.

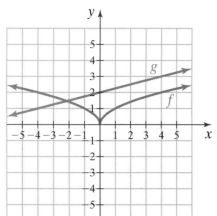

a. $(f - g)(-4)$ b. $(f - g)(-4)$

c. $(gg)(4)$ d. $\left(\dfrac{f}{g}\right)(4)$

14.

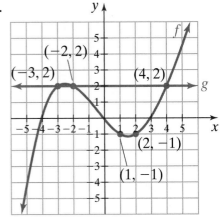

a. $(g + f)(4)$

b. $(g - f)(1)$

c. $(fg)(-3)$

d. $\left(\dfrac{f}{g}\right)(2)$

In Exercises 15–20, find the intersection of the given intervals and graph the set on a number line.

15. $(-\infty, \infty) \cap (-\infty, 3)$

16. $(-8, 5] \cap (-12, 3)$

17. $[0, \infty) \cap (-\infty, 4]$

18. $((-\infty, 1) \cup (1, \infty)) \cap (-\infty, \infty)$

19. $((-\infty, 0) \cup (0, \infty)) \cap [-3, \infty)$

20. $(-\infty, 10) \cap (-5, \infty) \cap (-\infty, 1]$

In Exercises 21–32, find a) $f + g$, b) $f - g$, c) fg, d) $\frac{f}{g}$, and the domain of each.

21. $f(x) = x^2 + 2, g(x) = x - 1$

22. $f(x) = x - 5, g(x) = x^2 - 4$

23. $f(x) = x + 5, g(x) = x^2 - 6x - 16$

24. $f(x) = \sqrt{x}, g(x) = x + 6$

25. $f(x) = \dfrac{1}{x}, g(x) = \sqrt{x - 1}$

26. $f(x) = \dfrac{2x + 1}{x + 3}, g(x) = \dfrac{x + 2}{3x + 5}$

27. $f(x) = \dfrac{x}{x^2 - 4}, g(x) = \dfrac{1}{x - 3}$

28. $f(x) = \dfrac{x - 4}{x^2 - x - 6}, g(x) = \dfrac{x + 5}{x^2 - 16}$

29. $f(x) = \sqrt{x + 2}, g(x) = \sqrt{2 - x}$

30. $f(x) = \sqrt{\dfrac{x + 1}{x + 2}}, g(x) = \sqrt{x - 3}$

31. $f(x) = \sqrt[3]{x + 1}, g(x) = \sqrt[3]{\dfrac{x - 1}{x + 2}}$

32. $f(x) = \sqrt{7 - x}, g(x) = \sqrt{x^2 - 2x - 15}$

In Exercises 33–44, let $f(x) = 3x + 1$, $g(x) = \dfrac{2}{x + 1}$, and $h(x) = \sqrt{x + 3}$.

33. Find the function $f \circ g$.

34. Find the function $g \circ f$.

35. Find the function $f \circ h$.

36. Find the function $g \circ h$.

37. Find the function $h \circ f$.

38. Find the function $h \circ g$.

39. Find the function $f \circ f$.

40. Find the function $g \circ g$.

41. Find the function $h \circ h$.

42. Find the function $f \circ g \circ h$.

43. Find the function $g \circ f \circ h$.

44. Find the function $h \circ f \circ g$.

In Exercises 45–56, evaluate the following composite functions given that
$f(x) = 3x + 1$, $g(x) = \dfrac{2}{x + 1}$, and $h(x) = \sqrt{x + 3}$.

45. $(f \circ g)(0)$

46. $(f \circ h)(6)$

47. $(g \circ f)(1)$

48. $(g \circ h)(-2)$

49. $(h \circ f)(0)$

50. $(h \circ g)(3)$

51. $(f \circ f)(-1)$

52. $(g \circ g)(4)$

53. $(h \circ h)(1)$

54. $(f \circ g \circ h)(-2)$

55. $(h \circ f \circ g)(2)$

56. $(g \circ f \circ h)(6)$

In Exercises 57 and 58, use the graph to evaluate each expression.

57. **a.** $(f \circ g)(1)$ **b.** $(g \circ f)(-1)$

    c. $(g \circ g)(0)$ **d.** $(f \circ f)(1)$

    e. $(f \circ g \circ f)(0)$ **f.** $(g \circ f \circ g)(0)$

58. **a.** $(f \circ g)(1)$ **b.** $(g \circ f)(-1)$

    c. $(g \circ g)(0)$ **d.** $(f \circ f)(1)$

    e. $(f \circ g \circ f)(0)$ **f.** $(g \circ f \circ g)(0)$

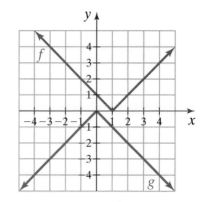

In Exercises 59–66, find the domain of $(f \circ g)(x)$ and $(g \circ f)(x)$.

59. $f(x) = x^2, g(x) = 2x - 1$

60. $f(x) = 3x - 5, g(x) = 2x^2 + 1$

61. $f(x) = x^2, g(x) = \sqrt{x}$

62. $f(x) = \dfrac{1}{x}, g(x) = x^2 - 4$

63. $f(x) = \dfrac{3}{x + 1}, g(x) = \dfrac{x}{x - 2}$

64. $f(x) = \dfrac{2x}{x - 3}, g(x) = \dfrac{x + 1}{x - 1}$

65. $f(x) = \sqrt{x}, g(x) = \dfrac{x + 3}{x - 2}$

66. $f(x) = \dfrac{1}{x - 2}, g(x) = \sqrt{4 - x}$

3.6 One-to-One Functions; Inverse Functions

THINGS TO KNOW

Before working through this section, be sure that you are familiar with the following concepts:

VIDEO   ANIMATION   INTERACTIVE

You Try It

1. Determining the Domain of a Function Given the Equation (Section 3.1)

You Try It

2. Determining the Domain and Range of a Function from Its Graph (Section 3.2)

3. Analyzing Piecewise-Defined Functions (Section 3.3)

4. Forming and Evaluating Composite Functions (Section 3.5)

OBJECTIVES

1 Understanding the Definition of a One-to-One Function

2 Determining Whether a Function Is One-to-One Using the Horizontal Line Test

3 Understanding and Verifying Inverse Functions

4 Sketching the Graphs of Inverse Functions

5 Finding the Inverse of a One-to-One Function

In this section, we study *inverse functions*. To illustrate an inverse function, let's consider the function $F(C) = \frac{9}{5}C + 32$. This function is used to convert a given temperature in degrees Celsius into its equivalent temperature in degrees Fahrenheit. For example,

$$F(25) = \frac{9}{5}(25) + 32 = 45 + 32 = 77$$

Thus, a temperature of 25°C corresponds to a temperature of 77°F. To convert a temperature of 77°F back into 25°C, we need a different function. The function that is used to accomplish this task is $C(F) = \frac{5}{9}(F - 32)$.

$$C(77) = \frac{5}{9}(77 - 32)$$

$$= \frac{5}{9}(45)$$

$$= 25$$

The function C is the *inverse* of the function F. In essence, these functions perform opposite actions (they "undo" each other). The first function converted 25°C into 77°F, whereas the second function converted 77°F back into 25°C.

The function F has an inverse function C. Later in this section, we examine a process for finding the inverse of a function, but keep the following in mind. We can find the inverse of many functions using this process, but that inverse will not always be a function. In this text, we are only interested in inverses that are functions, so we first develop a test to determine whether a function has an inverse *function*. When the word *inverse* is used throughout the remainder of this section, we assume that we are referring to the inverse that is a function.

OBJECTIVE 1:   UNDERSTANDING THE DEFINITION OF A ONE-TO-ONE FUNCTION

My video summary

First, we must define the concept of **one-to-one** functions.

DEFINITION   One-to-One Function

A function f is **one-to-one** if for any values $a \neq b$ in the domain of f, $f(a) \neq f(b)$.

This definition suggests that a function is one-to-one if for any two *different* input values (domain values), the corresponding output values (range values) must be different. An alternate definition says that if two range values are the same, $f(u) = f(v)$, then the domain values must be the same; that is, $u = v$.

Alternate Definition of a One-to-One Function

A function f is **one-to-one** if for any two range values $f(u)$ and $f(v)$, $f(u) = f(v)$ implies that $u = v$.

The function sketched in Figure 25a is one-to-one because for any two distinct x-values in the domain $(a \neq b)$, the function values or range values are not equal $(f(a) \neq f(b))$. In Figure 25b, we see that the function $y = g(x)$ is *not* one-to-one because we can easily find two different domain values that correspond to the same range value.

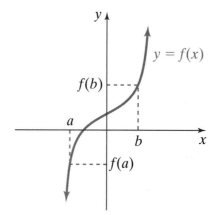

Figure 25a An example of a one-to-one function

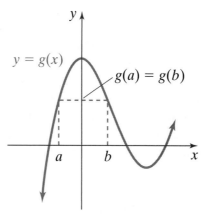

Figure 25b An example of a function that is not one-to-one

OBJECTIVE 2:   DETERMINING WHETHER A FUNCTION IS ONE-TO-ONE USING THE HORIZONTAL LINE TEST

My video summary

Notice in Figure 26 that the horizontal lines intersect the graph of $y = f(x)$ in at most one place, while we can find many horizontal lines that intersect the graph of $y = g(x)$ more than once. This gives us a visual example of how we can use horizontal lines to help us determine from the graph whether a function is one-to-one. Using horizontal lines to determine whether the graph of a function is one-to-one is known as the *horizontal line test*.

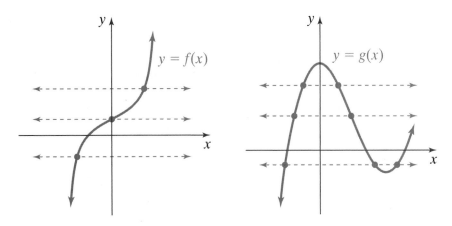

Figure 26 Drawing horizontal lines can help determine whether a graph represents a one-to-one function.

Horizontal Line Test

If every horizontal line intersects the graph of a function f at most once, then f is one-to-one.

My animation summary

 Example 1 Determine whether a Function Is One-to-One

Determine whether each function is one-to-one.

a.

b.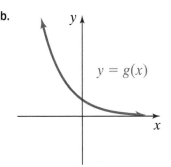

c. $f(x) = x^2 + 1$, $x \leq 0$

d. $f(x) = \begin{cases} 2x + 4 \text{ for } x \leq -1 \\ 2x - 6 \text{ for } x \geq 4 \end{cases}$

Solution The functions in parts b and c are one-to-one, whereas the functions in parts a and d are not one-to-one. Watch the animation to verify.

Work exercises 1–17 in this textbook or in the *MyMathLab*® Study Plan.

OBJECTIVE 3: UNDERSTANDING AND VERIFYING INVERSE FUNCTIONS

My video summary

We are now ready to ask ourselves the question, "Why should we be concerned with one-to-one functions?"

Answer: Every one-to-one function has an inverse function!

DEFINITION Inverse Function

Let f be a one-to-one function with domain A and range B. Then f^{-1} is the **inverse function of** f with domain B and range A. Furthermore, if $f(a) = b$, then $f^{-1}(b) = a$.

According to the definition of an inverse function, the domain of f is exactly the same as the range of f^{-1}, and the range of f is the same as the domain of f^{-1}. Figure 27 illustrates that if the function f assigns a number a to b, then the inverse function will assign b back to a. In other words, if the point (a, b) is an ordered pair on the graph of f, then the point (b, a) must be on the graph of f^{-1}.

Figure 27

 Do not confuse f^{-1} with $\frac{1}{f(x)}$. The negative 1 in f^{-1} is *not* an exponent!

My video summary

As with our opening example using an inverse function to convert a Fahrenheit temperature back into a Celsius temperature, inverse functions "undo" each other. For example, it can be shown that if $f(x) = x^3$, then the inverse of f is $f^{-1}(x) = \sqrt[3]{x}$. Note that

$$f(2) = (2)^3 = 8 \text{ and } f^{-1}(8) = \sqrt[3]{8} = 2$$

The function f takes the number 2 to 8, whereas f^{-1} takes 8 back to 2. Observe what happens if we look at the composition of f and f^{-1} and the composition of f^{-1} and f at specified values:

$$\left(f \circ f^{-1}\right)(8) = f\left(f^{-1}(8)\right) = f(2) = 8$$

$$\left(f^{-1} \circ f\right)(2) = f^{-1}\left(f(2)\right) = f^{-1}(8) = 2$$

Because of the "undoing" nature of inverse functions, we get the following **composition cancellation equations**:

Composition Cancellation Equations

$f(f^{-1}(x)) = x$ for all x in the domain of f^{-1} and

$f^{-1}(f(x)) = x$ for all x in the domain of f

These cancellation equations can be used to show whether two functions are inverses of each other. We can see from our example that if $f(x) = x^3$ and $f^{-1}(x) = \sqrt[3]{x}$, then

$$f(f^{-1}(x)) = f(\sqrt[3]{x}) = (\sqrt[3]{x})^3 = x \text{ and } f^{-1}(f(x)) = f^{-1}(x^3) = \sqrt[3]{x^3} = x$$

My interactive video summary

📺 Example 2 Verify Inverse Functions

Show that $f(x) = \frac{x}{2x + 3}$ and $g(x) = \frac{3x}{1 - 2x}$ are inverse functions using the composition cancellation equations.

Solution To show that f and g are inverses of each other, we must show that $(f \circ g)(x) = x$ and $(g \circ f)(x) = x$. Work through the interactive video to verify that both composition cancellation equations are satisfied.

Work exercises 18–24 in this textbook or in the *MyMathLab*® Study Plan.

OBJECTIVE 4:   SKETCHING THE GRAPHS OF INVERSE FUNCTIONS

If f is a one-to-one function, then we know that it must have an inverse function, f^{-1}. Given the graph of a one-to-one function f, we can obtain the graph of f^{-1} by simply interchanging the coordinates of each ordered pair that lies on the graph of f. In other words, for any point (a, b) on the graph of f, the point (b, a) must lie on the graph of f^{-1}. Notice in Figure 28 that the points (a, b) and (b, a) are symmetric about the line $y = x$. Therefore, the graph of f^{-1} is a reflection of the graph of f about the line $y = x$. Figure 29 shows the graph of $f(x) = x^3$ and $f^{-1}(x) = \sqrt[3]{x}$. You can see that if the functions have any points in common, they must lie along the line $y = x$.

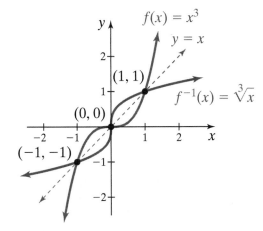

Figure 28 Graph of a one-to-one function and its inverse

Figure 29 Graph of $f(x) = x^3$ and $f^{-1}(x) = \sqrt[3]{x}$

My interactive video summary

📺 Example 3 Sketch the Graph of a One-to-One Function and Its Inverse

Sketch the graph of $f(x) = x^2 + 1$, $x \leq 0$, and its inverse. Also state the domain and range of f and f^{-1}.

Solution The graphs of f and f^{-1} are sketched in Figure 30. Notice how the graph of f^{-1} is a reflection of the graph of f about the line $y = x$. Also notice that the domain of f is the same as the range of f^{-1}. Likewise, the domain of f^{-1} is equivalent to the range of f. View the animation to see exactly how to sketch f and f^{-1}.

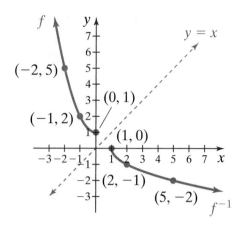

Domain of f: $(-\infty, 0]$ Domain of f^{-1}: $[1, \infty)$

Range of f: $[1, \infty)$ Range of f^{-1}: $(-\infty, 0]$

Figure 30 Graph of $f(x) = x^2 + 1$, $x \leq 0$, and its inverse

Using Technology

$$y_1 = (x^2 + 1)/(x \leq 0)$$

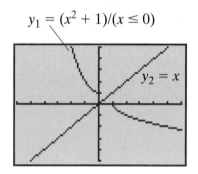

$y_2 = x$

Using a TI-83Plus, we can sketch the functions from Example 3 by letting $y_1 = (x^2 + 1)/(x \leq 0)$. We can draw the inverse function by typing the command **DrawInv** Y_1 in the calculator's main viewing window. The resultant graphs are shown here.

Work exercises 25–31 in this textbook or in the **MyMathLab**® Study Plan.

OBJECTIVE 5:   FINDING THE INVERSE OF A ONE-TO-ONE FUNCTION

We are now ready to find the inverse of a one-to-one function algebraically. We know that if a point (x, y) is on the graph of a one-to-one function, then the point (y, x) is on the graph of its inverse function. We can use this information to develop a process for finding the inverse of a function algebraically simply by switching x and y in the original function to produce its inverse function.

We use as a motivating example the function $f(x) = x^2 + 1$, $x \leq 0$, discussed in Example 3. To find the inverse of a one-to-one function, we follow the four-step process outlined here.

Step 1 **Change $f(x)$ to y:** $y = x^2 + 1$

Step 2 **Interchange x and y:** $x = y^2 + 1$

Step 3 **Solve for y:** $x - 1 = y^2$

                             $\pm\sqrt{x - 1} = y$

        (Because the domain of f is $(-\infty, 0]$, the range of f^{-1} must be $(-\infty, 0]$. Therefore, we must use the negative square root or $y = -\sqrt{x - 1}$.)

Step 4 **Change y to $f^{-1}(x)$:** $f^{-1}(x) = -\sqrt{x - 1}$

        Thus, the inverse of $f(x) = x^2 + 1$, $x \leq 0$, is $f^{-1}(x) = \sqrt{x - 1}$.

My animation summary

Example 4 Find the Inverse of a Function

Find the inverse of the function $f(x) = \frac{2x}{1 - 5x}$, and state the domain and range of f and f^{-1}.

Solution Work through the animation, and follow the four-step process to verify that $f^{-1}(x) = \frac{x}{5x + 2}$. The domain of f is $(-\infty, \frac{1}{5}) \cup (\frac{1}{5}, \infty)$, whereas the domain of f^{-1} is $(-\infty, -\frac{2}{5}) \cup (-\frac{2}{5}, \infty)$. Because the range of f must be the domain of f^{-1} and the range of f^{-1} must be the domain of f, we get the following result:

Domain of f: $(-\infty, \frac{1}{5}) \cup (\frac{1}{5}, \infty)$ $\qquad$ Domain of f^{-1}: $(-\infty, -\frac{2}{5}) \cup (-\frac{2}{5}, \infty)$

Range of f: $(-\infty, -\frac{2}{5}) \cup (-\frac{2}{5}, \infty)$ $\qquad$ Range of f^{-1}: $(-\infty, \frac{1}{5}) \cup (\frac{1}{5}, \infty)$

Work exercises 32−44 in this textbook or in the **MyMathLab** Study Plan.

Inverse Function Summary

1. f^{-1} exists if and only if the function f is one-to-one.

2. The domain of f is the same as the range of f^{-1}, and the range of f is the same as the domain of f^{-1}.

3. To verify that two one-to-one functions, f and g, are inverses of each other, we must use the composition cancellation equations to show that $f(g(x)) = g(f(x)) = x$.

4. The graph of f^{-1} is a reflection of the graph of f about the line $y = x$. That is, for any point (a, b) that lies on the graph of f, the point (b, a) must lie on the graph of f^{-1}.

5. To find the inverse of a one-to-one function, replace $f(x)$ with y, interchange the variables x and y, and solve for y. This is the function $f^{-1}(x)$.

3.6 Exercises

In Exercises 1–17, determine whether each function is one-to-one.

1. $f(x) = 3x - 1$

2. $f(x) = 2x^2$

3. $f(x) = (x - 1)^2, x \geq 1$

4. $f(x) = (x - 1)^2, x \geq -1$

5. $f(x) = \frac{1}{x} - 2$

6. $f(x) = 4\sqrt{x}$

7. $f(x) = -2|x|$

8. $f(x) = 2$

9. $f(x) = (x + 1)^3 - 2$

10. $f(x) = \begin{cases} x + 4 & \text{if } x < -1 \\ 2x + 2 & \text{if } x \geq \frac{1}{2} \end{cases}$

11. $f(x) = \begin{cases} -3x & \text{if } x \leq 0 \\ 2 - x & \text{if } x \geq 2 \end{cases}$

12.

13.

14.

15.

16.

17.

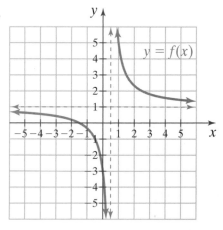

In Exercises 18–24, use the composition cancellation equations to verify that f and g are inverse functions.

18. $f(x) = \dfrac{3}{2}x - 4$ and $g(x) = \dfrac{2x + 8}{3}$

19. $f(x) = \dfrac{ax + b}{c}$ and $g(x) = \dfrac{cx - b}{a}$ for all real numbers a, b, and c such that $a \neq 0$, $c \neq 0$

20. $f(x) = (x - 1)^2$, $x \geq 1$, and $g(x) = \sqrt{x} + 1$

21. $f(x) = \dfrac{7}{x + 1}$ and $g(x) = \dfrac{7 - x}{x}$

22. $f(x) = \dfrac{x}{5 + 3x}$ and $g(x) = \dfrac{5x}{1 - 3x}$

23. $f(x) = 2\sqrt[3]{x - 1} + 3$ and $g(x) = \dfrac{(x - 3)^3}{8} + 1$

24. $f(x) = \dfrac{11x + 5}{4 - 7x}$ and $g(x) = \dfrac{4x - 5}{7x + 11}$

In Exercises 25–30, use the graph of f to sketch the graph of f^{-1}. Use the graphs to determine the domain and range of each function.

25.

26.

27.

28.

29.

30.

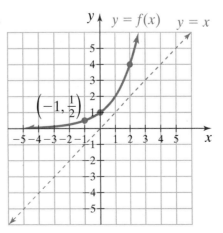

3.6 One-to-One Functions; Inverse Functions   157

In Exercise 31, refer to the graph of the function $y = f(x)$.

31.

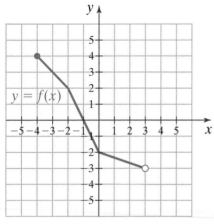

$y = f(x)$

a. What is the domain of f^{-1}?

b. What is the range of f^{-1}?

c. What is the y-intercept of f^{-1}?

d. Evaluate $f^{-1}(0)$.

e. Evaluate $f^{-1}(2)$.

f. Evaluate $f^{-1}(-2)$.

g. Evaluate $f^{-1}(3)$.

h. Evaluate $f^{-1}(4)$.

In Exercises 32–44, write an equation for the inverse function, and then state the domain and range of f and f^{-1}. (*Hint:* In Exercises 43 and 44, complete the square to solve for y.)

32. $f(x) = \dfrac{1}{3}x - 5$

33. $f(x) = \dfrac{3x + 9}{7}$

34. $f(x) = \sqrt[3]{2x - 3}$

35. $f(x) = 1 - \sqrt[5]{x + 4}$

36. $f(x) = -x^2 - 2,\ x \geq 0$

37. $f(x) = (x + 3)^2 - 5,\ x \leq -3$

38. $f(x) = \dfrac{3}{x}$

39. $f(x) = \dfrac{1 - x}{2x}$

40. $f(x) = \dfrac{8x - 1}{7 - 5x}$

41. $f(x) = \dfrac{1 - 4x}{11 + 13x}$

42. $f(x) = \dfrac{4x - 5}{8x + 4}$

43. $f(x) = x^2 + 4x - 1,\ x \geq -2$

44. $f(x) = -2x^2 + 6x + 8,\ x \leq \dfrac{3}{2}$

CHAPTER FOUR

Polynomial and Rational Functions

CHAPTER FOUR CONTENTS

4.1   Quadratic Functions

THINGS TO KNOW

Before working through this section, be sure that you are familiar with the following concepts:

VIDEO   ANIMATION   INTERACTIVE

You Try It
1. Solving Quadratic Equations by Factoring and the Zero Product Property (Section 1.4)

You Try It
2. Solving Quadratic Equations by Completing the Square (Section 1.4)

You Try It
3. Solving Quadratic Equations Using the Quadratic Formula (Section 1.4)

You Try It
4. Determining the Domain and Range of a Function From Its Graph (Section 3.2)

You Try It
5. Using Combinations of Transformations to Graph Functions (Section 3.4)

OBJECTIVES

1 Understanding the Definition of a Quadratic Function and Its Graph

2 Graphing Quadratic Functions Written in Standard Form

3 Graphing Quadratic Functions by Completing the Square

4 Graphing Quadratic Functions Using the Vertex Formula

5 Determining the Equation of a Quadratic Function Given Its Graph

OBJECTIVE 1: UNDERSTANDING THE DEFINITION OF A QUADRATIC FUNCTION AND ITS GRAPH

My video summary

In Section 1.4, we learned how to solve quadratic equations of the form $ax^2 + bx + c = 0$, $a \neq 0$. In this section, we learn about *quadratic functions*.

DEFINITION Quadratic Function

A **quadratic function** is a function of the form $f(x) = ax^2 + bx + c$, where a, b, and c are real numbers with $a \neq 0$. Every quadratic function has a "u-shaped" graph called a *parabola*.

The function $f(x) = x^2$ is a quadratic function with $a = 1$, $b = 0$, and $c = 0$ whose graph is shown in Figure 1a. The function $g(x) = -x^2$ is a quadratic function with $a = -1$, $b = 0$, and $c = 0$ whose graph is shown in Figure 1b. Notice that both graphs are parabolas.

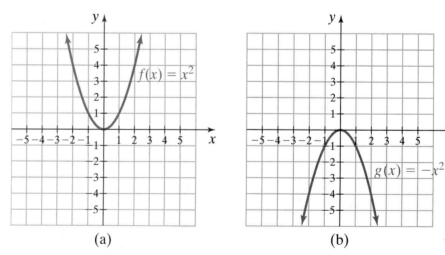

(a)   (b)

Figure 1 Graphs of $f(x) = x^2$ and $g(x) = -x^2$

A parabola either opens up or opens down, depending on the **leading coefficient**, a. If $a > 0$, as in Figure 1a, the parabola will "open up." If $a < 0$, as in Figure 1b, the parabola will "open down."

Example 1 Determine Whether the Graph of a Quadratic Function Opens Up or Opens Down

Without graphing, determine whether the graph of the quadratic function $f(x) = -3x^2 + 6x + 1$ opens up or opens down.

Solution Because the leading coefficient is $a = -3$, the graph of the quadratic function $f(x) = -3x^2 + 6x + 1$ must open down.

Using Technology

The graph of the quadratic function $f(x) = -3x^2 + 6x + 1$ is seen here using a graphing utility. You can see that the graph "opens down."

Work exercises 1–4 in this textbook or in the *MyMathLab*® Study Plan.

Characteristics of a Parabola

Before we learn how to sketch the graphs of other quadratic functions, it is important to identify the five basic characteristics of a parabola. Work through the following animation, and click on each characteristic to get a detailed description.

My animation summary

1. Vertex
2. Axis of symmetry
3. y-Intercept
4. x-Intercept(s) or real zeros
5. Domain and range

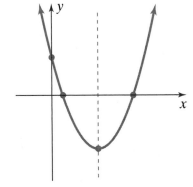

OBJECTIVE 2: GRAPHING QUADRATIC FUNCTIONS WRITTEN IN STANDARD FORM

My animation summary

In Section 3.4, Example 6, we sketched the graph of $f(x) = -2(x + 3)^2 - 1$ using **transformation techniques**. Watch the animation to see how to sketch the function $f(x) = -2(x + 3)^2 - 1$ using transformations. The graph of $f(x) = -2(x + 3)^2 - 1$ is sketched in Figure 2. Note that the domain is $(-\infty, \infty)$ and the range is $(-\infty, -1]$.

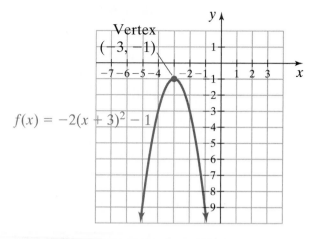

Figure 2 Graph of $f(x) = -2(x + 3)^2 - 1$

The function $f(x) = -2(x + 3)^2 - 1$ is an example of a quadratic function that is in *standard form* with vertex $(-3, -1)$.

Standard Form of a Quadratic Function

A quadratic function is in **standard form** if it is written as $f(x) = a(x - h)^2 + k$. The graph is a parabola with vertex (h, k). The parabola "opens up" if $a > 0$. The parabola "opens down" if $a < 0$.

My video summary

▣ **Example 2 Graph a Quadratic Function That Is in Standard Form**

Given that the quadratic function $f(x) = -(x - 2)^2 - 4$ is in standard form, address the following:

  a. What are the coordinates of the vertex?
  b. Does the graph "open up" or "open down"?
  c. What is the equation of the axis of symmetry?
  d. Find any x-intercepts.
  e. Find the y-intercept.

My video summary

f. Sketch the graph.

g. State the domain and range in interval notation.

Solution

a. The quadratic function $f(x) = -(x - 2)^2 - 4$ is in standard form. Therefore, we can find the vertex by determining the values of h and k.

$$f(x) = a(x - h)^2 + k \qquad \text{\textit{Quadratic function in standard form.}}$$

$$\boxed{a = -1} \quad \boxed{h = 2} \qquad \boxed{k = -4}$$

$$f(x) = -(x - 2)^2 - 4 \qquad \text{\textit{Given function.}}$$

Because $h = 2$ and $k = -4$, the vertex of this parabola is $(2, -4)$.

b. The leading coefficient is $a = -1 < 0$. Therefore, the parabola opens down.

c. The equation of the axis of symmetry is $x = 2$.

d. To find any x-intercepts, we must determine the real solution(s) to the equation $f(x) = 0$.

$$
\begin{aligned}
f(x) &= -(x - 2)^2 - 4 && \text{\textit{Write the original function.}} \\
0 &= -(x - 2)^2 - 4 && \text{\textit{Replace } } f(x) \text{ \textit{with 0 in }} f(x) = -(x - 2)^2 - 4. \\
(x - 2)^2 &= -4 && \text{\textit{Add }} (x - 2)^2 \text{ \textit{to both sides.}} \\
x - 2 &= \pm\sqrt{-4} && \text{\textit{Use the square root property.}} \\
x - 2 &= \pm 2i && \text{\textit{Recall that }} \sqrt{-4} = \sqrt{4} \cdot \sqrt{-1} = 2i. \\
x &= 2 \pm 2i && \text{\textit{Add 2 to both sides.}}
\end{aligned}
$$

Because the equation $f(x) = 0$ has no real solutions, the parabola has no x-intercepts.

e. The y-intercept is $f(0) = -(0 - 2)^2 - 4 = -4 - 4 = -8$.

f. The vertex is $(2, -4)$, the parabola opens down, the equation of the axis of symmetry is $x = 2$, and the y-intercept is -8. Using this information, we can sketch the parabola as shown in Figure 3.

g. The domain of f is the interval $(-\infty, \infty)$. The range of f is the interval $(-\infty, -4]$.

Watch the video to see each step worked out in detail.

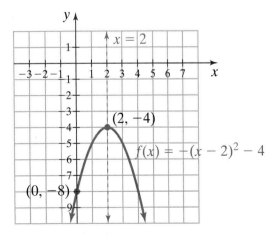

Figure 3 Graph of $f(x) = -(x - 2)^2 - 4$

Work exercises 5–10 in this textbook or in the **MyMathLab**® Study Plan.

OBJECTIVE 3: GRAPHING QUADRATIC FUNCTIONS BY COMPLETING THE SQUARE

We can see from Example 2 that if a quadratic function is in standard form $f(x) = a(x - h)^2 + k$, it is fairly straightforward to determine its graph. Therefore, to graph a quadratic function of the form $f(x) = ax^2 + bx + c$, we convert it into standard form by completing the square as in Example 3.

My video summary

Example 3 Write a Quadratic Function in Standard Form by Completing the Square and Sketch the Graph

Rewrite the quadratic function $f(x) = 2x^2 - 4x - 3$ in standard form, and then complete a) through g), as in Example 2.

Solution Our goal is to rewrite $f(x) = 2x^2 - 4x - 3$ in the form $f(x) = a(x - h)^2 + k$. To do this, we must complete the square.

$f(x) = 2x^2 - 4x - 3$	Write the original function.
$f(x) = (2x^2 - 4x) - 3$	Isolate the constant.
$f(x) = 2(x^2 - 2x) - 3$	Factor out the leading coefficient 2 from $2x^2 - 4x$.
$f(x) = 2(x^2 - 2x + 1) - 3 - 2$	Complete the square of $x^2 - 2x$. Note that we are adding $2 \cdot 1 = 2$, so we must subtract 2 as well!
$f(x) = 2(x - 1)^2 - 5$	Rewrite $(x^2 - 2x + 1)$ as a binomial squared.

Thus, the quadratic function $f(x) = 2x^2 - 4x - 3$ written in standard form is $f(x) = 2(x - 1)^2 - 5$. Watch the video to verify the following:

a. The vertex is $(1, -5)$.

b. The parabola opens up.

c. The equation of the axis of symmetry is $x = 1$.

d. The two x-intercepts are $x = 1 + \frac{\sqrt{10}}{2} \approx 2.5811$ and $x = 1 - \frac{\sqrt{10}}{2} \approx -0.5811$.

e. The y-intercept is -3.

f. The graph of $f(x) = 2x^2 - 4x - 3$ is sketched in Figure 4.

g. The domain of f is the interval $(-\infty, \infty)$. The range of f is the interval $[-5, \infty)$.

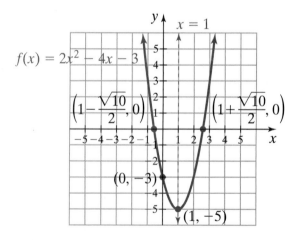

Figure 4 Graph of $f(x) = 2x^2 - 4x - 3$

Work exercises 11–20 in this textbook or in the *MyMathLab*® Study Plan.

OBJECTIVE 4:   GRAPHING QUADRATIC FUNCTIONS USING THE VERTEX FORMULA

My video summary

We can establish a "formula" for the vertex by completing the square on the quadratic function $f(x) = ax^2 + bx + c$. Work through the video to verify that the function $f(x) = ax^2 + bx + c$ is equivalent to $f(x) = a\left(x + \frac{b}{2a}\right)^2 + c - \frac{b^2}{4a}$.

We can now compare the function $f(x) = a\left(x + \frac{b}{2a}\right)^2 + c - \frac{b^2}{4a}$ to $f(x) = a(x - h)^2 + k$.

$$f(x) = a\left(x - \left(-\frac{b}{2a}\right)\right)^2 + \underbrace{c - \frac{b^2}{4a}}$$

$$\downarrow \qquad\qquad\qquad \downarrow \qquad \downarrow$$

$$f(x) = a(x - \qquad h)^2 + \quad k$$

We see that the coordinates of the vertex must be $\left(-\frac{b}{2a}, c - \frac{b^2}{4a}\right)$. Because the y-coordinate of the vertex can be found more easily by evaluating the function at the x-coordinate of the vertex than by memorizing the expression $c - \frac{b^2}{4a}$, we write the coordinates of the vertex as $\left(-\frac{b}{2a}, f\left(-\frac{b}{2a}\right)\right)$.

Formula for the Vertex of a Parabola

Given a quadratic function of the form $f(x) = ax^2 + bx + c$, $a \neq 0$, the vertex of the parabola is $\left(-\frac{b}{2a}, f\left(-\frac{b}{2a}\right)\right)$.

Example 4 Graph a Quadratic Function Using the Vertex Formula and Intercepts

Given the quadratic function $f(x) = -2x^2 - 4x + 5$, address the following:

a. Use the vertex formula to determine the vertex.
b. Does the graph "open up" or "open down"?
c. What is the equation of the axis of symmetry?
d. Find any x-intercepts.
e. Find the y-intercept.
f. Sketch the graph.
g. State the domain and range in interval notation.

Solution

a. The x-coordinate of the vertex is $-\frac{b}{2a} = -\frac{(-4)}{2(-2)} = -1$. The y-coordinate of the vertex is $f\left(-\frac{b}{2a}\right) = f(-1) = -2(-1)^2 - 4(-1) + 5 = -2 + 4 + 5 = 7$. Therefore, the vertex is $(-1, 7)$.

b. Because $a = -2 < 0$, the parabola must open down.

c. The axis of symmetry is $x = -1$.

d. To find the x-intercepts, we solve the quadratic equation $-2x^2 - 4x + 5 = 0$. Click here to verify that the x-intercepts are $x = -1 + \frac{\sqrt{14}}{2} \approx 0.8708$ and $x = -1 - \frac{\sqrt{14}}{2} \approx -2.8708$.

e. The y-intercept is $f(0) = 5$.

f. Plotting the vertex and all intercepts, we get the graph sketched in Figure 5.

g. The domain of f is the interval $(-\infty, \infty)$. The range of f is the interval $(-\infty, 7]$.

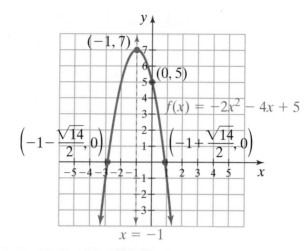

$$f(x) = -2x^2 - 4x + 5$$

$(-1, 7)$

$(0, 5)$

$\left(-1 - \dfrac{\sqrt{14}}{2}, 0\right)$ $\left(-1 + \dfrac{\sqrt{14}}{2}, 0\right)$

$x = -1$

Figure 5 Graph of $f(x) = -2x^2 - 4x + 5$

Using Technology

Maximum
X=-1   Y=7

Using a graphing utility and the maximum function, we can verify that the vertex is $(-1, 7)$.

Work exercises 21–30 in this textbook or in the *MyMathLab*® Study Plan.

OBJECTIVE 5: DETERMINING THE EQUATION OF A QUADRATIC FUNCTION GIVEN ITS GRAPH

Given the graph of a parabola, it may be necessary to find the equation of the quadratic function that it represents.

My video summary

▣ **Example 5 Determine the Equation of a Quadratic Function Given Its Graph**

Analyze the graph to address the following about the quadratic function it represents.

a. Is the leading coefficient positive or negative?

b. What is the value of h? What is the value of k?

c. What is the value of the leading coefficient a?

d. Write the equation of the function in standard form $f(x) = a(x - h)^2 + k$.

e. Write the equation of the function in the form $f(x) = ax^2 + bx + c$.

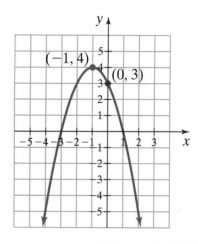

$(-1, 4)$ $(0, 3)$

My video summary

🎞 **Solution**

a. The leading coefficient is negative because the parabola opens down.

b. The value of h is the x-coordinate of the vertex, so $h = -1$. The value of k is the y-coordinate of the vertex, so $k = 4$.

c. Substitute the values of $h = -1$ and $k = 4$ into the standard form $f(x) = a(x - h)^2 + k$ to get $f(x) = a(x - (-1))^2 + 4$ or $f(x) = a(x + 1)^2 + 4$.

To determine the value of a, we can use the fact that the graph contains the point $(0, 3)$, and therefore, $f(0) = 3$.

$$f(x) = a(x + 1)^2 + 4 \qquad \text{\textit{Use the standard form with }} h = -1$$
$$\text{\textit{and }} k = 4.$$
$$f(0) = a(0 + 1)^2 + 4 \qquad \text{\textit{Evaluate }} f(0).$$
$$f(0) = 3 \qquad \text{\textit{The graph contains the point }} (0, 3).$$
$$a(0 + 1)^2 + 4 = 3 \qquad \text{\textit{Equate the two expressions for }} f(0).$$
$$a = -1 \qquad \text{\textit{Solve for }} a.$$

d. Substituting the values of $a = -1$, $h = -1$, and $k = 4$ into the equation $f(x) = a(x - h)^2 + k$, we get $f(x) = -(x + 1)^2 + 4$.

e. Writing the function in the form $f(x) = ax^2 + bx + c$, we get $f(x) = -x^2 - 2x + 3$. Watch the video to see each step worked out in detail.

Work exercises 31–36 in this textbook or in the _MyMathLab_® Study Plan.

4.1 Exercises

In Exercises 1–4, without graphing, determine whether the graph of each quadratic function opens up or opens down.

1. $f(x) = x^2 - 3$ **2.** $f(x) = -2x^2 + 4x + 1$ **3.** $f(x) = -4 + 5x^2 - 2x$

4. $f(x) = -\dfrac{2}{3}x^2 + x + 7$

In Exercises 5–10, given the quadratic function in standard form, address the following.

a. What are the coordinates of the vertex? b. Does the graph "open up" or "open down"?

c. What is the equation of the axis of symmetry? d. Find any x-intercepts.

e. Find the y-intercept. f. Sketch the graph.

g. State the domain and range.

5. $f(x) = (x - 2)^2 - 4$ **6.** $f(x) = -(x + 1)^2 - 9$ **7.** $f(x) = -2(x - 3)^2 + 2$

8. $f(x) = \dfrac{1}{2}(x + 2)^2 + 2$ **9.** $f(x) = -\dfrac{1}{4}(x - 4)^2 + 2$ **10.** $f(x) = 3\left(x + \dfrac{1}{3}\right)^2 - 4$

In Exercises 11–20, first rewrite the quadratic function in standard form by completing the square, and then address the following.

a. What are the coordinates of the vertex? b. Does the graph "open up" or "open down"?

c. What is the equation of the axis of symmetry? d. Find any x-intercepts.

e. Find the y-intercept. f. Sketch the graph.

g. State the domain and range in interval notation.

11. $f(x) = x^2 - 8x$ **12.** $f(x) = -x^2 - 4x + 12$ **13.** $f(x) = 3x^2 + 6x - 4$

14. $f(x) = 2x^2 - 5x - 3$ **15.** $f(x) = -x^2 + 2x - 6$ **16.** $f(x) = \frac{1}{2}x^2 + 6x + 1$

17. $f(x) = -\frac{1}{3}x^2 - 2x + 5$ **18.** $f(x) = -3x^2 + 7x + 5$ **19.** $f(x) = x^2 + \frac{8}{3}x - 1$

20. $f(x) = -\frac{1}{4}x^2 + 6x - 1$

In Exercises 21–30, use the quadratic function to address the following.

   a. Use the vertex formula to determine the vertex.   **b.** Does the graph "open up" or "open down"?

   c. What is the equation of the axis of symmetry?   **d.** Find any x-intercepts.

   e. Find the y-intercept.   **f.** Sketch the graph.

   g. State the domain and range in interval notation.

21. $f(x) = x^2 - 8x$ **22.** $f(x) = -x^2 - 4x + 8$ **23.** $f(x) = 3x^2 + 6x - 4$

24. $f(x) = 2x^2 - 5x - 3$ **25.** $f(x) = -x^2 + 2x - 6$ **26.** $f(x) = \frac{1}{2}x^2 + 6x + 1$

27. $f(x) = -\frac{1}{3}x^2 - 9x + 5$ **28.** $f(x) = -3x^2 + 7x + 5$ **29.** $f(x) = x^2 + \frac{8}{3}x - 1$

30. $f(x) = -\frac{1}{4}x^2 + 6x - 1$

In Exercises 31–36, analyze the graph to address the following about the quadratic function it represents.

   a. Is the leading coefficient positive or negative?   **b.** What is the value of h? What is the value of k?

   c. What is the value of the leading coefficient a?   **d.** Write the equation of the function in standard form

   e. Write the equation of the function in the form $f(x) = a(x - h)^2 + k.$

       $f(x) = ax^2 + bx + c.$

31.

32.

33.

34.

35.

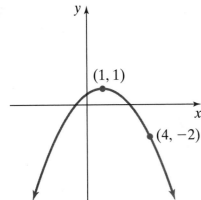

(1, 1)

(4, −2)

36.

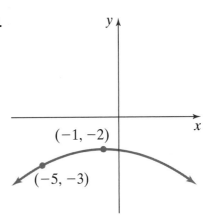

(−1, −2)

(−5, −3)

4.2 Applications and Modeling of Quadratic Functions

THINGS TO KNOW

Before working through this section, be sure that you are familiar with the following concept:

VIDEO   ANIMATION   INTERACTIVE

You Try It

1. Graphing Quadratic Functions Using the Vertex Formula (**Section 4.1**)

OBJECTIVES

1 Maximizing Projectile Motion Functions

2 Maximizing Functions in Economics

3 Maximizing Area Functions

My video summary

 With application problems involving functions, we are often interested in finding the maximum or minimum value of the function. For example, a professional golfer may want to control the trajectory of a struck golf ball. Thus, he may be interested in determining the maximum height of the ball given certain parameters such as swing velocity and club angle. An economist may want to minimize a cost function or maximize a revenue or profit function. A builder with a fixed amount of fencing may want to maximize an area function. In a calculus course, you learn how to maximize or minimize a wide variety of functions. In this section, we concentrate only on quadratic functions. Quadratic functions are relatively easy to maximize or minimize because we know a formula for finding the coordinates of the vertex. Recall that if $f(x) = ax^2 + bx + c$, $a \neq 0$, we know that the coordinates of the vertex are $\left(-\frac{b}{2a}, f\left(-\frac{b}{2a}\right)\right)$. If $a > 0$, the parabola *opens up* and has a minimum value at the vertex. If $a < 0$, the parabola *opens down* and has a maximum value at the vertex. See Figure 6.

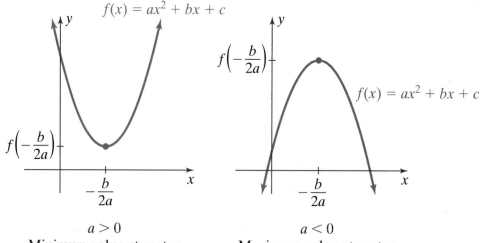

Figure 6   Minimum value at vertex   Maximum value at vertex

OBJECTIVE 1:   MAXIMIZING PROJECTILE MOTION FUNCTIONS

We start our discussion by looking at the projectile motion model that was introduced in Section 1.5. An object launched, thrown, or shot vertically into the air with an initial velocity of v_0 meters/second from an initial height of h_0 meters above the ground can be modeled by the function* $h(t) = -4.9t^2 + v_0 t + h_0$, where $h(t)$ is the height of the projectile t seconds after its departure.

Example 1   Launch a Toy Rocket

A toy rocket is launched with an initial velocity of 44.1 meters per second from a 1-meter-tall platform. The height h of the object at any time t seconds after launch is given by the function $h(t) = -4.9t^2 + 44.1t + 1$. How long after launch did it take the rocket to reach its maximum height? What is the maximum height obtained by the toy rocket?

Solution Because the function is quadratic with $a = -4.9 < 0$, we know that the graph is a parabola that opens down. Thus, the function obtains a maximum value at the vertex. The t-coordinate of the vertex is $t = -\dfrac{b}{2a} = -\dfrac{44.1}{2(-4.9)} = \dfrac{-44.1}{-9.8} = 4.5$ seconds. Therefore, the rocket reaches its maximum height 4.5 seconds after launch. The maximum height obtained by the rocket is $h(4.5) = -4.9(4.5)^2 + 44.1(4.5) + 1 = 100.225$ meters.

Maximum
X=4.5   Y=100.225

Using Technology

The graph of the quadratic function $h(t) = -4.9t^2 + 44.1t + 1$ is seen here using a graphing utility. Using the Maximum function, we can see that the maximum height is 100.225 meters when time is 4.5 seconds.

Work exercises 1–3 in this textbook or in the _MyMathLab_® Study Plan.

---

*Note that the leading coefficient of the projectile motion model is −4.9. This constant is derived using calculus and the acceleration of gravity on Earth, which is 9.8 meters per second per second. If the height was measured in feet, the leading coefficient of the projectile motion model would be −16, derived using calculus and acceleration of gravity on Earth, which is 32 feet per second per second.

Example 2 Horizontal and Vertical Projectile Motion

If an object is launched at an angle of 45 degrees from a 10-foot platform at 60 feet per second, it can be shown that the height of the object in feet is given by the quadratic function $h(x) = -\frac{32x^2}{(60)^2} + x + 10$, where x is the horizontal distance of the object from the platform. See Figure 7.

a. What is the height of the object when its horizontal distance from the platform is 20 feet? Round to two decimal places.
b. What is the horizontal distance from the platform when the object is at its maximum height?
c. What is the maximum height of the object?

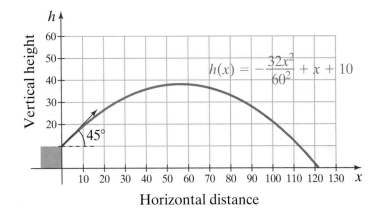

Figure 7

Solution

a. To find the height of the object when the horizontal distance from the platform is 20 feet, we must find the value of $h(20)$.

$$h(x) = -\frac{32x^2}{(60)^2} + x + 10 \qquad \textit{Write the original function.}$$

$$h(20) = -\frac{32(20)^2}{(60)^2} + 20 + 10 \qquad \textit{Substitute 20 for x.}$$

$$= -\frac{32(400)}{3,600} + 30 \qquad \textit{Simplify.}$$

$$= -\frac{12,800}{3,600} + 30$$

$$\approx 26.44$$

When the horizontal distance from the platform is 20 feet, the height of the object is approximately 26.44 feet.

b. To find the horizontal distance of the object when it is at its maximum height, we must find the x-coordinate of the vertex. The x-coordinate of the vertex is found using the vertex formula $x = -\frac{b}{2a}$. Since $a = -\frac{32}{60^2}$ and $b = 1$, we get

$$x = -\frac{1}{2\left(\dfrac{-32}{60^2}\right)} = \frac{-1}{\dfrac{-64}{3,600}} = -1 \cdot \frac{3,600}{-64} = 56.25 \text{ feet}$$

Thus, the object obtains its maximum height at a horizontal distance of 56.25 feet from the base of the platform.

4.2 Applications and Modeling of Quadratic Functions   **171**

c. From part b, the object obtains its maximum height at a horizontal distance of 56.25 feet from the base of the platform. Therefore, the maximum height is $h(56.25) = -\frac{32(56.25)^2}{(60)^2} + 56.25 + 10 = 38.125$ feet.

Work exercises 4–6 in this textbook or in the *MyMathLab*® Study Plan.

OBJECTIVE 2:   MAXIMIZING FUNCTIONS IN ECONOMICS

Revenue is defined as the dollar amount received by selling x items at a price of p dollars per item; that is, $R = xp$. For example, if a child sells 50 cups of lemonade at a price of \$.25 per cup, then the revenue generated is $R = \underbrace{(50)}_{x}\underbrace{(.25)}_{p} = \$12.50$.

A common model in economics states that as the quantity, x, increases, the price, p, tends to decrease. Likewise, if the quantity decreases, the price tends to increase.

My video summary

▣ Example 3 Maximize Revenue

Records can be kept on the price of shoes and the number of pairs sold in order to gather enough data to reasonably model shopping trends for a particular type of shoe. Demand functions of this type are often linear and can be developed using knowledge of the slope and equations of lines. Suppose that the marketing and research department of a shoe company determined that the price of a certain type of basketball shoe obeys the demand equation $p = -\frac{1}{50}x + 110$.

a. According to the demand equation, how much should the shoes sell for if 500 pairs of shoes are sold? 1,200 pairs of shoes?
b. What is the revenue if 500 pairs of shoes are sold? 1,200 pairs of shoes?
c. How many pairs of shoes should be sold in order to maximize revenue? What is the maximum revenue?
d. What price should be charged in order to maximize revenue?

Solution

a. The demand equation $p = -\frac{1}{50}x + 110$ is sketched in Figure 8. If 500 pairs of shoes are sold, the price should be $p(500) = -\frac{1}{50}(500) + 110 = -10 + 110 = \$100$. If 1,200 pairs of shoes are sold, the price should be $p(1,200) = -\frac{1}{50}(1,200) + 110 = -24 + 110 = \$86$.

Figure 8

b. Because $R = xp$, we can substitute the demand equation, $p = -\frac{1}{50}x + 110$, for p to obtain the function $R(x) = \underbrace{x(-\frac{1}{50}x + 110)}_{p}$ or $R(x) = -\frac{1}{50}x^2 + 110x$.

The graph of R is shown in Figure 9. The revenue generated by selling 500 pairs of shoes is $R(500) = \$50{,}000$, whereas the revenue generated by selling 1,200 pairs of shoes is $R(1{,}200) = \$103{,}200$.

Figure 9

c. $R(x) = -\frac{1}{50}x^2 + 110x$ is a quadratic function with $a = -\frac{1}{50}$ and $b = 110$. Because a is less than 0, the function has a *maximum* value at the vertex. Therefore, the value of x that produces the maximum revenue is

$$x = -\frac{b}{2a} = -\frac{110}{2\left(-\frac{1}{50}\right)} = \frac{-110}{-\frac{1}{25}} = 2{,}750 \text{ pairs of shoes.}$$

The maximum revenue is $R(2{,}750) = \$151{,}250$. See Figure 10.

Figure 10

d. Using the demand equation, the price that should be charged to maximize revenue when selling 2,750 pairs of shoes is $p(2{,}750) = -\frac{1}{50}(2{,}750) + 110 = \$55$.

Work exercises 7 and 8 in this textbook or in the *MyMathLab*® Study Plan.

Example 4 Maximize Profit

To sell x waterproof CD alarm clocks, WaterTime, LLC, has determined that the price in dollars must be $p = 250 - 2x$, which is the demand equation. Each clock costs \$2 to produce, with fixed costs of \$4,000 per month, producing the cost function of $C(x) = 2x + 4,000$.

a. Express the revenue R as a function of x.
b. Express the profit P as a function of x.
c. Find the value of x that maximizes profit. What is the maximum profit?
d. What is the price of the alarm clock that will maximize profit?

Solution

a. The equation for revenue is $R = xp$. Therefore, we can substitute the demand equation $p = 250 - 2x$ for p to obtain the function $R(x) = x\underbrace{(250 - 2x)}_{p}$ or $R(x) = -2x^2 + 250x$.

b. Profit is equal to revenue minus cost.

$$P(x) = R(x) - C(x)$$
$$= -2x^2 + 250x - (2x + 4,000)$$
$$= -2x^2 + 248x - 4,000$$

c. The value of x that will produce the maximum profit is

$$x = -\frac{b}{2a} = -\frac{248}{2(-2)} = \frac{-248}{-4} = 62 \text{ alarm clocks}$$

The maximum profit is $P(62) = -2(62)^2 + 248(62) - 4,000 = \$3,688$.

d. Using the demand equation $p = 250 - 2x$ and the fact that profit will be maximized when 62 alarm clocks are produced, we obtain a price of $p = 250 - 2(62) = \$126$.

Work exercises 9–12 in this textbook or in the MyMathLab® Study Plan.

My interactive video summary

Example 5 Maximize Revenue at a Country Club

A country club currently has 400 members who pay \$500 per month for membership dues. The country club's board members want to increase monthly revenue by *lowering* the monthly dues in hopes of attracting new members. A market research study has shown that for each \$1 decrease in the monthly membership price, two additional people will join the club. What price should the club charge to maximize revenue? What is the maximum revenue?

Solution We know that $R = xp$, where x is the number of members and p is the price of monthly membership. Currently, $R = (400)(\$500) = \$200,000$. If the price is reduced by \$1, two additional people will join the club, so the revenue generated would be $R = (402)(\$499) = \$200,598$. We can see that by reducing monthly dues to \$499, it can be anticipated that revenue will increase by \$598. We want to find the optimal price to charge in order to maximize revenue.

Because the number of members, x, depends on price, we can find an expression for x in terms of p:

$$x = \underbrace{400}_{\substack{\text{Number} \\ \text{of current} \\ \text{members}}} + \underbrace{2(500 - p)}_{\substack{\text{Additional two members} \\ \text{for each dollar decrease in} \\ \text{membership monthly dues}}}$$

We can now write the revenue as a function of p:

$$R = xp$$
$$R(p) = \underbrace{(400 + 2(500 - p))}_{x}p \text{ or}$$
$$R(p) = -2p^2 + 1,400p$$

The revenue is maximized when the price is $p = -\frac{b}{2a} = -\frac{1,400}{2(-2)} = \$350$.

The maximum revenue is $R(350) = -2(350)^2 + 1,400(350) = \$245,000$. Watch the interactive video to see this problem worked out using an alternate method!

Work exercises 13–20 in this textbook or in the *MyMathLab*® Study Plan.

OBJECTIVE 3:   MAXIMIZING AREA FUNCTIONS

Suppose you are asked to build a rectangular fence that borders a river but you have only 3,000 feet of fencing available. No fencing will be required along the side of the river. How many ways can you build this fence? As it turns out, there are infinitely many ways to construct such a fence.

My video summary

Although there are infinitely many ways in which you could build this fence, there is only one way to build this fence in order to *maximize* the area! What should the length and width of the fence be in order to maximize the area? Watch this video to see how to determine the length and width of the fence that will maximize the area.

Example 6 Maximize Area

Mark has 100 feet of fencing available to build a rectangular pen for his hens and roosters. He wants to separate the hens and roosters by dividing the pen into two equal areas. What should the length of the center partition be in order to maximize the area? What is the maximum area?

Solution Watch the video to verify that the length of the partition should be $16\frac{2}{3}$ feet. The maximum area is $416\frac{2}{3}$ square feet.

Work exercises 21–24 in this textbook or in the _MyMathLab_® Study Plan.

4.2 Exercises

1. A baseball player swings and hits a pop fly straight up in the air to the catcher. The height of the baseball in meters t seconds after it is hit is given by the quadratic function $h(t) = -4.9t^2 + 34.3t + 1$. How long does it take for the baseball to reach its maximum height? What is the maximum height obtained by the baseball?

2. An object is launched vertically in the air at 36.75 meters per second from a 10-meter-tall platform. Using the projectile motion model, determine how long it will take for the object to reach its maximum height. What is the maximum height?

3. A toy rocket is shot vertically into the air from a 5-foot-tall launching pad with an initial velocity of 112 feet per second. Suppose the height of the rocket in feet t seconds after being launched can be modeled by the function $h(t) = -16t^2 + v_0t + h_0$, where v_0 is the initial velocity of the rocket and h_0 is the initial height of the rocket. How long will it take for the rocket to reach its maximum height? What is the maximum height?

4. A football player punts a football at an inclination of 45 degrees to the horizontal at an initial velocity of 75 feet per second. The height of the football in feet is given by the quadratic function $h(x) = -\frac{32x^2}{(75)^2} + x + 4$, where x is the horizontal distance of the football from the point at which it was kicked. What was the horizontal distance of the football from the point at which it was kicked when the height of the ball is at a maximum? What is the maximum height of the football?

5. A golf ball is struck by a 60-degree golf club at an initial velocity of 88 feet per second. The height of the golf ball in feet is given by the quadratic function $h(x) = -\frac{16x^2}{(44)^2} + \frac{76.2}{44}x$, where x is the horizontal distance of the golf ball from the point of impact. What is the horizontal distance of the golf ball from the point of impact when the ball is at its maximum height? What is the maximum height obtained by the golf ball?

6. If a projectile is launched upward at an acute angle, then the height of the projectile at a horizontal distance of x units from the launch site is given by the quadratic function $h(x) = -\frac{1}{2}g\left(\frac{x}{w_0}\right)^2 + \left(\frac{v_0}{w_0}\right)x + h_o$, where g is acceleration due to gravity, w_0 is horizontal initial velocity, v_0 is vertical initial velocity, and h_0 is initial height of the object.

a. Find a formula describing the horizontal distance from the launch point where the projectile is at a maximum height in terms of g, w_0, and v_0.

b. Find a formula for the maximum height of the projectile in terms of g, v_0, and h_0.

c. Suppose $g = 32$, $w_0 = 64$, and $v_0 = 64$, use the formulas from parts a and b to find the horizontal distance from the launch point where the projectile is at a maximum height and find the maximum height.

7. The price p and the quantity x sold of a small flat screen television set obeys the demand equation $p = -.15x + 300$.

a. How much should be charged for the television set if there are 50 television sets in stock?

b. What quantity x will maximize revenue? What is the maximum revenue?

c. What price should be charged in order to maximize revenue?

8. The dollar price for a barrel of oil sold at a certain oil refinery tends to follow the demand equation $p = -\frac{1}{10}x + 72$, where x is the number of barrels of oil on hand (in millions).

a. How much should be charged for a barrel of oil if there are 4 million barrels on hand?

b. What quantity x will maximize revenue? What is the maximum revenue?

c. What price should be charged in order to maximize revenue?

In Exercises 9 and 10, use the fact that profit is defined as revenue minus cost or $P(x) = R(x) - C(x)$.

9. Rite-Cut riding lawnmowers obey the demand equation $p = -\frac{1}{20}x + 1{,}000$. The cost of producing x lawnmowers is given by the function $C(x) = 100x + 5{,}000$.

a. Express the revenue R as a function of x.

b. Express the profit P as a function of x.

c. Find the value of x that maximizes profit. What is the maximum profit?

d. What price should be charged in order to maximize profit?

10. The CarryItAll minivan, a popular vehicle among soccer moms, obeys the demand equation $p = -\frac{1}{40}x + 8{,}000$. The cost of producing x vans is given by the function $C(x) = 4{,}000x + 20{,}000$.

a. Express the revenue R as a function of x.

b. Express the profit P as a function of x.

c. Find the value of x that maximizes profit. What is the maximum profit?

d. What price should be charged in order to maximize profit?

11. Silver Scooter, Inc., finds that it costs $200 to produce each motorized scooter and that the fixed costs are $1,500 per day. The price is given by $p = 600 - 5x$, where p is the price in dollars at which exactly x scooters will be sold. Find the quantity of scooters that Silver Scooter, Inc., should produce and the price it should charge to maximize profit. Find the maximum profit.

12. Amy, the owner of Amy's Pottery, can produce china pitchers at a cost of $5 each. She estimates her price function to be $p = 17 - 5x$, where p is the price at which exactly x pitchers will be sold per week. Find the number of pitchers that she should produce and the price that she should charge in order to maximize profit. Also find the maximum profit.

13. Lewiston Golf and Country Club currently has 300 members who pay $450 per month for membership dues. The country club's board members want to increase monthly revenue by *lowering* the monthly dues in hopes of attracting new members. A market research study has shown that for each $1 decrease in the monthly membership price, three additional people will join the club. What price should the club charge to maximize revenue? What is the maximum revenue?

14. An orange grove produces a profit of $90 per tree when there are 1,200 trees planted. Because of overcrowding, the profit per tree (for every tree in the grove) is reduced by 5 cents per tree for each additional tree planted. How many trees should be planted in order to maximize the total profit of the orange grove? What is the maximum profit?

15. All-Nite Fitness Health Club currently charges its 1,500 clients monthly membership dues of $45. The board of directors decides to increase the monthly membership dues. Market research states that each $1 increase in dues will result in the loss of five clients. How much should the club charge each month to optimize the revenue from monthly dues?

16. The marketing department of a local shoe company found that approximately 600 pairs of running shoes are sold monthly when the price of each pair is $90. It was also observed that for each $1 reduction in price, an additional 20 pairs of running shoes are sold monthly. What price should the shoe store charge for a pair of running shoes in order to maximize revenue?

17. Pearson Travel, Inc., charges $10 for a trip to a concert if 30 people travel in a group. But for each 1 person above the 30-person level, the charge will be reduced by $.20. How many people will maximize the total revenue for the company?

18. Sunshine Bicycle Rental Company at a California resort rents 100 bicycles per day at a flat rate of $10 per day for each bicycle. For each $1 increase in the rate, 5 fewer bikes are rented. At what rate should the bicycles be rented to produce the maximum daily revenue? What is the maximum daily revenue?

19. Cherrypickers, Inc., estimates from past records that if 30 trees are planted per acre, each tree will yield an average of 50 pounds of cherries per season. If for each additional tree planted per acre the average yield per tree is reduced by 1 pound, how many trees should be planted per acre to obtain the maximum yield per acre?

20. An orchard contains 30 apple trees, each of which yields about 400 apples over the growing season. The owner plans to add more trees, but each new tree will reduce the average yield per tree by about 10 apples over the season. How many trees should be added to maximize the total yield of apples? What would be the maximum yield for the apple orchard for the season?

21. A farmer has 1,800 feet of fencing available to enclose a rectangular area bordering a river. If no fencing is required along the river, find the dimensions of the fence that will maximize the area. What is the maximum area?

22. Jim wants to build a rectangular parking lot along a busy street but has only 2,500 feet of fencing available. If no fencing is required along the street, find the maximum area of the parking lot.

23. A rancher has 5,000 feet of fencing available to enclose a rectangular area bordering a river. He wants to separate his cows and horses by dividing the enclosure into two equal areas. If no fencing is required along the river, find the length of the center partition that will yield the maximum area.

24. A 32-inch wire is to be cut. One piece is to be bent into the shape of a square, whereas the other piece is to be bent into the shape of a rectangle whose length is twice the width. Find the width of the rectangle that will minimize the total area.

4.3 The Graphs of Polynomial Functions

THINGS TO KNOW

Before working through this section, be sure that you are familiar with the following concepts:

VIDEO   ANIMATION   INTERACTIVE

You Try It

1. Determining the Intercepts of a Function (Section 3.2)

You Try It

2. Sketching the Graphs of the Basic Functions (Section 3.3)

You Try It

3. Using Transformations to Sketch Functions (Section 3.4)

OBJECTIVES

1 Understanding the Definition of a Polynomial Function

2 Sketching the Graphs of Power Functions

3 Determining the End Behavior of Polynomial Functions

4 Determining the Intercepts of a Polynomial Function

5 Determining the Real Zeros of Polynomial Functions and Their Multiplicities

6 Sketching the Graph of a Polynomial Function

7 Determining a Possible Equation of a Polynomial Function Given Its Graph

OBJECTIVE 1:   UNDERSTANDING THE DEFINITION OF A POLYNOMIAL FUNCTION

My video summary

📺 We have already studied linear functions of the form $f(x) = mx + b$, where $m \neq 0$, and quadratic functions of the form $f(x) = ax^2 + bx + c$, where $a \neq 0$. Both functions are examples of **polynomial functions**.

DEFINITION Polynomial Function

The function $f(x) = a_n x^n + a_{n-1}x^{n-1} + a_{n-2}x^{n-2} + \cdots + a_1 x + a_0$ is a polynomial function of degree n, where n is a nonnegative integer. The numbers $a_0, a_1, a_2, \ldots, a_n$ are called the **coefficients** of the polynomial function. The number a_n is called the **leading coefficient**, and a_0 is called the **constant coefficient**.

The important thing to remember from the definition is that the power (exponent) of each term must be an integer greater than or equal to zero. Although the coefficients can be any number (real or imaginary), we focus primarily on polynomial functions with real coefficients.

Example 1 Determine whether a Function Is a Polynomial Function

Determine which functions are polynomial functions. If the function is a polynomial function, identify the degree, the leading coefficient, and the constant coefficient.

a. $f(x) = \sqrt{3}x^3 - 2x^2 - \dfrac{1}{6}$

b. $g(x) = 4x^5 - 3x^3 + x^2 + \dfrac{7}{x} - 3$

c. $h(x) = \dfrac{3x - x^2 + 7x^4}{9}$

Solution

a. The function $f(x) = \sqrt{3}x^3 - 2x^2 - \frac{1}{6}$ is a polynomial function because all powers of x are integers greater than or equal to 0. The highest power is 3, so this polynomial is of degree 3. The leading coefficient is $\sqrt{3}$, and the constant coefficient is $-\frac{1}{6}$.

b. The function $g(x) = 4x^5 - 3x^3 + x^2 + \frac{7}{x} - 3$ is *not* a polynomial function because the term $\frac{7}{x}$ can be rewritten as $7x^{-1}$ and -1 is *not* greater than or equal to 0.

c. The function $h(x) = \frac{3x - x^2 + 7x^4}{9}$ is a polynomial function of degree 4. Dividing each term by 9 and reordering the terms, we can write this polynomial function as $h(x) = \frac{7}{9}x^4 - \frac{1}{9}x^2 + \frac{1}{3}x$. The leading coefficient is $\frac{7}{9}$, and the constant coefficient is 0.

Work exercises 1–6 in this textbook or in the *MyMathLab*® Study Plan.

OBJECTIVE 2:   SKETCHING THE GRAPHS OF POWER FUNCTIONS

The most basic polynomial functions are **monomial functions** of the form $f(x) = ax^n$. Polynomial functions of this form are called **power functions**. Figure 11 shows the graphs of power functions for $a = 1$ and for $n = 1, 2, 3, 4,$ and 5.

You can see from Figure 11 that the graph of $f(x) = x^4$ closely resembles the graph of $f(x) = x^2$. The difference in the two graphs is that the graph of $f(x) = x^4$ flattens out near the origin and gets steep quicker than the graph of $f(x) = x^2$. Likewise, the graph of $f(x) = x^5$ resembles the graph of $f(x) = x^3$. In general if n is even, the graph of $f(x) = x^n$ will resemble the graph of $f(x) = x^2$. If n is odd, the graph of $f(x) = x^n$ will resemble the graph of $f(x) = x^3$.

Because we know the basic shape of $f(x) = x^n$, we can use the transformation techniques discussed in Section 3.4 to sketch more complicated polynomial functions. Click here for a quick review of the transformation techniques or refer to Section 3.4.

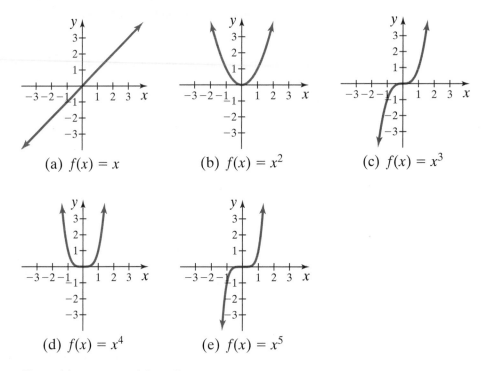

(a) $f(x) = x$ (b) $f(x) = x^2$ (c) $f(x) = x^3$

(d) $f(x) = x^4$ (e) $f(x) = x^5$

Figure 11 Graphs of $f(x) = x^n$, where $n = 1, 2, 3, 4,$ and 5.

My interactive video summary

Example 2 Use Transformations to Sketch Polynomial Functions

Use transformations to sketch the following functions:

a. $f(x) = -x^6$ b. $f(x) = (x + 1)^5 + 2$ c. $f(x) = 2(x - 3)^4$

Solution

a. The graph of $f(x) = -x^6$ is obtained by reflecting the graph of $y = x^6$ about the x-axis. See Figure 12(a).

b. Beginning with the graph of $y = x^5$, we can obtain the graph of $f(x) = (x + 1)^5 + 2$ by horizontally shifting $y = x^5$ to the *left* one unit and then *up* two units. See Figure 12(b).

c. To sketch the graph of $f(x) = 2(x - 3)^4$, we start with the graph of $y = x^4$. We then horizontally shift the graph to the *right* three units and vertically stretch the graph by a factor of 2. See Figure 12(c).

Watch the interactive video for a detailed solution.

My video summary

(a)

(b)

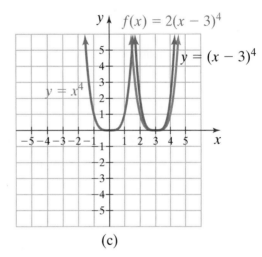
(c)

Figure 12

Work exercises 7–15 in this textbook or in the *MyMathLab*® Study Plan.

OBJECTIVE 3: DETERMINING THE END BEHAVIOR OF POLYNOMIAL FUNCTIONS

My video summary

To begin to sketch the graph of a more complicated polynomial function, it is necessary to understand the behavior of the graph as the x-coordinate gets large in the positive direction (approaches infinity) and as the x-coordinate gets small in the negative direction (approaches negative infinity). The nature of the graph of a polynomial function for large values of x in the positive and negative direction is known as the **end behavior**. The end behavior of the graph of a polynomial function depends on the leading term, that is, the term of highest degree. The graph of a polynomial function $f(x) = a_n x^n + a_{n-1}x^{n-1} + a_{n-2}x^{n-2} + \cdots + a_1 x + a_0$ has the same end behavior as the graph of the power function $f(x) = a_n x^n$.

To illustrate end behavior, consider the polynomial function $f(x) = x^5 - 2x^4 - 6x^3 + 8x^2 + 5x - 6$ sketched in Figure 13a and compare it to the graph of $f(x) = x^5$ sketched in Figure 13b. Notice that the ends of both graphs show similar behavior. The right-hand ends of both graphs increase without bound; that is, the y-values approach infinity as the x-values approach infinity. The left-hand ends of both graphs approach negative infinity as the x-values approach negative infinity.

4.3 The Graphs of Polynomial Functions   181

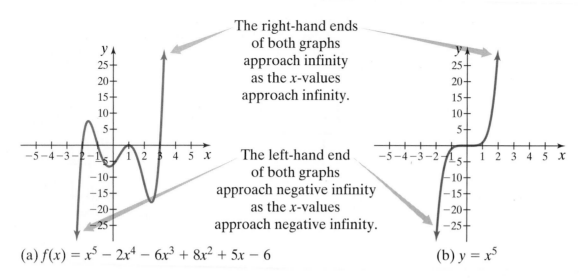

The right-hand ends of both graphs approach infinity as the *x*-values approach infinity.

The left-hand end of both graphs approach negative infinity as the *x*-values approach negative infinity.

(a) $f(x) = x^5 - 2x^4 - 6x^3 + 8x^2 + 5x - 6$ (b) $y = x^5$ **Figure 13**

To determine the end behavior of a polynomial function $f(x) = a_n x^n + a_{n-1}x^{n-1} + a_{n-2}x^{n-2} + \cdots + a_1 x + a_0$, we look at the leading term $a_n x^n$ and follow a two-step process.

Two-Step Process for Determining the End Behavior of a Polynomial Function $f(x) = a_n x^n + a_{n-1}x^{n-1} + a_{n-2}x^{n-2} + \cdots + a_1 x + a_0$

Step 1 Determine the sign of the leading coefficient a_n.

If $a_n > 0$, the right-hand behavior "finishes up."

If $a_n < 0$, the right-hand behavior "finishes down."

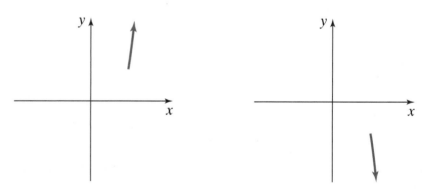

Step 2 Determine the degree.

If the degree n is odd, the graph has opposite left-hand and right-hand end behavior; that is, the graph "starts" and "finishes" in *opposite* directions.

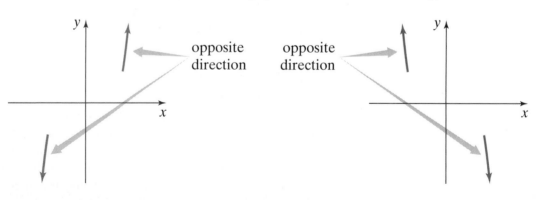

opposite direction   opposite direction

$a_n > 0$, odd degree $a_n < 0$, odd degree

If the degree n is even, the graph has the same left-hand and right-hand end behavior; that is, the graph "starts" and "finishes" in the *same* direction.

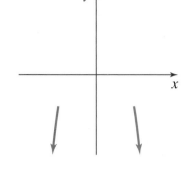

$a_n > 0$, even degree $a_n < 0$, even degree

▣ Example 3 Use End Behavior to Determine the Degree and Leading Coefficient

Use the end behavior of each graph to determine whether the degree is even or odd and whether the leading coefficient is positive or negative.

a.

b.

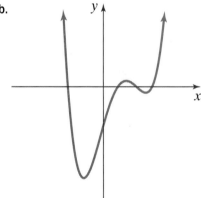

Solution

a. The leading coefficient is *negative* because the right-hand end behavior "ends down." The degree is *odd* because the graph has opposite left-hand and right-hand end behavior.

b. The leading coefficient is *positive* because the right-hand end behavior "ends up." The degree is *even* because the graph has the same left-hand and right-hand end behavior.

Watch the video to see a more detailed solution.

Work exercises 16–21 in this textbook or in the *MyMathLab*® Study Plan.

OBJECTIVE 4:   DETERMINING THE INTERCEPTS OF A POLYNOMIAL FUNCTION

Now that we know how to determine the end behavior of a polynomial function, it is time to find out what happens between the ends. We start by trying to locate the **intercepts** of the graph. Every polynomial function, $y = f(x)$, has a y-intercept that can be found by evaluating $f(0)$. Locating the x-intercepts is not that easy of a task. Recall that the number $x = c$ is called a **zero** of a function f if $f(c) = 0$. If c is a real number, then c is an x-intercept (see Section 3.2). Therefore, to find the x-intercepts of a polynomial function $y = f(x)$, we must find the real solutions of the equation $f(x) = 0$.

Example 4 Find the Intercepts of a Polynomial Function

Find the intercepts of the polynomial function $f(x) = x^3 - x^2 - 4x + 4$.

Solution The polynomial function $f(x) = x^3 - x^2 - 4x + 4$ has a y-intercept at $f(0) = (0)^3 - (0)^2 - 4(0) + 4 = 4$. Notice that the y-intercept is the same as the constant coefficient. In general, every polynomial function of the form $f(x) = a_n x^n + a_{n-1} x^{n-1} + a_{n-2} x^{n-2} + \cdots + a_1 x + a_0$ has a y-intercept at the constant coefficient a_0.

To find the x-intercepts of $f(x) = x^3 - x^2 - 4x + 4$, we must find the real solutions of the equation $x^3 - x^2 - 4x + 4 = 0$. This equation can be solved by factoring and using the **zero product property**.

$$x^3 - x^2 - 4x + 4 = 0 \quad \textit{Write the equation } f(x) = 0.$$
$$x^2(x - 1) - 4(x - 1) = 0 \quad \textit{Factor by grouping.}$$
$$(x - 1)(x^2 - 4) = 0 \quad \textit{Factor out the common factor of } (x - 1).$$
$$(x - 1)(x - 2)(x + 2) = 0 \quad \textit{Factor } (x^2 - 4) \textit{ using difference of squares.}$$

By the **zero product property**, $x - 1 = 0$, $x - 2 = 0$, or $x + 2 = 0$. Thus, the zeros of $f(x) = x^3 - x^2 - 4x + 4$ are $x = 1$, $x = 2$, and $x = -2$. Because the zeros are all real numbers, $x = 1$, $x = 2$, and $x = -2$ are the x-intercepts. By determining the end behavior and plotting the intercepts, we begin to get an idea of what the graph of $f(x) = x^3 - x^2 - 4x + 4$ looks like. See Figure 14.

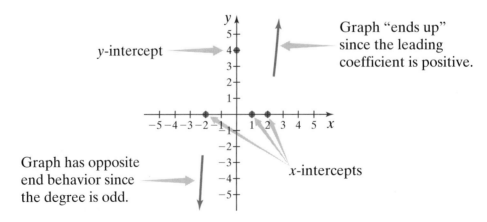

y-intercept

Graph "ends up" since the leading coefficient is positive.

Graph has opposite end behavior since the degree is odd.

x-intercepts

Figure 14

Work exercises 22–25 in this textbook or in the **MyMathLab**® Study Plan.

OBJECTIVE 5:   DETERMINING THE REAL ZEROS OF POLYNOMIAL FUNCTIONS
AND THEIR MULTIPLICITIES

My video summary

In Example 4, we were able to factor the polynomial function $f(x) = x^3 - x^2 - 4x + 4$ into $f(x) = (x - 1)(x - 2)(x + 2)$. Once the polynomial function was in factored form, we were able to determine the three zeros $x = 1$, $x = 2$, and $x = -2$. In general, if $(x - c)$ is a factor of a polynomial function, then $x = c$ is a zero. The converse of the previous statement is also true. (If f is a polynomial function and $x = c$ is a zero, then $(x - c)$ is a factor.)

For example, consider the polynomial functions $f(x) = (x - 1)^2$ and $g(x) = (x - 1)^3$. Both functions have a real zero at $x = 1$ because $(x - 1)$ is a factor of both polynomials. For the function $f(x) = (x - 1)^2$, $x = 1$ is called a zero of multiplicity 2 because the factor $(x - 1)$ occurs two times in the factorization of f. Likewise, for $g(x) = (x - 1)^3$, $x = 1$ is a zero of multiplicity 3. Notice in Figure 15 that the graph of $f(x) = (x - 1)^2$ is *tangent* to (touches) the x-axis at $x = 1$, whereas the graph of $g(x) = (x - 1)^3$ *crosses* the x-axis at $x = 1$. The graph of a polynomial function will touch the x-axis at a real zero of even multiplicity and will cross the x-axis at a real zero of odd multiplicity.

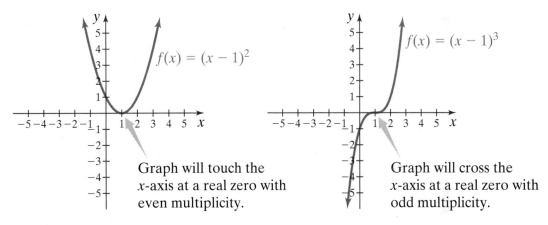

Graph will touch the x-axis at a real zero with even multiplicity.

Graph will cross the x-axis at a real zero with odd multiplicity.

Figure 15

Shape of the Graph of a Polynomial Function Near a Zero of Multiplicity k

Suppose c is a real zero of a polynomial function f of multiplicity k (where k is a positive integer); that is, $(x - c)^k$ is a factor of f. Then the shape of the graph of f near c is as follows:

If $k > 1$ is even, then the graph touches the x-axis at c.

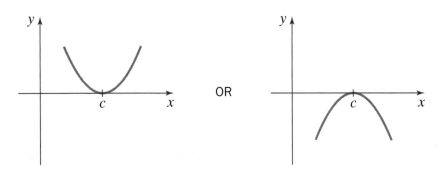

OR

If $k \geq 1$ is odd, then the graph crosses the x-axis at c.

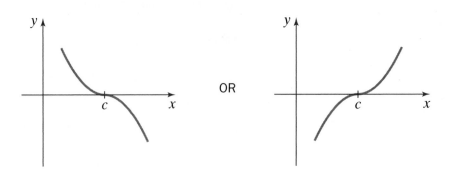

OR

Example 5 Determine the Real Zeros (and Their Multiplicities) of a Polynomial Function

Find all real zeros of $f(x) = x(x^2 - 1)(x - 1)$. Determine the multiplicities of each zero, and decide whether the graph touches or crosses at each zero.

Solution Completely factor the polynomial function. We can use difference of squares to factor $x^2 - 1$.

$$f(x) = x\underbrace{(x^2 - 1)}(x - 1) \qquad \textit{Write the original function.}$$
$$f(x) = x(x + 1)(x - 1)(x - 1) \qquad \textit{Factor using difference of squares.}$$
$$f(x) = x(x + 1)(x - 1)^2 \qquad \textit{Rewrite } (x - 1)(x - 1) \textit{ as } (x - 1)^2.$$

Because the first factor, x, is the same as $(x - 0)$, this implies that $x = 0$ is a zero of multiplicity 1. The next factor, $(x + 1)$, means that $x = -1$ is a zero of multiplicity 1.
The final factor, $(x - 1)^2$, means that $x = 1$ is a zero of multiplicity 2.
The graph of $f(x) = x(x + 1)(x - 1)^2$ must cross the x-axis at $x = 0$ and $x = -1$ and must touch the x-axis at $x = 1$. See the following calculator graph.

Using Technology

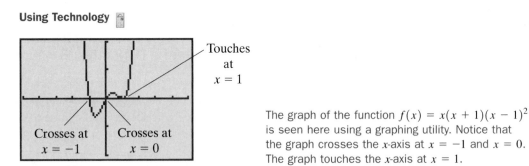

Touches at $x = 1$

Crosses at $x = -1$

Crosses at $x = 0$

The graph of the function $f(x) = x(x + 1)(x - 1)^2$ is seen here using a graphing utility. Notice that the graph crosses the x-axis at $x = -1$ and $x = 0$. The graph touches the x-axis at $x = 1$.

Work exercises 26–33 in this textbook or in the *MyMathLab*® Study Plan.

OBJECTIVE 6: SKETCHING THE GRAPH OF A POLYNOMIAL FUNCTION

If we are able to find the zeros of a polynomial function, we will be able to sketch an approximate graph. Earlier in this section, we were able to factor the polynomial function $f(x) = x^3 - x^2 - 4x + 4$ into $f(x) = (x - 1)(x - 2)(x + 2)$.

Note that the three zeros $x = 1$, $x = 2$, and $x = -2$ have odd multiplicities. Therefore, the graph of f must cross the x-axis at each x-intercept. By determining the end behavior of this function and by plotting the x-intercepts and y-intercept, we can begin to sketch the graph as seen in Figure 16.

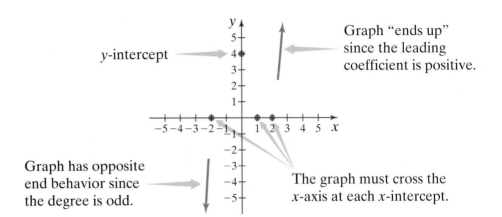

Figure 16

To complete the graph, we can create a table of values to plot additional points by choosing a **test value** between each of the zeros. Here, we choose $x = -1$ and $x = 1.5$. The approximate graph is sketched in Figure 17.

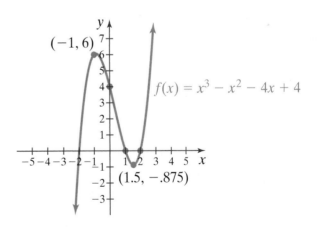

Figure 17

We see from the graph in Figure 17 that there are two **turning points**. A polynomial of degree n has at most $n - 1$ turning points. A turning point in which the graph changes from increasing to decreasing is also called a **relative maximum**. A turning point in which the graph changes from decreasing to increasing is called a **relative minimum**.

Without the use of calculus (or a graphing utility), there is no way to determine the precise coordinates of the relative minima and relative maxima. Using a graphing utility, we can use the Maximum and Minimum feature to determine the coordinates of the relative maxima as follows.

Using Technology

Most graphing calculators have a Maximum and Minimum feature that approximates the coordinates of a relative maximum or a relative minimum. The graphs above show the approximate coordinates of the relative maximum and relative minimum of $f(x) = x^3 - x^2 - 4x + 4$. Using calculus, it can be shown that the exact coordinates are $\left(\frac{1 - \sqrt{13}}{3}, \frac{70 + 26\sqrt{13}}{27}\right)$ and $\left(\frac{1 + \sqrt{13}}{3}, \frac{70 - 26\sqrt{13}}{27}\right)$, respectively.

Note: Some texts use the terms *local maximum* and *local minimum* to describe turning points.

We can summarize the sketching procedure with the following four-step process.

Four-Step Process for Sketching the Graph of a Polynomial Function

Step 1: Determine the end behavior.

Step 2: Plot the y-intercept $f(0) = a_0$.

Step 3: Completely factor f to find all real zeros and their **multiplicities**.

Step 4: Choose a test value between each real zero and sketch the graph.

(Remember that without calculus, there is no way to determine the precise coordinates of the turning points.)

My interactive video summary

Example 6 Use the Four-Step Process to Sketch the Graph of a Polynomial Function

Use the four-step process to sketch the graphs of the following polynomial functions:

   a. $f(x) = -2(x + 2)^2(x - 1)$ **b.** $f(x) = x^4 - 2x^3 - 3x^2$

Solution The graphs of each polynomial are sketched in Figure 18. Work through the interactive video to see how to sketch these polynomial functions.

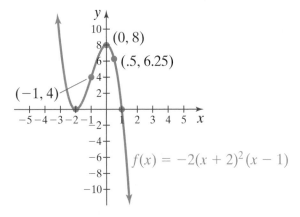

Figure 18 (a) Graph of $f(x) = -2(x + 2)^2(x - 1)$

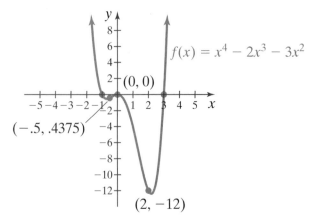

Figure 18 (b) Graph of $f(x) = x^4 - 2x^3 - 3x^2$

Work exercises 34–44 in this textbook or in the *MyMathLab* ® Study Plan.

OBJECTIVE 7: DETERMINING A POSSIBLE EQUATION OF A POLYNOMIAL FUNCTION GIVEN ITS GRAPH

Now that we have identified the characteristics of a polynomial function and can create an approximate sketch of its graph, we should be able to analyze the graph of a polynomial function and describe features of its equation. The end behavior of the graph gives us information about the degree and leading coefficient, the *y*-intercept gives us the value of the constant coefficient, and the behavior of the graph at the *x*-intercepts gives us information about the multiplicity of the zeros. We can use this information to identify possible equations that the graph of a polynomial function might represent.

My interactive video summary

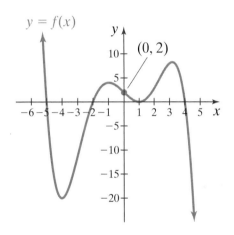 **Example 7** **Determine a Possible Equation of a Polynomial Function Given Its Graph**

Analyze the graph to address the following about the polynomial function it represents.

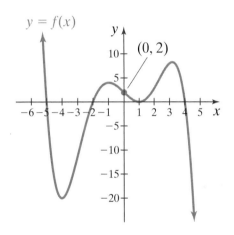

a. Is the degree of the polynomial function even or odd?
b. Is the leading coefficient positive or negative?
c. What is the value of the constant coefficient?
d. Identify the real zeros, and state the multiplicity of each.
e. Select from this list a possible function that could be represented by this graph.

$$\text{i. } f(x) = -\frac{1}{20}(x + 5)(x + 2)(x - 1)^2(x - 4)$$

$$\text{ii. } f(x) = -\frac{1}{800}(x + 5)^2(x + 2)^2(x - 1)(x - 4)^2$$

$$\text{iii. } f(x) = \frac{1}{20}(x + 5)(x + 2)(x - 1)^2(x - 4)$$

$$\text{iv. } f(x) = -\frac{1}{10}(x + 5)(x + 2)(x - 1)^2(x - 4)$$

Solution

a. The degree of the polynomial function is odd because the graph "starts" and "finishes" in opposite directions.

b. The leading coefficient is negative because the right-hand behavior "finishes down."

c. The *y*-intercept is 2, so $f(0) = 2$.

d. The real zeros are -5, -2, 1, and 4. The zeros that have odd multiplicity are -5, -2, and 4 because the graph crosses the *x*-axis at these zeros. The only zero of even multiplicity is 1 because the graph touches the *x*-axis at $x = 1$.

e. The best choice is item i. Watch the interactive video for a more detailed explanation.

Work exercises 45–47 in this textbook or in the *MyMathLab*® Study Plan.

4.3 Exercises

In Exercises 1–6, determine whether the function is a polynomial function. For each function that is a polynomial, identify the degree, the leading coefficient, and the constant coefficient.

1. $f(x) = \dfrac{1}{3}x^7 - 8x^3 + \sqrt{5}x - 8$

2. $g(x) = \dfrac{8x^3 - 5x + 3}{7x^2}$

3. $h(x) = 2x^{\frac{2}{3}} - 8x^2 + 5$

4. $f(x) = \dfrac{3x - 5x^3}{11}$

5. $G(x) = 7x^4 + 1 + 3x - 9x^5$

6. $f(x) = 12$

In Exercises 7–15, use the graph of a power function and transformations to sketch the graph of each polynomial function.

7. $f(x) = -x^5$

8. $f(x) = x^6 + 1$

9. $f(x) = -x^5 - 2$

10. $f(x) = \dfrac{1}{2}(x - 1)^4$

11. $f(x) = 2(x + 1)^5$

12. $f(x) = -(x - 2)^6$

13. $f(x) = (x - 2)^5 - 3$

14. $f(x) = -(x + 2)^4 + 1$

15. $f(x) = -(x - 1)^5 + 2$

In Exercises 16–21, use the end behavior of the graph of the polynomial function to (a) determine whether the degree is even or odd and (b) determine whether the leading coefficient is positive or negative.

16.

17.

18.

19.

20.

21.

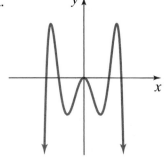

In Exercises 22–25, find the intercepts of each polynomial function.

22. $f(x) = x^2 - 3x - 10$

23. $f(x) = x^3 + 3x^2 - 16x - 48$

24. $f(x) = (x^2 + 1)\left(x - \dfrac{1}{2}\right)(x + 6)$

25. $f(x) = (x^2 - 9)(2 - x)$

In Exercises 26–33, determine the real zeros of each polynomial and their multiplicities. Then decide whether the graph touches or crosses the x-axis at each zero.

26. $f(x) = (x - 2)^2(x + 1)^3$

27. $f(x) = -(x + 1)(x + 3)^2$

28. $f(x) = x(x - 3)(x - 4)^4$

29. $f(x) = 3(x^2 - 2)^2$

30. $f(x) = -5(x - 2)(x^2 - 4)^3$

31. $f(x) = x^2(x^2 - x - 3)$

32. $f(x) = (x^2 - 3)(x + 1)^3$

33. $f(x) = -(x^2 - 7)^2(x + \sqrt{7})^3$

In Exercises 34–44, sketch each polynomial function using the four-step process.

34. $f(x) = x^2(x - 4)$

35. $f(x) = -x(x + 2)^2$

36. $f(x) = x^2(x + 2)(x - 2)$

37. $f(x) = -2(x - 1)^2(x + 1)$

38. $f(x) = -\dfrac{1}{2}x(x^2 - 4)^2$

39. $f(x) = x^4 + 2x^3 - 3x^2$

40. $f(x) = -2x^3 + 6x^2 + 8x$

41. $f(x) = x^3 + x^2 - 4x - 4$

42. $f(x) = -2x^3 - 4x^2 + 2x + 4$

43. $f(x) = x^4 + 2x^3 - 9x^2 - 18x$

44. $f(x) = -x^5 + 3x^4 + 9x^3 - 27x^2$

45. Analyze the graph to address the following about the polynomial function it represents.

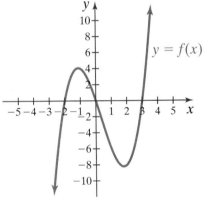

a. Is the degree of the polynomial function even or odd?

b. Is the leading coefficient positive or negative?

c. What is the value of the constant coefficient?

d. Identify the real zeros, and state the multiplicity of each.

e. Select from this list a possible function that could be represented by this graph.

    i. $f(x) = -x(x + 2)(x - 3)$

    ii. $f(x) = x(x + 2)(x - 3)$

    iii. $f(x) = -x^3(x + 2)(x - 3)$

    iv. $f(x) = (x + 2)(x - 3)$

46. Analyze the graph to address the following about the polynomial function it represents.

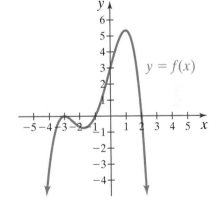

a. Is the degree of the polynomial function even or odd?

b. Is the leading coefficient positive or negative?

c. What is the value of the constant coefficient?

d. Identify the real zeros, and state the multiplicity of each.

e. Select from this list a possible function that could be represented by this graph.

    i. $f(x) = \dfrac{1}{4}(x + 3)(x + 1)^2(x - 2)^2$

    ii. $f(x) = -\dfrac{1}{18}(x + 3)^3(x + 1)(x - 2)$

    iii. $f(x) = -\dfrac{1}{6}(x - 3)^2(x - 1)(x + 2)$

    iv. $f(x) = -\dfrac{1}{6}(x + 3)^2(x + 1)(x - 2)$

47. Analyze the graph to address the following about the polynomial function it represents.

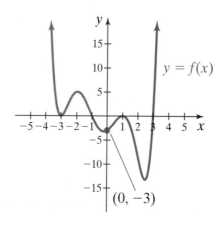

$y = f(x)$

$(0, -3)$

a. Is the degree of the polynomial function even or odd?

b. Is the leading coefficient positive or negative?

c. What is the value of the constant coefficient?

d. Identify the real zeros, and state the multiplicity of each.

e. Select from this list a possible function that could be represented by this graph.

i. $f(x) = \dfrac{1}{6}(x + 3)^2(x + 1)(x - 1)^2(x - 3)$

ii. $f(x) = \dfrac{1}{9}(x + 3)^2(x + 1)(x - 1)^2(x - 3)$

iii. $f(x) = \dfrac{1}{3}(x + 3)^2(x + 1)(x - 1)^2(x - 3)$

iv. $f(x) = \dfrac{1}{12}(x + 3)^2(x + 1)(x - 1)^2(x - 3)$

4.4 Synthetic Division; The Remainder and Factor Theorems

THINGS TO KNOW

Before working through this section, be sure that you are familiar with the following concepts:

VIDEO   ANIMATION   INTERACTIVE

 You Try It

1. Using Long Division to Divide Polynomials (Section R.4)

 You Try It

2. Simplifying Powers of i (Section 1.3)

 You Try It

3. Determining the Intercepts of a Function (Section 3.2)

You Try It

4. Using Combinations of Transformations to Sketch Functions (**Section 3.4**)

 You Try It

5. Sketching the Graph of a Polynomial Function (Section 4.3)

192   **Chapter 4**   Polynomial and Rational Functions

OBJECTIVES

1 Using the Division Algorithm

2 Using Synthetic Division

3 Using the Remainder Theorem

4 Using the Factor Theorem

5 Sketching the Graph of a Polynomial Function

OBJECTIVE 1:   USING THE DIVISION ALGORITHM

Long division of polynomials was discussed in Section R.4. For instance, we saw in Example 7 of Section R.4 that if we divide $3x^4 + x^3 + 7x + 4$ by $x^2 - 1$, we get

$$\frac{3x^4 + x^3 + 7x + 4}{x^2 - 1} = 3x^2 + x + 3 + \frac{8x + 7}{x^2 - 1}$$

Watch the video to see each step of the long division process. We see from this division that the quotient is $3x^2 + x + 3$ and the remainder is $8x + 7$. Letting $f(x) = 3x^4 + x^3 + 7x + 4$, $d(x) = x^2 - 1$, $q(x) = 3x^2 + x + 3$, and $r(x) = 8x + 7$, then

$$\frac{3x^4 + x^3 + 7x + 4}{x^2 - 1} = 3x^2 + x + 3 + \frac{8x + 7}{x^2 - 1} \text{ is equivalent to the equation}$$

$$\frac{f(x)}{d(x)} \quad = \quad q(x) \quad + \quad \frac{r(x)}{d(x)}.$$

If we multiply both sides of these equations by the divisor $d(x)$, we get

$$(x^2 - 1)\frac{3x^4 + x^3 + 7x + 4}{x^2 - 1} = (x^2 - 1)(3x^2 + x + 3) + \frac{8x + 7}{x^2 - 1}(x^2 - 1),$$

which is equivalent to

$$d(x)\frac{f(x)}{d(x)} \quad = \quad d(x)q(x) \quad + \quad \frac{r(x)}{d(x)}d(x) \quad \text{or}$$

$3x^4 + x^3 + 7x + 4 = (x^2 - 1)(3x^2 + x + 3) + (8x + 7)$, which is equivalent to

$$f(x) = d(x)q(x) + r(x).$$

The equation $f(x) = d(x)q(x) + r(x)$ is used to check that the long division was done properly. The original polynomial should equal the product of the divisor and the quotient plus the remainder. This process is known as the **division algorithm**.

Division Algorithm

If $f(x)$ and $d(x)$ are polynomial functions with $d(x) \neq 0$ and the degree of $d(x)$ is less than or equal to the degree of $f(x)$, then unique polynomial functions $q(x)$ and $r(x)$ exist such that

$$f(x) = d(x)q(x) + r(x)$$

---

or equivalently
$$\frac{f(x)}{d(x)} = q(x) + \frac{r(x)}{d(x)}$$

where the remainder $r(x) = 0$ or is of degree less than the degree of $d(x)$.

Example 1 Use the Division Algorithm

Given the polynomials $f(x) = 4x^4 - 3x^3 + 5x - 6$ and $d(x) = x - 2$, find polynomial functions $q(x)$ and $r(x)$, and write $f(x)$ in the form of $f(x) = d(x)q(x) + r(x)$.

Solution We use long division inserting the term $0x^2$ into the dividend as a placeholder to make sure that all columns of the long division process line up properly.

$$
\require{enclose}
\begin{array}{r}
4x^3 + 5x^2 + 10x + 25 \\
x - 2 \enclose{longdiv}{4x^4 - 3x^3 + 0x^2 + 5x - 6} \\
\underline{4x^4 - 8x^3} \\
5x^3 + 0x^2 \\
\underline{5x^3 - 10x^2} \\
10x^2 + 5x \\
\underline{10x^2 - 20x} \\
25x - 6 \\
\underline{25x - 50} \\
44
\end{array}
$$

The process is complete because 44 is of lesser degree than the divisor, $x - 2$. Therefore, $q(x) = 4x^3 + 5x^2 + 10x + 25$ and $r(x) = 44$, so

$$4x^4 - 3x^3 + 5x - 6 = (x - 2)(4x^3 + 5x^2 + 10x + 25) + 44$$

Notice in Example 1 that the remainder, $r(x)$, was a constant. We know that the remainder must always be of degree less than the degree of the divisor. Therefore, if a polynomial $f(x)$ is divided by a first-degree polynomial, then the remainder must be a constant.

Corollary to the Division Algorithm

If a polynomial $f(x)$ of degree greater than or equal to 1 is divided by a polynomial $d(x)$ where $d(x)$ is of degree 1, then there exists a unique polynomial function $q(x)$ and a constant r such that

$$f(x) = d(x)q(x) + r$$

OBJECTIVE 2:   USING SYNTHETIC DIVISION

My video summary

If a polynomial $f(x)$ is divided by $d(x) = x - c$, then we can use a "shortcut" method" called **synthetic division** to find the quotient, $q(x)$, and the remainder, r. To illustrate synthetic division, we do a side-by-side comparison of long division and synthetic division using the polynomial function of Example 1.

My video summary

Long Division

$$\begin{array}{r}
4x^3 + 5x^2 + 10x + 25 \\
x - 2\overline{)4x^4 - 3x^3 + 0x^2 + 5x - 6} \\
\underline{4x^4 - 8x^3} \\
5x^3 + 0x^2 \\
\underline{5x^3 - 10x^2} \\
10x^2 + 5x \\
\underline{10x^2 - 20x} \\
25x - 6 \\
\underline{25x - 50} \\
44
\end{array}$$

Synthetic Division

$$\begin{array}{r|rrrrr}
2 & 4 & -3 & 0 & 5 & -6 \\
& & 8 & 10 & 20 & 50 \\
\hline
& 4 & 5 & 10 & 25 & 44
\end{array}$$

You can see how much more compact the synthetic division process is compared to long division. This is how it works.

Step 1 The constant coefficient, c, of the divisor $d(x) = x - c$ is written to the far left, whereas all coefficients of the polynomial $f(x)$ are written inside the symbol $\underline{\quad\quad}$. Once again, be sure to include the 0 for $0x^2$.

This is c in $x - c$. ⟶ $2\lfloor 4 \quad -3 \quad 0 \quad 5 \quad -6$

These are the coefficients of $f(x) = 4x^4 - 3x^3 + 5x - 6$.

Step 2 Bring down the leading coefficient 4.

$$\begin{array}{r|rrrrr}
2 & 4 & -3 & 0 & 5 & -6 \\
& \downarrow \\
\hline
& 4
\end{array}$$

Step 3 Multiply c times the leading coefficient that was just brought down. In this case, we multiply $2 \cdot 4$. The product (8 in this case) is written in the next column in the second row.

Multiply $2 \cdot 4 = 8$
$$\begin{array}{r|rrrrr}
2 & 4 & -3 & 0 & 5 & -6 \\
& \downarrow & 8 \\
\hline
& 4
\end{array}$$

Step 4 Add down the column and write the sum (5 in this case) in the bottom row.

$$\begin{array}{r|rrrrr}
2 & 4 & -3 & 0 & 5 & -6 \\
& & 8 \\
\hline
& 4 & 5
\end{array}$$

We now repeat this process by multiplying $c = 2$ times the value in the last row, always adding down the columns.

Multiply $2 \cdot 5 = 10$
$$\begin{array}{r|rrrrr}
2 & 4 & -3 & 0 & 5 & -6 \\
& & 8 & 10 \\
\hline
& 4 & 5
\end{array}$$

Add $0 + 10 = 10$
$$\begin{array}{r|rrrrr}
2 & 4 & -3 & 0 & 5 & -6 \\
& & 8 & 10 \\
\hline
& 4 & 5 & 10
\end{array}$$

Multiply $2 \cdot 10 = 20$
$$\begin{array}{r|rrrrr}
2 & 4 & -3 & 0 & 5 & -6 \\
& & 8 & 10 & 20 \\
\hline
& 4 & 5 & 10
\end{array}$$

Add $5 + 20 = 25$
$$\begin{array}{r|rrrrr}
2 & 4 & -3 & 0 & 5 & -6 \\
& & 8 & 10 & 20 \\
\hline
& 4 & 5 & 10 & 25
\end{array}$$

Multiply $2 \cdot 25 = 50$
$$\begin{array}{r|rrrrr}
2 & 4 & -3 & 0 & 5 & -6 \\
& & 8 & 10 & 20 & 50 \\
\hline
& 4 & 5 & 10 & 25
\end{array}$$

Add $-6 + 50 = 44$
$$\begin{array}{r|rrrrr}
2 & 4 & -3 & 0 & 5 & -6 \\
& & 8 & 10 & 20 & 50 \\
\hline
& 4 & 5 & 10 & 25 & 44
\end{array}$$

4.4 Synthetic Division; The Remainder and Factor Theorems   195

The last row now represents the quotient and the remainder. The last entry of the bottom row (44 in this case) is the remainder. The other numbers of the last row represent the coefficients of the quotient.

$$
\begin{array}{r|rrrr}
2 & 4 & -3 & 0 & 5 & -6 \\
  &   & 8 & 10 & 20 & 50 \\
\hline
  & 4 & 5 & 10 & 25 & 44
\end{array}
$$

remainder $r = 44$

coefficients of $q(x) = 4x^3 + 5x^2 + 10x + 25$

You can watch the video to see this example worked out. Whenever possible, synthetic division is used throughout the rest of this chapter.

CAUTION Synthetic division can only be used when the divisor $d(x)$ has the form $(x - c)$.

My video summary

▦ Example 2 Use Synthetic Division to Find the Quotient and Remainder

Use synthetic division to divide $f(x)$ by $(x - c)$, and then write $f(x)$ in the form $f(x) = (x - c)q(x) + r$ for $f(x) = -2x^4 + 3x^3 + 7x^2 - x + 5$ divided by $x + 1$.

Solution The divisor $x + 1$ or $x - (-1)$ is of the form $x - c$ with $c = -1$ so we can use synthetic division. We start by writing $c = -1$ on the far left, putting the coefficients of f on the inside and bringing down the -2:

$$
\begin{array}{r|rrrrr}
-1 & -2 & 3 & 7 & -1 & 5 \\
   & \downarrow \\
\hline
   & -2
\end{array}
$$

Using a series of multiplications and additions, we can complete the synthetic division process:

$$
\begin{array}{r|rrrrr}
-1 & -2 & 3 & 7 & -1 & 5 \\
   & \downarrow & 2 & -5 & -2 & 3 \\
\hline
   & -2 & 5 & 2 & -3 & 8
\end{array}
$$

Watch the video to see each step. We can see from the last entry of the bottom row that the remainder is 8. Thus,

$$q(x) = -2x^3 + 5x^2 + 2x - 3 \quad \text{and} \quad r = 8, \text{ so}$$
$$
\begin{aligned}
f(x) &= -2x^4 + 3x^3 + 7x^2 - x + 5 \\
     &= (x + 1)(-2x^3 + 5x^2 + 2x - 3) + 8
\end{aligned}
$$

Work exercises 1–10 in this textbook or in the *MyMathLab*® Study Plan.

OBJECTIVE 3:   USING THE REMAINDER THEOREM

In Example 2, we see that the remainder is 8 when $f(x) = -2x^4 + 3x^3 + 7x^2 - x + 5$ is divided by $x + 1$. If we evaluate f at $x = -1$, we get

$$f(-1) = -2(-1)^4 + 3(-1)^3 + 7(-1)^2 - (-1) + 5$$
$$= -2(1) + 3(-1) + 7(1) + 1 + 5$$
$$= -2 - 3 + 7 + 1 + 5 = 8$$

It is no coincidence that $f(-1)$ is equal to the remainder. If we rewrite $f(x) = -2x^4 + 3x^3 + 7x^2 - x + 5$ in the form of $f(x) = (x + 1)(-2x^3 + 5x^2 + 2x - 3) + 8$ as in Example 2, we see that

$$f(-1) = \underbrace{(-1 + 1)(-2(-1)^3 + 5(-1)^2 + 2(-1) - 3)}_{0} + 8$$

$$= \qquad\qquad 0 \qquad\qquad\qquad + 8 = 8$$

Therefore, $f(-1) = 8$, and we can conclude that if a polynomial is divided by $x - c$, then the remainder is the value of $f(c)$. This is known as the **remainder theorem**.

Remainder Theorem

If a polynomial $f(x)$ is divided by $x - c$, then the remainder is $f(c)$.

Click here to see the proof of the remainder theorem.

My video summary

▦ Example 3 Use the Remainder Theorem

Use the remainder theorem to find the remainder when $f(x)$ is divided by $x - c$.

a. $f(x) = 5x^4 - 8x^2 + 3x - 1; x - 2$ **b.** $f(x) = 3x^3 + 5x^2 - 5x - 6; x + 2$

Solution

a. We could use long division or synthetic division to find the remainder, but the remainder theorem says that the remainder must be $f(2)$.

$$f(2) = 5(2)^4 - 8(2)^2 + 3(2) - 1 = 80 - 32 + 6 - 1 = 53$$

Thus, the remainder is 53.

b. To find the remainder when $f(x) = 3x^3 + 5x^2 - 5x - 6$ is divided by $x + 2 = x - (-2)$, we must evaluate $f(-2)$.

$$f(-2) = 3(-2)^3 + 5(-2)^2 - 5(-2) - 6$$
$$= -24 + 20 + 10 - 6 = 0$$

Therefore, the remainder is 0.

Work exercises 11–16 in this textbook or in the *MyMathLab*® Study Plan.

OBJECTIVE 4:   USING THE FACTOR THEOREM

In Example 3, we see that when $f(x) = 3x^3 + 5x^2 - 5x - 6$ is divided by $x + 2$, then the remainder is 0. The remainder of 0 implies that $x + 2$ divides $f(x)$ *evenly*, and therefore, $x + 2$ is a factor of f. In fact, by the remainder theorem, if $f(c) = 0$, then the remainder when a polynomial $f(x)$ is divided by $x - c$ must be zero. Conversely, if $x - c$ is a factor of $f(x)$, then $f(c)$ must equal zero. The last two statements can be rewritten as the following important theorem known as the **factor theorem**.

Factor Theorem

The polynomial $x - c$ is a factor of the polynomial $f(x)$ if and only if $f(c) = 0$.

My video summary

Example 4 Use the Factor Theorem to Determine whether $x - c$ Is a Factor of $f(x)$

Determine whether $x + 3$ is a factor of $f(x) = 2x^3 + 7x^2 + 2x - 3$.

Solution By the factor theorem, $x + 3$ is a factor of $f(x)$ if and only if $f(-3) = 0$. We can use synthetic division to determine the value of $f(-3)$.

$$
\begin{array}{r|rrrr}
-3 & 2 & 7 & 2 & -3 \\
   & \downarrow & -6 & -3 & 3 \\
\hline
   & 2 & 1 & -1 & 0
\end{array}
\quad \longleftarrow f(-3) = 0
$$

The remainder $f(-3)$ is equal to zero, and therefore, $x + 3$ is a factor of $f(x)$. Furthermore, using the bottom row of the synthetic division, we can rewrite $f(x)$ as $f(x) = (x + 3)(2x^2 + x - 1)$.

Work exercises 17–26 in this textbook or in the MyMathLab® Study Plan.

OBJECTIVE 5:   SKETCHING THE GRAPH OF A POLYNOMIAL FUNCTION

In Example 4, we show that $x + 3$ is a factor of $f(x) = 2x^3 + 7x^2 + 2x - 3$. Using synthetic division and the division algorithm, we were able to rewrite $f(x)$ as $f(x) = (x + 3)(2x^2 + x - 1)$. Because the quadratic function $2x^2 + x - 1$ can also be factored, we can now completely factor $f(x)$ as

$$f(x) = 2x^3 + 7x^2 + 2x - 3 \qquad \text{\textit{Write the original polynomial.}}$$

$$f(x) = (x + 3)(2x^2 + x - 1) \qquad \text{\textit{Because } } f(-3) = 0, \, x + 3 \text{ \textit{is a factor}}$$
$$\text{\textit{(See Example 4).}}$$

$$f(x) = (x + 3)(2x - 1)(x + 1) \qquad \text{\textit{Factor } } 2x^2 + x - 1 \text{ \textit{as }} (2x - 1)(x + 1).$$

$$f(x) = 2(x + 3)\left(x - \frac{1}{2}\right)(x + 1) \qquad \text{\textit{Factor }} (2x - 1) \text{ \textit{as }} 2\left(x - \frac{1}{2}\right).$$

The polynomial $f(x) = 2(x + 3)\left(x - \frac{1}{2}\right)(x + 1)$ is now said to be written in **completely factored form**. Because the polynomial is in completely factored form, we can follow the four-step process discussed in Section 4.3 to sketch the graph of $f(x)$. The graph of $f(x) = 2(x + 3)\left(x - \frac{1}{2}\right)(x + 1)$ is sketched in Figure 19. Click here to see the video of how to use the four-step process to sketch this polynomial.

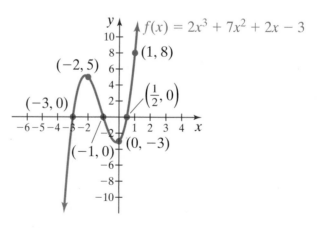

Figure 19 Graph of $f(x) = 2x^3 + 7x^2 + 2x - 3$

My video summary

■ **Example 5 Factor Completely and Sketch the Graph of a Polynomial Function**

Given that $x = 2$ is a zero of $f(x) = x^3 - 6x + 4$, completely factor f and sketch its graph.

Solution We begin by using the synthetic division process to divide $f(x) = x^3 - 6x + 4$ by $x - 2$, remembering to use a 0 for the coefficient of the x^2-term.

$$
\begin{array}{r|rrrr}
2 & 1 & 0 & -6 & 4 \\
  & \downarrow & 2 & 4 & -4 \\
\hline
  & 1 & 2 & -2 & 0
\end{array}
$$

We see from the synthetic division process that we can rewrite $f(x)$ as $f(x) = (x - 2)(x^2 + 2x - 2)$. Because the quadratic $x^2 + 2x - 2$ does not factor with integer coefficients, we can use the quadratic formula or complete the square to find the zeros of $x^2 + 2x - 2$. The zeros of $x^2 + 2x - 2$ are $x = -1 + \sqrt{3} \approx .732$ and $x = -1 - \sqrt{3} \approx -2.732$. Because $x = -1 + \sqrt{3}$ is a zero of $x^2 + 2x - 2$, this implies that $x - (-1 + \sqrt{3})$ is a factor of $x^2 + 2x - 2$. Similarly, $x - (-1 - \sqrt{3})$ is also a factor of $x^2 + 2x - 2$.

Thus, the complete factorization of f is $f(x) = (x - 2)(x - (-1 + \sqrt{3}))$ $(x - (-1 - \sqrt{3}))$. Once again, we can now use the four-step process to sketch the graph seen in Figure 20. Watch the video to see this example worked out in detail.

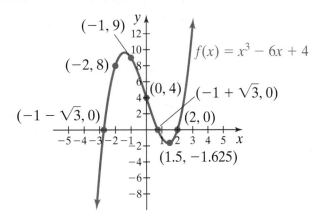

Figure 20 Graph of $f(x) = x^3 - 6x + 4$

Work exercises 27–34 in this textbook or in the *MyMathLab*® Study Plan.

4.4 Exercises

In Exercises 1–10, use synthetic division to divide $f(x)$ by $x - c$, and then write $f(x)$ in the form $f(x) = (x - c)q(x) + r$.

1. $f(x) = x^3 - 3x^2 + x - 5;\ x - 1$

2. $f(x) = x^3 + 5x^2 - 3x - 1;\ x + 1$

3. $f(x) = 6x^3 - 2x + 3;\ x - 2$

4. $f(x) = 2x^3 - x^2 + 4;\ x + 3$

5. $f(x) = x^4 + x^3 - x^2 - x - 4;\ x - 2$

6. $f(x) = -2x^4 + 3x^3 - x^2 + 7x + 2;\ x + 1$

7. $f(x) = -x^4 - x^3 - 5x;\ x + 4$

8. $f(x) = 3x^5 - x^2 - 2;\ x + 1$

9. $f(x) = x^3 - 1;\ x - 1$

10. $f(x) = x^5 + 1;\ x + 1$

In Exercises 11–16, use synthetic division and the remainder theorem to find the remainder when $f(x)$ is divided by $x - c$.

11. $f(x) = 2x^3 - 3x^2 - 5x - 7;\ x - 1$

12. $f(x) = x^4 + 3x^3 - 8x^2 - x + 2;\ x + 1$

13. $f(x) = x^5 - 2x^2 + x + 6;\ x - 3$

14. $f(x) = -4x^3 + 2x - 1;\ x + 3$

15. $f(x) = x^3 - 3x^2 + x - 5;\ x - i$

16. $f(x) = x^4 - x^3 - 2x^2 - x - 3;\ x + i$

In Exercises 17–26, use synthetic division and the factor theorem to determine whether $x - c$ is a factor of $f(x)$.

17. $f(x) = 2x^3 - 5x^2 + 7x - 4; x - 1$

18. $f(x) = 2x^3 - 5x^2 + 7x - 4; x + 1$

19. $f(x) = x^5 + x^4 - 3x^3 - 3x^2 - 4x - 4; x - 2$

20. $f(x) = x^5 + x^4 - 3x^3 - 3x^2 - 4x - 4; x + 2$

21. $f(x) = 2x^3 + 5x^2 - 8x - 5; x - \dfrac{1}{2}$

22. $f(x) = 2x^3 + 5x^2 - 8x - 5; x + \dfrac{1}{2}$

23. $f(x) = x^4 + x^3 + x^2 - 2x - 6; x - \sqrt{2}$

24. $f(x) = x^4 + x^3 + x^2 - 2x - 6; x + \sqrt{2}$

25. $f(x) = 3x^4 - x^3 - x^2 - x - 4; x - i$

26. $f(x) = x^3 + x^2 + 3x - 5; x - (1 + 2i)$

In Exercises 27–34, find the remaining zeros of $f(x)$ given that c is a zero. Then rewrite $f(x)$ in completely factored form and sketch its graph.

27. $f(x) = x^3 + 2x^2 - x - 2; c = 1$ is a zero

28. $f(x) = -x^3 + 5x^2 - 3x - 9; c = -1$ is a zero

29. $f(x) = 2x^3 + 7x^2 + 2x - 3; c = \dfrac{1}{2}$ is a zero

30. $f(x) = \dfrac{1}{2}x^3 - \dfrac{3}{2}x - 1; c = 2$ is a zero

31. $f(x) = -\dfrac{1}{3}x^3 - \dfrac{4}{3}x^2 - \dfrac{1}{3}x + 2; c = -3$ is a zero

32. $f(x) = 3x^3 + 11x^2 + 8x - 4; c = -2$ is a zero of multiplicity 2

33. $f(x) = 4x^4 + 5x^3 - 3x^2 - 5x - 1; c = -1$ is a zero of multiplicity 2

34. $f(x) = 3x^5 - 5x^4 - 2x^3 + 6x^2 - x - 1; c = 1$ is a zero of multiplicity 3

4.5 The Zeros of Polynomial Functions; The Fundamental Theorem of Algebra

THINGS TO KNOW

Before working through this section, be sure that you are familiar with the following concepts:

VIDEO   ANIMATION   INTERACTIVE

 1. Sketching the Graph of a Polynomial Function (Section 4.3)
You Try It

 2. Using the Remainder Theorem (Section 4.4)
You Try It

 3. Using the Factor Theorem (Section 4.4)
You Try It

 4. Sketching the Graph of a Polynomial Function (Section 4.4)
You Try It

OBJECTIVES

1 Using the Rational Zeros Theorem

2 Finding the Zeros of a Polynomial Function

3 Solving Polynomial Equations

4 Using the Complex Conjugate Pairs Theorem

5 Using the Intermediate Value Theorem

6 Sketching the Graphs of Polynomial Functions

The goal of this section is to be able to find the zeros of polynomial functions, or equivalently, we are interested in solving polynomial equations of the form $f(x) = 0$. To begin our discussion, we must take a look at a very important theorem that will help us create a list of potential zeros of a polynomial function.

OBJECTIVE 1:   USING THE RATIONAL ZEROS THEOREM

In Example 5 of Section 4.4, we were able to find all real zeros of $f(x) = x^3 - 6x + 4$ given that $x = 2$ was a zero. You may want to work through this **video** to review this example. Because we knew that $x = 2$ was a zero, we were able to use the **factor theorem** and synthetic division to rewrite f as $f(x) = (x - 2)(x^2 + 2x - 2)$. We then solved the quadratic equation $x^2 + 2x - 2 = 0$ to find the other two zeros.

But what if we were not given the fact that $x = 2$ was a zero? How do we start looking for the zeros? Is there a systematic way to determine possible zeros of a polynomial function? The answer to this question is yes *if* we are given a polynomial with **integer** coefficients. If a polynomial has integer coefficients, then we are able to create a list of the **potential** rational zeros using the **rational zeros theorem**.

Rational Zeros Theorem

Let f be a polynomial function of the form $f(x) = a_n x^n + a_{n-1} x^{n-1} + a_{n-2} x^{n-2} + \cdots + a_1 x + a_0$ of degree $n \geq 1$, where each **coefficient** is an integer. If $\frac{p}{q}$ is a **rational zero** of f (where $\frac{p}{q}$ is written in lowest terms), then p must be a factor of the constant coefficient, a_0, and q must be a factor of the leading coefficient, a_n.

⬦ CAUTION
The rational zeros theorem can provide us with only a list of possible rational zeros. It *does not* guarantee that the polynomial will have a zero from the list. Simply stated, if a polynomial with integer coefficients has a rational zero, then it must be on the list created using this theorem. Example 1 illustrates how this theorem works.

Example 1 Use the Rational Zeros Theorem

Use the rational zeros theorem to determine the potential rational zeros of each polynomial function.

$$f(x) = 4x^4 - 7x^3 + 9x^2 - x - 10$$

Solution Because f has integer coefficients, we can use the rational zeros theorem to create a list of **potential** rational zeros. The factors of the constant coefficient, -10, are $p = \pm 1, \pm 2, \pm 5, \pm 10$, whereas the factors of the leading coefficient, 4, are $q = \pm 1, \pm 2, \pm 4$. Listing all possibilities of $\frac{p}{q}$, we get

$$\frac{p}{q} \colon \pm 1, \pm 2, \pm 5, \pm 10, \pm \frac{1}{2}, \pm \frac{1}{4}, \pm \frac{5}{2}, \pm \frac{5}{4}$$

Thus, if f has a rational zero, then it must appear on this list.

Work exercises 1–6 in this textbook or in the *MyMathLab*® Study Plan.

OBJECTIVE 2: FINDING THE ZEROS OF A POLYNOMIAL FUNCTION

Before we begin to find the zeros of polynomial functions, we should know how many zeros to expect. The fundamental theorem of algebra says that every polynomial of degree $n \geq 1$ has at least one complex zero.

Fundamental Theorem of Algebra

Every polynomial function of degree $n \geq 1$ has at least one complex zero.

Suppose $f(x)$ is a polynomial function of degree $n \geq 1$. By the fundamental theorem of algebra, f has at least one complex zero, call it c_1. By the factor theorem, $x - c_1$ is a factor of $f(x)$, and it follows that we can rewrite $f(x)$ as $f(x) = (x - c_1)q_1(x)$, where $q_1(x)$ is another polynomial. If $q_1(x)$ is of degree 1 or more, then we repeat the process again and rewrite $f(x)$ as $f(x) = (x - c_1)(x - c_2)q_2(x)$, where c_2 is a zero of $q_1(x)$. If $f(x)$ is a degree n polynomial, then we can repeat this process a total of n times to rewrite $f(x)$ as $f(x) = a(x - c_1)(x - c_2)(x - c_3) \cdots (x - c_n)$, where $c_1, c_2, c_3 \ldots c_n$ are zeros of $f(x)$ and a is the leading coefficient.

    Therefore, every polynomial of degree n has n **complex zeros** and can be written in **completely factored form**. Note that some of the zeros could be the same and need to be counted each time, as indicated by the following theorem.

Number of Zeros Theorem

Every polynomial of degree n has n **complex zeros** provided each zero of multiplicity greater than 1 is counted accordingly.

My video summary

▣ Example 2 Find Zeros of a Polynomial Function

Find all **complex zeros** of $f(x) = 6x^4 + 13x^3 + 61x^2 + 8x - 10$, and rewrite $f(x)$ in completely factored form.

Solution This is a degree 4 polynomial so there must be four complex zeros. The factors of the constant coefficient, -10, are $p = \pm 1, \pm 2, \pm 5, \pm 10$. The factors of the leading coefficient, 6, are $q = \pm 1, \pm 2, \pm 3, \pm 6$. By the rational zeros theorem, the list of possible rational zeros are $\frac{p}{q}$: $\pm 1, \pm 2, \pm 5, \pm 10, \pm \frac{1}{2}, \pm \frac{1}{3}, \pm \frac{1}{6}, \pm \frac{2}{3}, \pm \frac{5}{2}, \pm \frac{5}{3}, \pm \frac{5}{6}, \pm \frac{10}{3}$.

    We go through this list evaluating f at each number on the list or using synthetic division until we find a zero. Remember, if $f\left(\frac{p}{q}\right) = 0$, then $\frac{p}{q}$ is a zero. Eventually getting to $-\frac{1}{2}$ and using synthetic division, we see that we obtain a zero.

$$
\begin{array}{r|rrrrr}
-\frac{1}{2} & 6 & 13 & 61 & 8 & -10 \\
& \downarrow & -3 & -5 & -28 & 10 \\
\hline
& 6 & 10 & 56 & -20 & 0
\end{array}
$$
    $\longleftarrow f\left(-\frac{1}{2}\right) = 0$, so $-\frac{1}{2}$ is a zero.

Thus, we can rewrite f as $f(x) = \left(x + \frac{1}{2}\right)(6x^3 + 10x^2 + 56x - 20)$. We now try to find a zero of the polynomial $q_1(x) = 6x^3 + 10x^2 + 56x - 20$. We again test $-\frac{1}{2}$ because we may have a repeated zero.

$$
\begin{array}{r|rrrr}
-\frac{1}{2} & 6 & 10 & 56 & -20 \\
& \downarrow & -3 & -\frac{7}{2} & -\frac{105}{4} \\
\hline
& 6 & 7 & \frac{105}{2} & -\frac{185}{4}
\end{array}
$$
    $\longleftarrow q_1\left(-\frac{1}{2}\right) \neq 0$, so $-\frac{1}{2}$ is not a zero of $q_1(x)$.

The remainder when $q_1(x) = 6x^3 + 10x^2 + 56x - 20$ is divided by $x + \frac{1}{2}$ is $-\frac{185}{4}$. Therefore, we know that $-\frac{1}{2}$ is *not* a repeated zero.

We next try $\frac{1}{3}$:

$$\frac{1}{3} \begin{array}{|rrrr} 6 & 10 & 56 & -20 \\ \downarrow & 2 & 4 & 20 \\ \hline 6 & 12 & 60 & 0 \end{array} \longleftarrow q_1\left(\frac{1}{3}\right) = 0, \text{ so } \frac{1}{3} \text{ is a zero}$$

Because the remainder when $q_1(x) = 6x^3 + 10x^2 + 56x - 20$ is divided by $x - \frac{1}{3}$ is zero implies that $\frac{1}{3}$ is a zero. Using the bottom row of the synthetic division, we see that $f(x) = \left(x + \frac{1}{2}\right)(6x^3 + 10x^2 + 56x - 20) = \left(x + \frac{1}{2}\right)\left(x - \frac{1}{3}\right)(6x^2 + 12x + 60)$. Factoring out a 6 from $6x^2 + 12x + 60$, we can rewrite f as $f(x) = 6\left(x + \frac{1}{2}\right)\left(x - \frac{1}{3}\right)(x^2 + 2x + 10)$.

We must now solve the quadratic equation $x^2 + 2x + 10 = 0$ to find the remaining two zeros. Solving either by **completing the square** or by using the **quadratic formula**, we get $x = -1 \pm 3i$. Thus, the four complex zeros are $-\frac{1}{2}, \frac{1}{3}$, $-1 + 3i$, and $-1 - 3i$. The completely factored form of this polynomial is $f(x) = 6\left(x + \frac{1}{2}\right)\left(x - \frac{1}{3}\right)(x - (-1 + 3i))(x - (-1 - 3i))$.

Note: Once we factor a polynomial into the product of linear factors and a quadratic function of the form $f(x) = (x - c_1)(x - c_2)(x - c_3) \cdots (x - c_k)(ax^2 + bx + c)$, we need simply to solve the quadratic equation $ax^2 + bx + c = 0$ to find the remaining two zeros. For example, if we want to find all zeros of $g(x) = 2x^3 - 3x^2 + 4x - 3$, we need only try to find *one* zero from the list created by the rational zeros theorem and then find the two zeros of the remaining quadratic function. See if you can find the zeros of $g(x) = 2x^3 - 3x^2 + 4x - 3$. Watch the video to see if you are correct!

Work exercises 7–15 in this textbook or in the *MyMathLab*® Study Plan.

OBJECTIVE 3:   SOLVING POLYNOMIAL EQUATIONS

Solving polynomial equations of the form $f(x) = 0$ is equivalent to finding the zeros of $f(x)$. For example, the solutions to the equation $6x^4 + 13x^3 + 61x^2 + 8x - 10 = 0$ are exactly the zeros of the polynomial function $f(x) = 6x^4 + 13x^3 + 61x^2 + 8x - 10$. Using the result from Example 2, we conclude that the real solutions of this equation are $-\frac{1}{2}$ and $\frac{1}{3}$, whereas the two nonreal solutions are $-1 + 3i$ and $-1 - 3i$.

My video summary

▣ Example 3 Solve a Polynomial Equation

Solve $5x^5 - 9x^4 + 23x^3 - 35x^2 + 12x + 4 = 0$.

Solution Let $f(x) = 5x^5 - 9x^4 + 23x^3 - 35x^2 + 12x + 4$. The solutions to this equation must be the zeros of f. Using the rational zeros theorem, we see that the potential rational zeros are $\pm 1, \pm 2, \pm 4, \pm \frac{1}{5}, \pm \frac{2}{5}, \pm \frac{4}{5}$. Our goal is to attempt to find three zeros from this list and then solve the remaining quadratic to find the last two zeros. Watch the video to verify that the rational zeros of f are $x = 1$ (of multiplicity 2) and $-\frac{1}{5}$. Using synthetic division and factoring out the leading coefficient, we can rewrite f as $f(x) = 5(x - 1)^2\left(x + \frac{1}{5}\right)(x^2 + 4)$. The solutions to the quadratic equation $x^2 + 4 = 0$ are $\pm 2i$. Thus, the equation $5x^5 - 9x^4 + 23x^3 - 35x^2 + 12x + 4 = 0$ is equivalent to $5(x - 1)^2\left(x + \frac{1}{5}\right)(x - 2i)(x + 2i) = 0$. Therefore, the solutions are $\left\{1, -\frac{1}{5}, 2i, -2i\right\}$.

Work exercises 16–21 in this textbook or in the *MyMathLab*® Study Plan.

OBJECTIVE 4:   USING THE COMPLEX CONJUGATE PAIRS THEOREM

In Example 2, the two nonreal zeros of $f(x) = 6x^4 + 13x^3 + 61x^2 + 8x - 10$ are $-1 + 3i$ and $-1 - 3i$. In Example 3, $2i$ and $-2i$ are the two nonreal solutions of $5x^5 - 9x^4 + 23x^3 - 35x^2 + 12x + 4 = 0$. Notice that the nonreal zeros in both examples are **complex conjugates** of each other. This is no coincidence! In fact, the following theorem explains that complex zeros must occur in pairs.

Complex Conjugate Pairs Theorem

If $a + bi$ is a zero of a polynomial function with real coefficients, then the complex conjugate $a - bi$ is also a zero.

⬦ **CAUTION**   To use the complex conjugate pairs theorem, it is imperative that the coefficients of the polynomial are real numbers. For example, the polynomial function $f(x) = x - i$ has a constant coefficient i. Note that $f(i) = i - i = 0$, and therefore, i is a zero of the polynomial function $f(x) = x - i$. However, the complex conjugate, $-i$, is not a zero because $f(-i) = -i - i = -2i$.

My interactive video summary

▶ **Example 4 Use the Complex Conjugate Pairs Theorem**

Find the equation of a polynomial function f with real coefficients that satisfies the given conditions:

a. Fourth-degree polynomial function such that 1 is a zero of multiplicity 2 and $2 - i$ is also a zero.

b. Fifth-degree polynomial function sketched in Figure 21 given that $1 + 3i$ is a zero.

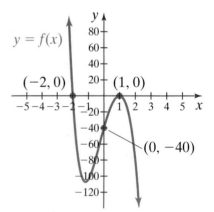

Figure 21

Solution

a. By the complex conjugate pairs theorem, because $2 - i$ is a zero, this implies that $2 + i$ is also a zero. The four zeros of f are 1, 1, $2 - i$, and $2 + i$. Using the factor theorem, we can write f as $f(x) = a(x - 1)(x - 1)(x - (2 - i))(x - (2 + i))$, where a is a constant. We can actually let a equal any number here because there are infinitely many polynomial functions that meet the stated conditions, and they all differ only by the leading coefficient. So, when no specific information is given to allow the exact leading coefficient to be determined, we choose a equal to 1 just for the sake of simplicity. Letting $a = 1$ and multiplying the factors together, we get $f(x) = x^4 - 6x^3 + 14x^2 - 14x + 5$. Click here to see this multiplication or watch the interactive video.

b. Because we are looking for a fifth-degree polynomial, there must be five zeros (including multiplicities). From the graph, we see that $x = -2$ is a zero with *odd* multiplicity because the graph crosses the x-axis at $x = -2$. Similarly, $x = 1$ is a zero with even multiplicity because the graph is tangent to the x-axis at $x = 1$. By the complex conjugate pairs theorem, $1 + 3i$ and $1 - 3i$ are zeros. Thus, the completely factored form of f is $f(x) = a(x + 2)^m (x - 1)^n (x - (1 + 3i))(x - (1 - 3i))$, where m is odd and n is even.

Because f is a fifth-degree polynomial, m must equal 1 and n must equal 2. Therefore, $f(x) = a(x + 2)(x - 1)^2 (x - (1 + 3i))(x - (1 - 3i))$.

Multiplying the factors, we get $f(x) = a(x^5 - 2x^4 + 7x^3 + 8x^2 - 34x + 20)$. Click here to see this multiplication or watch the interactive video. Unlike in part a, we cannot randomly assign a value for the leading coefficient because specific information has been given. To determine the value of a, we can use the fact that the graph contains the point $(0, -40)$ and, therefore, $f(0) = -40$.

$f(x) = a(x^5 - 2x^4 + 7x^3 + 8x^2 - 34x + 20)$	*Write the polynomial function.*
$f(0) = a(0^5 - 2(0)^4 + 7(0)^3 + 8(0)^2 - 34(0) + 20)$	*Evaluate $f(0)$.*
$f(0) = a(20)$	*Simplify.*
$f(0) = -40$	*The graph contains the point $(0, -40)$.*
$a(20) = -40$	*Equate the two expressions for $f(0)$.*
$a = -2$	*Solve for a.*

Thus, $f(x) = -2(x^5 - 2x^4 + 7x^3 + 8x^2 - 34x + 20)$ or $f(x) = -2x^5 + 4x^4 - 14x^3 - 16x^2 + 68x - 40$.

Work exercises 22–27 in this textbook or in the *MyMathLab*® Study Plan.

OBJECTIVE 5:   USING THE INTERMEDIATE VALUE THEOREM

A direct consequence of the complex conjugate pairs theorem is the fact that every *odd degree* polynomial with real coefficients has at least one real zero. This is true because complex zeros always occur in pairs; thus, the number of nonreal zeros of a polynomial with real coefficients is always *even*. Therefore, if f is a polynomial function of odd degree, then at least one of its zeros must be real. For example, the polynomial function $f(x) = x^3 + 2x - 1$ is of odd degree, so we know that there exists at least one real zero. By the rational zeros theorem, the only possible rational zeros are ± 1. But

$$f(1) = 1 + 2 - 1 = 2 \text{ and}$$

$$f(-1) = -1 - 2 - 1 = -4$$

Neither 1 nor -1 are zeros of f, so there are no real **rational zeros**. Therefore, the real zero(s) of f must be **irrational**. So how do we find the irrational zero(s)? The following theorem will help us approximate the irrational zeros to as many decimal places as we want.

4.5 The Zeros of Polynomial Functions; The Fundamental Theorem of Algebra   **205**

Intermediate Value Theorem

Let f be a polynomial function with $a < b$. If $f(a)$ and $f(b)$ have opposite signs, then there exists at least one real zero strictly between a and b.

Although the proof of the intermediate value theorem involves calculus, you can see from Figure 22 that if $f(a)$ is negative and $f(b)$ is positive, then the graph of f *must* cross the x-axis somewhere between a and b.

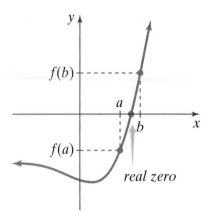

Figure 22 If $f(a)$ is negative and $f(b)$ is positive, then there is a real zero somewhere between a and b.

My video summary

🎬 **Example 5 Use the Intermediate Value Theorem to Approximate a Real Zero**

Find the real zero of $f(x) = x^3 + 2x - 1$ correct to two decimal places.

Solution Because f is of odd degree, there is at least one real zero. To approximate this zero, we must first find values of a and b such that $f(a)$ and $f(b)$ have opposite signs.

Note that for $a = 0$, $f(a) = f(0) = -1$. We must now find a number b such that $f(b)$ is positive. Using trial and error, we see that for $b = 1$, $f(b) = f(1) = 2$. Therefore, by the intermediate value theorem, there exists at least one real zero strictly between 0 and 1. We now evaluate the function at the decimal values .1, .2, .3, . . . until we obtain another sign change:

$$f(.1) = -.7990$$
$$f(.2) = -.5920$$
$$f(.3) = -.3730$$
$$\left.\begin{array}{l} f(.4) = -.1360 \\ f(.5) = .1250 \end{array}\right\} \quad \textit{Sign change!}$$

Notice that $f(.4)$ and $f(.5)$ have opposite signs. Therefore, by the intermediate value theorem, there exists a zero strictly between .4 and .5. We repeat the process again, this time evaluating the function at .41, .42, .43, . . . until we obtain another sign change.

$$\left.\begin{array}{l} f(.45) = -.0089 \\ f(.46) = .01734 \end{array}\right\} \quad \textit{Sign change!}$$

Because $f(.45)$ is closer to zero than $f(.46)$, we can conclude that the irrational real zero of $f(x) = x^3 + 2x - 1$ approximated to two decimal places is .45. Using a graphing utility, we can see that the zero carried out to eight decimal places is .45339765.

Using Technology

$$f(x) = x^3 + 2x - 1$$

Use a graphing utility and the Zero feature to approximate the real zero of $f(x) = x^3 + 2x - 1$.

Work exercises 28–33 in this textbook or in the *MyMathLab*® Study Plan.

OBJECTIVE 6: **SKETCHING THE GRAPHS OF POLYNOMIAL FUNCTIONS**

Now that we have a procedure for finding the zeros of a polynomial function, we can modify the four-step process that was discussed in Section 4.3 and put it all together to sketch the graphs of polynomial functions of the form $f(x) = a_n x^n + a_{n-1}x^{n-1} + a_{n-2}x^{n-2} + \cdots + a_1 x + a_0$.

Steps for Sketching the Graphs of Polynomial Functions

Step 1: Determine the end behavior.

Step 2: Plot the y-intercept $f(0) = a_0$.

Step 3: Use the rational zeros theorem, the factor theorem and synthetic division, or the intermediate value theorem to find all zeros and completely factor f.

Step 4: Choose a test value between each real zero and complete the graph.

My video summary

▦ **Example 6 Sketch a Polynomial Function**

Sketch the graph of $f(x) = 2x^5 - 5x^4 - 2x^3 + 7x^2 - 4x + 12$.

Solution Work through the video to see how to sketch this polynomial function. The polynomial is sketched in Figure 23.

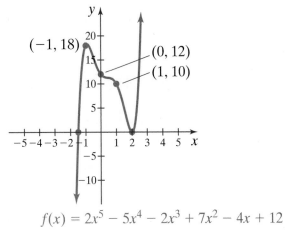

$$f(x) = 2x^5 - 5x^4 - 2x^3 + 7x^2 - 4x + 12$$

Figure 23 Graph of $f(x) = 2x^5 - 5x^4 - 2x^3 + 7x^2 - 4x + 12$

Work exercises 34–39 in this textbook or in the *MyMathLab*® Study Plan.

4.5 Exercises

In Exercises 1–6, use the rational zeros theorem to determine the potential rational zeros of each polynomial function. Do *not* find the zeros.

1. $f(x) = x^4 - 2$

2. $f(x) = x^3 - 3x + 16$

3. $f(x) = x^5 - 2x^4 + 8x^3 - x^2 + 2x - 12$

4. $f(x) = 3x^4 - 7x^2 - x + 8$

5. $f(x) = x^4 + 4 + 3x - 9x^5$

6. $f(x) = 5x^3 - 27x^6 + 10x^3 + 16 - 4x$

In Exercises 7–15, find all complex zeros of the given polynomial function, and write the polynomial in completely factored form.

7. $f(x) = 4x^3 - 5x^2 - 23x + 6$

8. $f(x) = 2x^3 + 7x^2 - 6x - 21$

9. $f(x) = -2x^3 - 7x^2 - 36x + 20$

10. $f(x) = -6x^4 + 19x^3 - 15x^2 - 3x + 5$

11. $f(x) = 3x^4 + 4x^3 + 73x^2 - 54x - 26$

12. $f(x) = -2x^5 - 11x^4 - 19x^3 - 8x^2 + 7x + 5$

13. $f(x) = 8x^5 + 36x^4 + 70x^3 - 5x^2 - 68x + 24$

14. $f(x) = 63x^5 + 33x^4 - 125x^3 - 67x^2 - 2x + 2$

15. $f(x) = 5x^6 - 18x^5 + 29x^4 - 32x^3 + 27x^2 - 14x + 3$

In Exercises 16–21, solve each polynomial equation in the complex numbers.

16. $x^3 - 1 = 0$

17. $x^4 - 1 = 0$

18. $x^3 - 4x^2 + x + 6 = 0$

19. $2x^4 + x^3 - 4x^2 + 11x - 10 = 0$

20. $3x^4 + 4x^3 + 73x^2 - 54x - 26 = 0$

21. $12x^4 + 56x^3 + 3x^2 - 13x - 2 = 0$

In Exercises 22–27, form a polynomial function $f(x)$ with real coefficients using the given information.

22. Form a third-degree polynomial function with real coefficients such that $5 + i$ and 2 are zeros.

23. Form a fourth-degree polynomial function with real coefficients such that $3 - 2i$ is a zero and 1 is a zero of multiplicity 2.

24. Form a fifth-degree polynomial function with real coefficients such that $3i$, $1 - 5i$, and -1 are zeros and $f(0) = -3$.

25. Form the third-degree polynomial function with real coefficients sketched here given that $2i$ is a zero.

26. Form the fifth-degree polynomial function with real coefficients sketched here given that $1 + i$ is a zero.

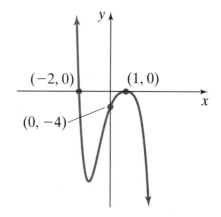

27. Form the fourth-degree polynomial function with real coefficients sketched here given that $-1 - 2i$ is a zero.

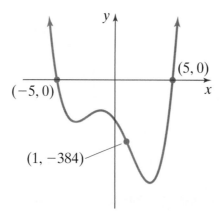

In Exercises 28–30, use the intermediate value theorem to show that the polynomial has a real zero on the given interval.

28. $f(x) = x^3 + 2x - 5; [1, 2]$ **29.** $f(x) = 3x^4 - x^3 + 8x^2 + x - 1; [0, 1]$ **30.** $f(x) = 2x^5 + 5x + 1; [-1, 0]$

In Exercises 31–33, use the intermediate value theorem to approximate the one real zero of each polynomial. Round your answer to two decimal places.

31. $f(x) = x^3 + 2x - 5$ **32.** $f(x) = -x^3 + x^2 + x - 1$ **33.** $f(x) = 2x^5 + 5x + 1$

In Exercises 34–39, determine the end behavior, plot the *y*-intercept, find and plot all real zeros, and plot at least one test value between each intercept. Then connect the points with a smooth curve.

34. $f(x) = x^3 + x^2 - x - 1$ **35.** $f(x) = 4x^3 - 5x^2 - 23x + 6$

36. $f(x) = -x^3 - 5x^2 - 3x + 9$ **37.** $f(x) = -6x^4 + 19x^3 - 15x^2 - 3x + 5$

38. $f(x) = 2x^5 + 11x^4 + 18x^3 + 8x^2 - 4x - 3$ **39.** $f(x) = -6x^4 + x^3 + 53x^2 - 3x - 105$

4.6 Rational Functions and Their Graphs

THINGS TO KNOW

Before working through this section, be sure that you are familiar with the following concepts:

VIDEO   ANIMATION   INTERACTIVE

 1. Determining the Domain of a Function Given the Equation (**Section 3.1**)
You Try It

 2. Determining Whether a Function Is Even, Odd, or Neither (**Section 3.2**)
You Try It

 3. Using Combinations of Transformations to Graph Functions (**Section 3.4**)
You Try It

OBJECTIVES

1 Finding the Domain and Intercepts of Rational Functions

2 Identifying Vertical Asymptotes

3 Identifying Horizontal Asymptotes

4 Using Transformations to Sketch the Graphs of Rational Functions

5 Sketching Rational Functions Having Removable Discontinuities

6 Identifying Slant Asymptotes

7 Sketching Rational Functions

In this section, we investigate the properties and graphs of rational functions.

DEFINITION Rational Function

A rational function is a function of the form $f(x) = \frac{g(x)}{h(x)}$, where g and h are **polynomial functions** such that $h(x) \neq 0$.

Note: If $h(x) = c$, where c is a real number, then we will consider the function $f(x) = g(x)/h(x) = g(x)/c$ to be a polynomial function.

4.6 Rational Functions and Their Graphs   **209**

OBJECTIVE 1:   FINDING THE DOMAIN AND INTERCEPTS OF RATIONAL FUNCTIONS

Because polynomial functions are defined for all values of x, it follows that rational functions are defined for all values of x *except* those for which the denominator $h(x)$ is equal to zero. If $f(x)$ has a y-intercept, it can be found by evaluating $f(0)$, provided that $f(0)$ is defined. If $f(x)$ has any x-intercepts, they can be found by solving the equation $g(x) = 0$ (provided that g and h do not share a common factor).

My video summary

▦ Example 1 Find the Domain and Intercepts of a Rational Function

Let $f(x) = \dfrac{x - 4}{x^2 + x - 6}$.

a. Determine the domain of f.
b. Determine the y-intercept (if any).
c. Determine any x-intercepts.

Solution Factor the denominator, and rewrite f as $f(x) = \dfrac{x - 4}{(x + 3)(x - 2)}$.

a. The domain of f is the set of all real numbers except those for which the denominator is equal to zero.

   The denominator, $(x + 3)(x - 2)$, is equal to zero when $x = -3$ or $x = 2$. Therefore, the domain is the set of all real numbers except -3 and 2 or $\{x \mid x \neq -3,\ x \neq 2\}$. In interval notation, the domain is $(-\infty, -3) \cup (-3, 2) \cup (2, \infty)$.

b. The y-intercept is $f(0) = \dfrac{-4}{-6} = \dfrac{2}{3}$.

c. To find any x-intercepts, we must solve the equation $x - 4 = 0$. The solution to this equation is $x = 4$. Thus, the only x-intercept is 4.

Watch the video to see the solution to Example 1 worked out in detail.

Work exercises 1–6 in this textbook or in the *MyMathLab*® Study Plan.

OBJECTIVE 2:   IDENTIFYING VERTICAL ASYMPTOTES

To begin our discussion of the graphs of rational functions, we first look at a basic function that was introduced in Section 3.3, the reciprocal function, $f(x) = \frac{1}{x}$. The function $f(x) = \frac{1}{x}$ is defined everywhere except at $x = 0$; therefore, the domain of $f(x) = \frac{1}{x}$ is $(-\infty, 0) \cup (0, \infty)$. You can see from the graph of $f(x) = \frac{1}{x}$ sketched in Figure 24 that as the values of x get closer to 0 from the right side of 0, the

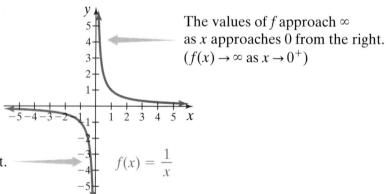

The values of f approach ∞ as x approaches 0 from the right. $(f(x) \to \infty$ as $x \to 0^{+})$

The values of f approach $-\infty$ as x approaches 0 from the left. $(f(x) \to -\infty$ as $x \to 0^{-})$

Figure 24 Graph of $f(x) = \frac{1}{x}$

values of $f(x)$ increase without bound. Mathematically, we say, "as x approaches zero from the right, $f(x)$ approaches infinity." In symbols, this is written as

$$f(x) \rightarrow \infty \quad \text{as} \quad x \rightarrow 0^+$$

The "+" symbol indicates that we are only looking at values of x that are located to the *right* of zero. Similarly, as x gets closer to 0 from the left side of 0, the values of $f(x)$ approach negative infinity. Symbolically, we can write the following:

$$f(x) \rightarrow -\infty \quad \text{as} \quad x \rightarrow 0^-$$

The "−" symbol indicates that we are only looking at values of x that are located to the *left* of zero.

   The graph of the function $f(x) = \frac{1}{x}$ gets closer and closer to the line $x = 0$ (the *y*-axis) as x gets closer and closer to 0 from either side of 0, but the graph never touches the line $x = 0$. The line $x = 0$ is said to be a **vertical asymptote** of the graph of $f(x) = \frac{1}{x}$. Many rational functions have vertical asymptotes. The vertical asymptotes occur when the graph of a function approaches positive or negative infinity as x approaches some finite number a as stated in the following definition.

DEFINITION Vertical Asymptote

The vertical line $x = a$ is a **vertical asymptote** of a function $y = f(x)$ if *at least* one of the following occurs:

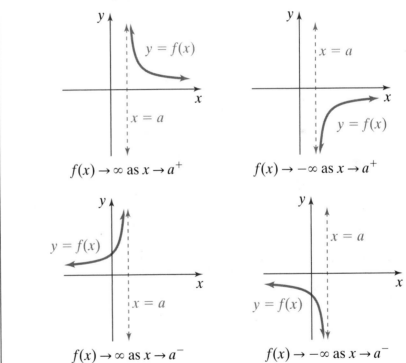

$f(x) \rightarrow \infty$ as $x \rightarrow a^+$ 　　　 $f(x) \rightarrow -\infty$ as $x \rightarrow a^+$

$f(x) \rightarrow \infty$ as $x \rightarrow a^-$ 　　　 $f(x) \rightarrow -\infty$ as $x \rightarrow a^-$

A rational function of the form $f(x) = \frac{g(x)}{h(x)}$, where $g(x)$ and $h(x)$ have no common factors, will have a vertical asymptote at $x = a$ if $h(a) = 0$.

![CAUTION] It is essential to cancel any common factors before locating the vertical asymptotes.

The situation that arises when $g(x)$ and $h(x)$ share a common factor is discussed in objective 5 of this section.

My video summary

▣ Example 2 Find the Vertical Asymptotes of a Rational Function

Find the vertical asymptotes (if any) of the function $f(x) = \frac{x-3}{x^2 + x - 6}$, and then sketch the graph near the vertical asymptotes.

Solution Factor the denominator of $f(x)$: $f(x) = \frac{x-3}{(x+3)(x-2)}$.

Because the numerator and denominator have no common factors, f will have vertical asymptotes when the denominator, $(x + 3)(x - 2)$, is equal to zero or when $x = -3$ and $x = 2$. Therefore, the equations of the two vertical asymptotes are $x = -3$ and $x = 2$. To sketch the graph near the asymptotes, we must determine whether $f \to \infty$ or $f \to -\infty$ on either side of the vertical asymptote. To do this, we evaluate the sign of each factor of f by choosing a test value "close to" $x = -3$ and $x = 2$ from the left side and from the right side of $x = -3$ and $x = 2$.

Vertical Asymptote	$x = -3$		$x = 2$	
Approach From	Left	Right	Left	Right
Test Value	$x = -3.1$	$x = -2.9$	$x = 1.9$	$x = 2.1$
Substitute Test Value into $f(x) = \dfrac{x-3}{(x+3)(x-2)}$	$f(-3.1) =$ $\dfrac{(-3.1-3)}{(-3.1+3)(-3.1-2)} =$ $\dfrac{(-)}{(-)(-)} = -$	$f(-2.9) =$ $\dfrac{(-2.9-3)}{(-2.9+3)(-2.9-2)} =$ $\dfrac{(-)}{(+)(-)} = +$	$f(1.9) =$ $\dfrac{(1.9-3)}{(1.9+3)(1.9-2)} =$ $\dfrac{(-)}{(+)(-)} = +$	$f(2.1) =$ $\dfrac{(2.1-3)}{(2.1+3)(2.1-2)} =$ $\dfrac{(-)}{(+)(+)} = -$
Result	Sign is negative. $f \to -\infty$	Sign is positive. $f \to \infty$	Sign is positive. $f \to \infty$	Sign is negative. $f \to -\infty$

◆ CAUTION If there is an x-intercept near the vertical asymptote, it is essential to choose a test value that is between the x-intercept and the vertical asymptote.

We see from the table that as the values of x approach -3 from the left side of -3, the values of f approach $-\infty$. Similarly, as the values of x approach -3 from the right side of -3, the values of f approach ∞. As x approaches 2 from the left side of 2, the values of f approach ∞. Finally, as the values of x approach 2 from the right side of 2, the values of f approach $-\infty$. Notice that we chose 2.1 as our test value for the case when x approaches 2 from the right. If we would have picked a test value greater than 3, our conclusions would have been incorrect. This is because $x = 3$ is an x-intercept and the graph to the right of 3 is above the x-axis. It is absolutely crucial to choose test values that are strictly between the x-intercept and the vertical asymptote.

Watch the video to see this entire example worked out in detail. The graph of f near both vertical asymptotes is sketched in Figure 25.

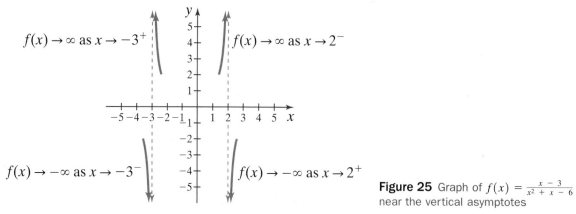

$f(x) \to \infty$ as $x \to -3^{+}$

$f(x) \to \infty$ as $x \to 2^{-}$

$f(x) \to -\infty$ as $x \to -3^{-}$

$f(x) \to -\infty$ as $x \to 2^{+}$

Figure 25 Graph of $f(x) = \frac{x-3}{x^2 + x - 6}$ near the vertical asymptotes

My video summary

Example 3 Find the Vertical Asymptotes of a Rational Function

Find the vertical asymptotes (if any) of the function $f(x) = \frac{x+3}{x^2+x-6}$, and then sketch the graph near the vertical asymptotes.

Solution We start by factoring the denominator: $f(x) = \frac{x+3}{(x+3)(x-2)}$. Notice that the numerator and denominator share a common factor of $x + 3$. Canceling the common factor, we get $f(x) = \frac{x+3}{(x+3)(x-2)} = \frac{1}{x-2}$ for $x \neq -3$. After cancellation, the only zero in the denominator is $x = 2$. Thus, the line $x = 2$ is the only vertical asymptote of the graph of f. To sketch the graph near the asymptote, we must determine the sign of the denominator as x approaches 2 from the left side and the right side of $x = 2$.

Vertical Asymptote	$x = 2$	
Approach from	**Left**	**Right**
Test Value	$x = 1.9$	$x = 2.1$
Substitute Test Value into $f(x) = \dfrac{1}{x-2}$ for $x \neq -3$.	$f(1.9) =$ $\dfrac{1}{(1.9-2)} =$ $\dfrac{1}{(-)} = -$	$f(2.1) =$ $\dfrac{1}{(2.1-2)} =$ $\dfrac{1}{(+)} = +$
Result	Sign is negative. $f \to -\infty$	Sign is positive. $f \to \infty$

We see that as the values of x approach 2 from the left side of 2, the values of f approach $-\infty$. Similarly, as the values of x approach 2 from the right side of 2, the values of f approach ∞. Watch the video to see this entire example worked out in detail. A rough sketch of the graph of f near the vertical asymptote can be seen in Figure 26. We sketch the complete graph of f in Example 6.

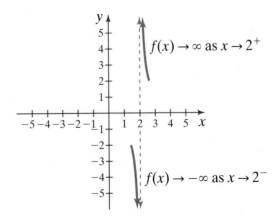

Figure 26 Graph of $f(x) = \frac{x+3}{x^2+x-6}$ near the vertical asymptote

Work exercises 7–12 in this textbook or in the *MyMathLab*® Study Plan.

4.6 Rational Functions and Their Graphs   213

OBJECTIVE 3:   IDENTIFYING HORIZONTAL ASYMPTOTES

My video summary

The graph of $f(x) = \frac{1}{x}$ is sketched again in Figure 27. The graph shows that as the values of x increase without bound (as $x \to \infty$), the values of f approach 0 ($f(x) \to 0$). Similarly, as the values of x decrease without bound (as $x \to -\infty$), the values of f approach 0 ($f(x) \to 0$). In this case, the line $y = 0$ is called a **horizontal asymptote** of the graph of $f(x) = \frac{1}{x}$.

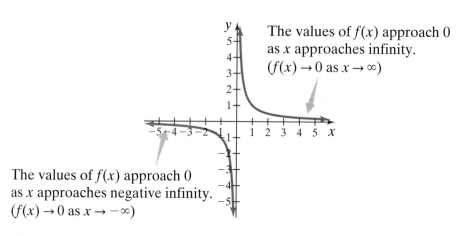

The values of $f(x)$ approach 0 as x approaches infinity. ($f(x) \to 0$ as $x \to \infty$)

The values of $f(x)$ approach 0 as x approaches negative infinity. ($f(x) \to 0$ as $x \to -\infty$)

Figure 27 The line $y = 0$ is a horizontal asymptote of the graph of $f(x) = \frac{1}{x}$.

DEFINITION Horizontal Asymptote

A horizontal line $y = H$ is a **horizontal asymptote** of a function f if the values of $f(x)$ approach some fixed number H as the values of x approach ∞ or $-\infty$.

  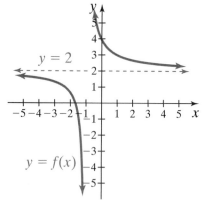

The line $y = -1$ is a horizontal asymptote because the values of $f(x)$ approach -1 as x approaches $-\infty$.

The line $y = 3$ is a horizontal asymptote because the values of $f(x)$ approach 3 as x approaches $\pm\infty$.

The line $y = 2$ is a horizontal asymptote because the values of $f(x)$ approach 2 as x approaches $\pm\infty$.

Properties of Horizontal Asymptotes of Rational Functions

• Although a rational function can have many vertical asymptotes, it can have at most one horizontal asymptote.

- The graph of a rational function will never intersect a vertical asymptote but may intersect a horizontal asymptote.
- A rational function $f(x) = \frac{g(x)}{h(x)}$ that is written in lowest terms (all common factors of the numerator and denominator have been canceled) will have a horizontal asymptote whenever the degree of $h(x)$ is greater than or equal to the degree of $g(x)$.

The third property listed here tells us that a rational function (written in lowest terms) will have a horizontal asymptote whenever the degree of the denominator is greater than or equal to the degree of the numerator. If the degree of the denominator is less than the degree of the numerator, then the rational function will not have a horizontal asymptote. We now summarize a technique for finding the horizontal asymptotes of a rational function.

Finding Horizontal Asymptotes of a Rational Function

Let $f(x) = \frac{g(x)}{h(x)} = \frac{a_n x^n + a_{n-1}x^{n-1} + a_{n-2}x^{n-2} + \cdots + a_1 x + a_0}{b_m x^m + b_{m-1}x^{m-1} + b_{m-2}x^{m-2} + \cdots + b_1 x + b_0}$, $a_n \neq 0$, $b_m \neq 0$, where

f is written in lowest terms, n is the degree of g, and m is the degree of h.

- If $m > n$, then $y = 0$ is the horizontal asymptote.
- If $m = n$, then the horizontal asymptote is $y = \frac{a_n}{b_m}$, the ratio of the leading coefficients.
- If $m < n$, then there are no horizontal asymptotes.

My interactive video summary

▶ Example 4 Find the Horizontal Asymptotes of a Rational Function

Find the horizontal asymptote of the graph of each rational function or state that one does not exist.

a. $f(x) = \dfrac{x}{x^2 - 4}$

b. $f(x) = \dfrac{4x^2 - x + 1}{1 - 2x^2}$

c. $f(x) = \dfrac{2x^3 + 3x^2 - 2x - 2}{x - 1}$

Solution

a. $f(x) = \dfrac{x}{x^2 - 4}$

The degree of the denominator is *greater than* the degree of the numerator. Therefore, the graph has a horizontal asymptote whose equation is $y = 0$ (the x-axis). Click here to see the graph of f or work through the interactive video.

b. $f(x) = \dfrac{4x^2 - x + 1}{1 - 2x^2}$

The degree of the denominator is *equal* to the degree of the numerator. Thus, the graph has a horizontal asymptote. Note that the leading coefficient of the polynomial in the numerator is 4. The leading coefficient of the polynomial in the denominator is -2 because -2 is the coefficient of the term with the highest degree. Thus, the equation of the horizontal asymptote is $y = \frac{4}{-2}$ or $y = -2$. Click here to see the graph of f or work through the interactive video.

c. $f(x) = \dfrac{2x^3 + 3x^2 - 2x - 2}{x - 1}$

The degree of the denominator is *less than* the degree of the numerator. Therefore, the graph has no horizontal asymptotes. Click here to see the graph of f or work through the interactive video.

Work exercises 13–20 in this textbook or in the *MyMathLab*® Study Plan.

OBJECTIVE 4:   USING TRANSFORMATIONS TO SKETCH THE GRAPHS OF RATIONAL FUNCTIONS

We now introduce another basic rational function, $f(x) = \frac{1}{x^2}$. Like the reciprocal function, the domain of $f(x) = \frac{1}{x^2}$ is $(-\infty, 0) \cup (0, \infty)$. The denominator, x^2, is always greater than zero, and this implies that the range includes all values of y greater than zero. The graph has a y-axis vertical asymptote ($x = 0$) and an x-axis horizontal asymptote ($y = 0$). Following are the graphs of $f(x) = \frac{1}{x}$ and $f(x) = \frac{1}{x^2}$, along with some important properties of each. Knowing these two basic graphs will help us sketch more complicated rational functions.

Graphs of $f(x) = \frac{1}{x}$ and $f(x) = \frac{1}{x^2}$

   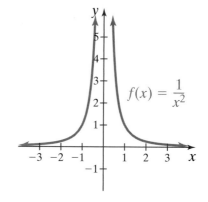

Properties of the graph of $f(x) = \frac{1}{x}$

Domain: $(-\infty, 0) \cup (0, \infty)$

Range: $(-\infty, 0) \cup (0, \infty)$

No intercepts

Vertical asymptote: $x = 0$

Horizontal asymptote: $y = 0$

Odd function $f(-x) = -f(x)$

The graph is symmetric about the origin.

Properties of the graph of $f(x) = \frac{1}{x^2}$

Domain: $(-\infty, 0) \cup (0, \infty)$

Range: $(0, \infty)$

No intercepts

Vertical asymptote: $x = 0$

Horizontal asymptote: $y = 0$

Even function $f(x) = f(-x)$

The graph is symmetric about the y-axis.

Sometimes we are able to use transformations to sketch the graphs of rational functions. Click here for a quick review of the transformations that are discussed in Section 3.4.

Example 5 Use Transformations to Sketch the Graph of a Rational Function

Use transformations to sketch the graph of $f(x) = \frac{-2}{(x+3)^2} + 1$.

Solution Because $x + 3$ is raised to the second power in the denominator, start with the graph of $y = \frac{1}{x^2}$. We can obtain the graph of $f(x) = \frac{-2}{(x+3)^2} + 1$ using the following sequence of transformations:

1. Horizontally shift the graph of $y = \frac{1}{x^2}$ to the left three units to obtain the graph of $y = \frac{1}{(x+3)^2}$.

2. Vertically stretch the graph of $y = \frac{1}{(x+3)^2}$ by a factor of 2 to obtain the graph of $y = \frac{2}{(x+3)^2}$.

3. Reflect the graph of $y = \frac{2}{(x+3)^2}$ about the x-axis to obtain the graph of $y = \frac{-2}{(x+3)^2}$.

4. Vertically shift the graph of $y = \frac{-2}{(x+3)^2}$ up one unit to obtain the final graph of $f(x) = \frac{-2}{(x+3)^2} + 1$.

You can see from the final graph in Figure 28 that the horizontal asymptote is now shifted up one unit to $y = 1$. The x-intercepts of $-3 - \sqrt{2}$ and $-3 + \sqrt{2}$ can be found by setting $f(x) = 0$ and solving for x. The y-intercept is $f(0) = \frac{7}{9}$. Watch the video to see each step worked out in detail.

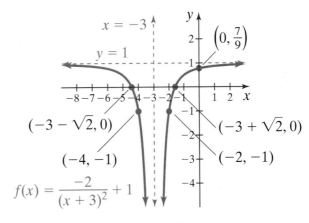

Figure 28 Graph of $f(x) = \frac{-2}{(x+3)^2} + 1$

Work exercises 21–32 in this textbook or in the *MyMathLab*® Study Plan.

OBJECTIVE 5: **SKETCHING RATIONAL FUNCTIONS HAVING REMOVABLE DISCONTINUITIES**

A rational function $f(x) = \frac{g(x)}{h(x)}$ may sometimes have a "hole" in its graph. In calculus, these holes are called **removable discontinuities**. For a rational function to have a removable discontinuity, $g(x)$ and $h(x)$ must share a common factor as in Example 6.

My video summary

▦ Example 6 Sketch a Rational Function Having a Removable Discontinuity

Sketch the graph of $f(x) = \frac{x^2 - 1}{x + 1}$, and find the coordinates of all removable discontinuities.

Solution The domain of f is $\{x \mid x \neq -1\}$. Factoring the numerator, we can rewrite f as $f(x) = \frac{(x + 1)(x - 1)}{x + 1}$. Canceling the common factor of $x + 1$, we get $f(x) = x - 1$ for $x \neq -1$. Notice that the domain of this simplified function is still all real numbers *except* $x = -1$. Because there are no longer any factors containing the variable x in the denominator of the simplified function, there are no vertical asymptotes.

The graph of $f(x) = \frac{x^2 - 1}{x + 1}$ behaves exactly like the graph of the simplified function $f(x) = x - 1$ for $x \neq -1$. This graph will look like the graph of $y = x - 1$ except we must remove the point on the graph when $x = -1$. We can find this point by evaluating $y = x - 1$ at $x = -1$. We get $y = (-1) - 1 = -2$. Thus, the removable discontinuity is the point $(-1, -2)$. The graph no longer looks like a typical rational function because the entire denominator canceled leaving only $f(x) = x - 1$ for $x \neq -1$. In Figure 29, the graph is a line with a "hole" at the point $(-1, -2)$.

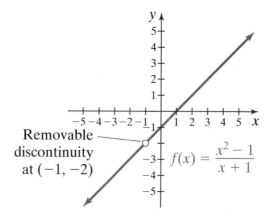

Removable discontinuity at $(-1, -2)$

$f(x) = \frac{x^2 - 1}{x + 1}$

Figure 29 Graph of $f(x) = \frac{x^2 - 1}{x + 1}$

It is possible to have a rational function with a common factor in the numerator and denominator whose graph still looks like the graph of a rational function but has a "hole" in it, as in Example 7.

My video summary

▦ Example 7 Sketch a Rational Function Having a Removable Discontinuity

Sketch the graph of $f(x) = \frac{x + 3}{x^2 + x - 6}$, and find the coordinates of all removable discontinuities.

Solution Notice that this is the function that was discussed in Example 3. Factoring the denominator, we get $f(x) = \frac{x + 3}{(x + 3)(x - 2)}$. The domain of $f(x)$ is all real numbers, except $x = 2$ and $x = -3$. Canceling the common factor of $x + 3$, we get $f(x) = \frac{1}{x - 2}$ for $x \neq -3$. Therefore, the graph of $f(x) = \frac{x + 3}{(x + 3)(x - 2)}$ will behave exactly like the graph of $f(x) = \frac{1}{x - 2}$ for $x \neq -3$. Notice that this graph will look like the graph of $y = \frac{1}{x - 2}$, except we must remember that $x = -3$ is *not* in the domain! There will be a removable discontinuity at the point where $x = -3$. We can find this point by evaluating $y = \frac{1}{x - 2}$ at $x = -3$. We get $y = \frac{1}{(-3) - 2} = -\frac{1}{5}$. Using transformations, we can shift the graph of $y = \frac{1}{x}$ to the right two units to obtain the graph of $y = \frac{1}{x-2}$, remembering to remove the point $\left(-3, -\frac{1}{5}\right)$. See Figure 30.

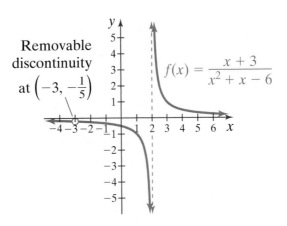

Figure 30 Graph of $f(x) = \frac{x+3}{x^2+x-6}$

> **CAUTION** Be careful when using graphing devices. Most graphing calculators will not show removable discontinuities!

Using Technology

The graph of $y_1 = \frac{x+3}{x^2+x-6}$ is shown using a graphing utility. The graph does not suggest that the function is undefined at $x = -3$. However, the table of values does illustrate this fact.

Work exercises 33–38 in this textbook or in the *MyMathLab* Study Plan.

OBJECTIVE 6:   IDENTIFYING SLANT ASYMPTOTES

Recall that the graph of a rational function $f(x) = \frac{g(x)}{h(x)}$, where f is written in lowest terms has a horizontal asymptote whenever the degree of $h(x)$ is greater than or equal to the degree of $g(x)$. But what happens if the degree of $h(x)$ is less than the degree of $g(x)$? If the degree of $h(x)$ is exactly 1 less than the degree of $g(x)$, then there will be a **slant asymptote**.

By the **division algorithm**, there exist polynomial functions $q(x)$ and $r(x)$ such that

$$\frac{g(x)}{h(x)} = q(x) + \frac{r(x)}{h(x)}$$

where the degree of $r(x)$ is less than the degree of $h(x)$. Because the degree of $h(x)$ is 1 less than the degree of $g(x)$, this means that the polynomial function $q(x)$ must be a **linear function**. Thus, we can rewrite $f(x)$ as

$$f(x) = mx + b + \frac{r(x)}{h(x)}$$

Because the degree of $r(x)$ is less than the degree of $h(x)$, we know that as $x \to \pm\infty$, $\frac{r(x)}{h(x)} \to 0$. Therefore, as $x \to \pm\infty$, the graph of $f(x) = mx + b + \frac{r(x)}{h(x)}$ approaches the line $y = mx + b$. This line is the **slant asymptote**.

My interactive video summary

Example 8 Find the Slant Asymptote of a Rational Function

Find the slant asymptote of $f(x) = \frac{2x^2 + 3x - 2}{x - 1}$.

Solution The numerator and denominator have no common factors, and the degree of the denominator is exactly 1 less than the degree of the numerator. Thus, the function has a slant asymptote. Using **synthetic division** or **long division**, we can rewrite f as $f(x) = 2x + 5 + \frac{3}{x - 1}$ so the slant asymptote is the line $y = 2x + 5$.

The graph of $f(x) = \frac{2x^2 + 3x - 2}{x - 1}$ from Example 8 is sketched in Figure 31. You may work through this interactive video to see how to sketch this function. Notice in Figure 31 that the y-intercept is $f(0) = 2$. The x-intercepts are $x = \frac{1}{2}$ and $x = -2$, which are the solutions to the equation $2x^2 + 3x - 2 = 0$. The vertical asymptote is the line $x = 1$, and the slant asymptote is the line $y = 2x + 5$.

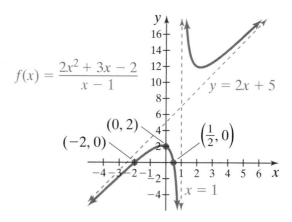

Figure 31 Graph of $f(x) = \frac{2x^2 + 3x - 2}{x - 1}$

Work exercises 39–42 in this textbook or in the MyMathLab® Study Plan.

OBJECTIVE 7:   SKETCHING RATIONAL FUNCTIONS

Given a rational function of the form $f(x) = \frac{g(x)}{h(x)}$, we can sketch the graph of f by carefully following the nine-step procedure outlined here.

Steps for Graphing Rational Functions of the Form $f(x) = \frac{g(x)}{h(x)}$

Step 1: Find the domain.

Step 2: If $g(x)$ and $h(x)$ have common factors, cancel all common factors determining the x-coordinates of any removable discontinuities and rewrite f in lowest terms.

Step 3: Check for symmetry.

If $f(-x) = -f(x)$, then the graph of $f(x)$ is *odd* and thus symmetric about the origin.

If $f(x) = f(-x)$, then the graph of $f(x)$ is *even* and thus symmetric about the *y*-axis.

Step 4: Find the *y*-intercept by evaluating $f(0)$.

Step 5: Find the *x*-intercepts by finding the zeros of the numerator of f, being careful to use the new numerator if a common factor has been removed.

Step 6: Find the vertical asymptotes by finding the zeros of the denominator of f, being careful to use the new denominator if a common factor has been removed. Use test values to determine the behavior of the graph on each side of the vertical asymptotes.

Step 7: Determine whether the graph has any horizontal or slant asymptotes.

Step 8: Plot points, choosing values of *x* between each intercept and values of *x* on either side of all vertical asymptotes.

Step 9: Complete the sketch.

My video summary

▦ Example 9 Sketch a Rational Function

Sketch the graph of $f(x) = \frac{x^3 + 2x^2 - 9x - 18}{x^3 + 6x^2 + 5x - 12}$.

Solution Watch the video to see each of the following steps worked out in detail:

1. We can use the rational zeros theorem and synthetic division to factor and rewrite f as $f(x) = \frac{(x + 3)(x - 3)(x + 2)}{(x + 3)(x - 1)(x + 4)}$. The domain is $\{x \mid x \neq -4, x \neq -3, x \neq 1\}$.

2. Because the numerator and denominator share a common factor of $x + 3$, there will be a removable discontinuity at $x = -3$. The function simplifies to $f(x) = \frac{(x - 3)(x + 2)}{(x - 1)(x + 4)}$ for $x \neq -3$ or $f(x) = \frac{x^2 - x - 6}{x^2 + 3x - 4}$ for $x \neq -3$. The removable discontinuity is the point $\left(-3, -\frac{3}{2}\right)$.

3. This function is neither odd nor even because $f(-x) \neq -f(x)$ and $f(x) \neq f(-x)$. Therefore, the graph is not symmetric about the origin or the *y*-axis.

4. The *y*-intercept is $f(0) = \frac{3}{2}$.

5. The *x*-intercepts occur when $(x - 3)(x + 2) = 0$. The *x*-intercepts are $x = -2$ and $x = 3$.

6. The equations of the two vertical asymptotes are $x = -4$ and $x = 1$. Using test values, we can determine the behavior of the graph near the vertical asymptotes.

   As $x \rightarrow -4^-$, $f(x) \rightarrow \infty$. The graph approaches positive infinity as *x* approaches -4 from the left.

   As $x \rightarrow -4^+$, $f(x) \rightarrow -\infty$. The graph approaches negative infinity as *x* approaches -4 from the right.

   As $x \rightarrow 1^-$, $f(x) \rightarrow \infty$. The graph approaches positive infinity as *x* approaches 1 from the left.

   As $x \rightarrow 1^+$, $f(x) \rightarrow -\infty$. The graph approaches negative infinity as *x* approaches 1 from the right.

7. Because the degree of the denominator is equal to the degree of the numerator, the graph of f has a horizontal asymptote at $y = \frac{1}{1} = 1$.

8. We can evaluate the function at several values of x. Note that $f(-5) = 4$, $f\left(\frac{1}{2}\right) = \frac{25}{9}$, and $f(2) = -\frac{2}{3}$. Hence, the graph must pass through the points $(-5, 4)$, $\left(\frac{1}{2}, \frac{25}{9}\right)$, and $\left(2, -\frac{2}{3}\right)$.

9. The completed graph is sketched in Figure 32. Watch the video to see each step worked out in detail.

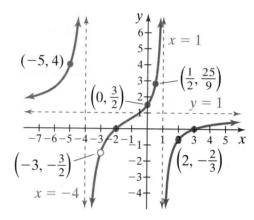

Figure 32 Graph of $f(x) = \frac{x^3 + 2x^2 - 9x - 18}{x^3 + 6x^2 + 5x - 12}$

The graph seen in Figure 32 does not have any symmetry. Some rational functions, however, do have symmetry so it is important not to skip step 3. For example, the rational function $f(x) = \frac{x}{x^2 - 4}$ is symmetric about the origin because $f(-x) = -f(x)$. See if you can follow the nine-step graphing process to sketch this graph. Click here to see a video of how to sketch this function.

Work exercises 43–54 in this textbook or in the *MyMathLab*® Study Plan.

4.6 Exercises

In Exercises 1–6, for each rational function, answer parts a to c.
  a. Determine the domain.
  b. Determine the y-intercept (if any).
  c. Determine any x-intercepts.

1. $f(x) = \dfrac{x}{x^2 - 1}$

2. $f(x) = \dfrac{x}{x^2 + 1}$

3. $f(x) = \dfrac{x^2 - 1}{x}$

4. $f(x) = \dfrac{3x - 1}{x + 2}$

5. $f(x) = \dfrac{x^2 + x - 12}{x^2 + x}$

6. $f(x) = \dfrac{x^3 - 2x^2 - 5x + 6}{x^2 - 4x - 5}$

In Exercises 7–12, find all vertical asymptotes, and create a rough sketch of the graph near each asymptote.

7. $f(x) = \dfrac{3}{x + 1}$

8. $f(x) = \dfrac{x}{x - 2}$

9. $f(x) = \dfrac{x - 1}{x^2 - 1}$

10. $f(x) = \dfrac{x + 3}{x^2 - 6x + 8}$

11. $f(x) = \dfrac{x^2 - 2x - 15}{x^2 + x - 2}$

12. $f(x) = \dfrac{x^2 - 9}{x^3 - 2x^2 - 5x + 6}$

In Exercises 13–20, find the equation of all horizontal asymptotes (if any) of each rational function.

13. $f(x) = \dfrac{3}{x + 1}$

14. $f(x) = \dfrac{x^2 - 1}{x + 1}$

15. $f(x) = \dfrac{3x - 1}{5x + 2}$

16. $f(x) = \dfrac{2x^3 - x^2 + 1}{x^2 + 1}$

17. $f(x) = \dfrac{3x^2 - 4x^3}{7x^3 + x^2 - x + 1}$

18. $f(x) = \dfrac{9x^2}{4 - x^2}$

19. $f(x) = \dfrac{1}{x^3 + x^2 - 1}$

20. $f(x) = \dfrac{3x^2 - 2x - 9}{11x^2 + x - 2}$

In Exercises 21–32, use transformations of $y = \frac{1}{x}$ or $y = \frac{1}{x^2}$ to sketch each rational function. Label all intercepts, and find the equation of all asymptotes.

21. $f(x) = \dfrac{1}{x} - 2$

22. $f(x) = \dfrac{1}{x^2} + 2$

23. $f(x) = \dfrac{1}{x - 1}$

24. $f(x) = \dfrac{1}{(x - 1)^2}$

25. $f(x) = \dfrac{1}{x + 1} - 3$

26. $f(x) = \dfrac{1}{(x + 3)^2} - 2$

27. $f(x) = \dfrac{-1}{x - 2} - 3$

28. $f(x) = \dfrac{-1}{(x + 1)^2} + 2$

29. $f(x) = \dfrac{2}{x + 1}$

30. $f(x) = \dfrac{2}{(x - 3)^2}$

31. $f(x) = \dfrac{-2}{x + 1} - 1$

32. $f(x) = \dfrac{-2}{(x - 1)^2} + 3$

In Exercises 33–38, identify the coordinates of all removable discontinuities, and sketch the graph of each rational function. Label all intercepts, and find the equations of all asymptotes.

33. $f(x) = \dfrac{x^2 - 9}{x - 3}$

34. $f(x) = \dfrac{x^2 + 3x - 4}{x + 4}$

35. $f(x) = \dfrac{x - 1}{x^2 - x}$

36. $f(x) = \dfrac{x + 2}{x^2 - x - 6}$

37. $f(x) = \dfrac{x + 2}{(x + 2)(x - 1)^2}$

38. $f(x) = \dfrac{x - 1}{x^3 + 3x^2 - 4}$

In Exercises 39–42, find the slant asymptote of the graph of each rational function.

39. $f(x) = \dfrac{2x^2 - 1}{x}$

40. $f(x) = \dfrac{x^2 - 2x + 5}{x + 1}$

41. $f(x) = \dfrac{x^3 - 1}{x^2 + x}$

42. $f(x) = \dfrac{2x^4 - 1}{x^3 - x}$

In Exercises 43–54, follow the nine-step graphing strategy to sketch the graph of each rational function. Be sure to label all intercepts and find the equations of all asymptotes. Also label any removable discontinuities.

43. $f(x) = \dfrac{2x + 4}{x - 1}$

44. $f(x) = \dfrac{2x + 4}{x^2 - x - 2}$

45. $f(x) = \dfrac{x^2 - 1}{x^2 + 1}$

46. $f(x) = \dfrac{2x^2 - 5x - 3}{x + 2}$

47. $f(x) = \dfrac{1}{x^2 - 9}$

48. $f(x) = \dfrac{x}{x^2 - 9}$

49. $f(x) = \dfrac{x^2 + 2x - 3}{x + 1}$

50. $f(x) = \dfrac{2x^2 + 4x - 6}{x^2 - 4x + 3}$

51. $f(x) = \dfrac{x^3 - 3x^2 - x + 3}{x^2 - x - 6}$

52. $f(x) = \dfrac{x^3 - 7x + 6}{x^3 - 5x^2 + 2x + 8}$

53. $f(x) = \dfrac{2x^3 - 3x^2 - 3x + 2}{x^3 - 3x^2 - 6x + 8}$

54. $f(x) = \dfrac{2x^3 - 3x^2 - 3x + 2}{x^2 - 9}$

CHAPTER FIVE

Exponential and Logarithmic Functions and Equations

CHAPTER FIVE CONTENTS

5.1   Exponential Functions

THINGS TO KNOW

Before working through this section, be sure that you are familiar with the following concepts:

VIDEO   ANIMATION   INTERACTIVE

You Try It

1. Using Combinations of Transformations to Graph Functions (Section 3.4)

You Try It

2. Determining whether a Function Is One-to-One Using the Horizontal Line Test (Section 3.6)

OBJECTIVES

1 Understanding the Characteristics of Exponential Functions

2 Sketching the Graphs of Exponential Functions Using Transformations

3 Solving Exponential Equations by Relating the Bases

4 Solving Applications of Exponential Functions

OBJECTIVE 1: UNDERSTANDING THE CHARACTERISTICS OF EXPONENTIAL FUNCTIONS

Many natural phenomena and real life applications can be modeled using exponential functions. Before we define the exponential function, it is important to remember how to manipulate exponential expressions because this skill is necessary when solving certain equations involving exponents. In Section R.3, expressions of the form b^r were evaluated for **rational numbers** r. For example,

$$3^2 = 9, \quad 4^{-2} = \frac{1}{4^2} = \frac{1}{16}, \quad \text{and} \quad 27^{-2/3} = \frac{1}{27^{2/3}} = \frac{1}{(\sqrt[3]{27})^2} = \frac{1}{(3)^2} = \frac{1}{9}$$

In this section, we extend the meaning of b^r to include all **real** values of r by defining the exponential function $f(x) = b^x$.

DEFINITION Exponential Function

An **exponential function** is a function of the form $f(x) = b^x$, where x is any real number and $b > 0$ such that $b \neq 1$.

   The constant, b, is called the base of the exponential function.

Notice in the definition that the base, b, must be positive and must not equal 1. If $b = 1$, then the function $f(x) = 1^x$ is equal to 1 for all x and is hence equivalent to the constant function $f(x) = 1$. If b were negative, then $f(x) = b^x$ would not be defined for all real values of x. For example, if $b = -4$, then $f\left(\frac{1}{2}\right) = (-4)^{1/2} = \sqrt{-4} = 2i$, which is *not* a positive real number.

My video summary

   Before we generalize the graph of $f(x) = b^x$, we create a table of values and sketch the graph of $y = b^x$ for $b = 2, 3, \frac{1}{2}$ and $\frac{1}{3}$. See Table 1.

Table 1

x	$y = 2^x$	$y = 3^x$	$y = \left(\frac{1}{2}\right)^x$	$y = \left(\frac{1}{3}\right)^x$
-2	$2^{-2} = \frac{1}{2^2} = \frac{1}{4}$	$3^{-2} = \frac{1}{3^2} = \frac{1}{9}$	$\left(\frac{1}{2}\right)^{-2} = 2^2 = 4$	$\left(\frac{1}{3}\right)^{-2} = 3^2 = 9$
-1	$2^{-1} = \frac{1}{2^1} = \frac{1}{2}$	$3^{-1} = \frac{1}{3^1} = \frac{1}{3}$	$\left(\frac{1}{2}\right)^{-1} = 2^1 = 2$	$\left(\frac{1}{3}\right)^{-1} = 3^1 = 3$
0	$2^0 = 1$	$3^0 = 1$	$\left(\frac{1}{2}\right)^0 = 1$	$\left(\frac{1}{3}\right)^0 = 1$
1	$2^1 = 2$	$3^1 = 3$	$\left(\frac{1}{2}\right)^1 = \frac{1}{2}$	$\left(\frac{1}{3}\right)^1 = \frac{1}{3}$
2	$2^2 = 4$	$3^2 = 9$	$\left(\frac{1}{2}\right)^2 = \frac{1}{4}$	$\left(\frac{1}{3}\right)^2 = \frac{1}{9}$

You can see from the graphs sketched in Figure 1 that all four functions intersect the y-axis at the point $(0, 1)$. This is true because $b^0 = 1$ for all nonzero values of b. For values of $b > 1$, the graph of $y = b^x$ increases rapidly as the values of x approach positive infinity ($b^x \to \infty$ as $x \to \infty$). In fact, the larger the base, the faster the graph will grow. Also, for $b > 1$, the graph of $y = b^x$ decreases quickly, approaching 0 as the values of x approach negative infinity ($b^x \to 0$ as $x \to -\infty$). Thus, the x-axis (the line $y = 0$) is a **horizontal asymptote**.

   However, for $0 < b < 1$, the graph decreases quickly, approaching the horizontal asymptote $y = 0$ as the values of x approach positive infinity ($b^x \to 0$ as $x \to \infty$), whereas the graph increases rapidly as the values of x approach negative infinity ($b^x \to \infty$ as $x \to -\infty$). The preceding statements, along with some other characteristics of the graphs of exponential functions, are outlined on the following page.

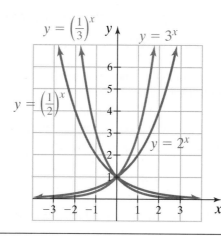

Figure 1 Graphs of $y = 2^x$, $y = 3^x$, $y = \left(\frac{1}{2}\right)^x$, and $y = \left(\frac{1}{3}\right)^x$

Characteristics of Exponential Functions

For $b > 0$, $b \neq 1$, the exponential function with base b is defined by $f(x) = b^x$. The domain of $f(x) = b^x$ is $(-\infty, \infty)$, and the range is $(0, \infty)$. The graph of $f(x) = b^x$ has one of the following two shapes:

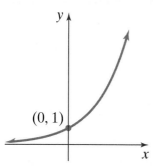

$$f(x) = b^x, b > 1$$

The graph intersects the y-axis at $(0, 1)$.

$$b^x \to \infty \text{ as } x \to \infty$$
$$b^x \to 0 \text{ as } x \to -\infty$$

The line $y = 0$ is a **horizontal asymptote**.

The function is **one-to-one**.

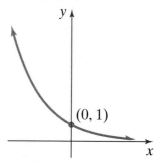

$$f(x) = b^x, 0 < b < 1$$

The graph intersects the y-axis at $(0, 1)$.

$$b^x \to 0 \text{ as } x \to \infty$$
$$b^x \to \infty \text{ as } x \to -\infty$$

The line $y = 0$ is a **horizontal asymptote**.

The function is **one-to-one**.

My video summary

Example 1 Sketch the Graph of an Exponential Function

Sketch the graph of $f(x) = \left(\frac{2}{3}\right)^x$.

Solution Because the base of the exponential function is $\frac{2}{3}$, which is between 0 and 1, the graph must approach the x-axis as the value of x approaches positive infinity. The graph intersects the y-axis at $(0, 1)$. We can find a few more points by choosing some negative and positive values of x:

$$f(-2) = \left(\frac{2}{3}\right)^{-2} = \left(\frac{3}{2}\right)^2 = \frac{9}{4}$$

$$f(-1) = \left(\frac{2}{3}\right)^{-1} = \left(\frac{3}{2}\right)^1 = \frac{3}{2}$$

$$f(1) = \left(\frac{2}{3}\right)^1 = \frac{2}{3}$$

$$f(2) = \left(\frac{2}{3}\right)^2 = \frac{4}{9}$$

We can complete the graph by connecting the points with a smooth curve. See Figure 2.

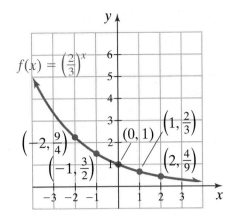

Figure 2 Graph of $f(x) = \left(\frac{2}{3}\right)^x$

Work exercises 1–5 in this textbook or in the *MyMathLab*® Study Plan.

Example 2 Determine an Exponential Function Given the Graph

Find the exponential function $f(x) = b^x$ whose graph is given as follows.

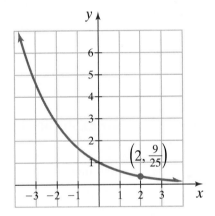

Solution From the point $\left(2, \frac{9}{25}\right)$, we see that $f(2) = \frac{9}{25}$. Thus,

$$f(x) = b^x \qquad \textit{Write the exponential function } f(x) = b^x.$$
$$f(2) = b^2 \qquad \textit{Evaluate } f(2).$$
$$f(2) = \frac{9}{25} \qquad \textit{The graph contains the point } \left(2, \frac{9}{25}\right).$$
$$b^2 = \frac{9}{25} \qquad \textit{Equate the two expressions for } f(2).$$

Therefore, we are looking for a constant b such that $b^2 = \frac{9}{25}$. Using the **square root property**, we get

$$\sqrt{b^2} = \pm\sqrt{\frac{9}{25}}$$

$$b = \pm\frac{3}{5}$$

By definition of an exponential function, $b > 0$; thus, $b = \frac{3}{5}$. Therefore, this is the graph of $f(x) = \left(\frac{3}{5}\right)^x$.

Work exercises 6–12 in this textbook or in the *MyMathLab*® Study Plan.

5.1 Exponential Functions   **227**

(EBook Screens   5.1-1–5.1-32)

OBJECTIVE 2: SKETCHING THE GRAPHS OF EXPONENTIAL FUNCTIONS USING TRANSFORMATIONS

Often we can use the **transformation techniques** that are discussed in Section 3.4 to sketch the graph of exponential functions. For example, the graph of $f(x) = 3^x - 1$ can be obtained by vertically shifting the graph of $y = 3^x$ down one unit. You can see in Figure 3 that the y-intercept of $f(x) = 3^x - 1$ is $(0, 0)$ and the horizontal asymptote is the line $y = -1$.

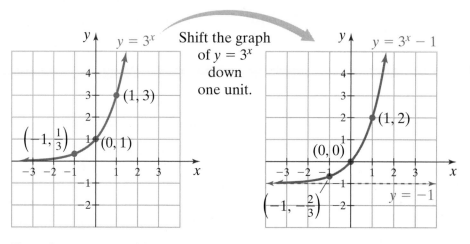

Figure 3 The graph of $f(x) = 3^x - 1$ can be obtained by vertically shifting the graph of $y = 3^x$ down one unit.

My video summary

▣ Example 3 Use Transformations to Sketch an Exponential Function

Use transformations to sketch the graph of $f(x) = -2^{x+1} + 3$.

Solution Starting with the graph of $y = 2^x$, we can obtain the graph of $f(x) = -2^{x+1} + 3$ through a series of three transformations:

1. Horizontally shift the graph of $y = 2^x$ to the left one unit, producing the graph of $y = 2^{x+1}$.

2. Reflect the graph of $y = 2^{x+1}$ about the x-axis, producing the graph of $y = -2^{x+1}$.

3. Vertically shift the graph of $y = -2^{x+1}$ up three units, producing the graph of $f(x) = -2^{x+1} + 3$.

The graph of $f(x) = -2^{x+1} + 3$ is shown in Figure 4. Watch the video to see every step.

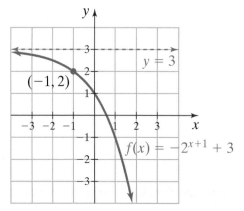

Figure 4 Graph of $f(x) = -2^{x+1} + 3$

Notice that the graph of $f(x) = -2^{x+1} + 3$ in Figure 4 has a y-intercept. We can find the y-intercept by evaluating $f(0) = -2^{0+1} + 3 = -2 + 3 = 1$. Also notice that the graph has an x-intercept. Can you find it? Recall that to find an x-intercept, we need to set $f(x) = 0$ and solve for x.

$$f(x) = -2^{x+1} + 3 \qquad \textit{Write the original function.}$$
$$0 = -2^{x+1} + 3 \qquad \textit{Substitute 0 for } f(x).$$
$$-3 = -2^{x+1} \qquad \textit{Subtract 3 from both sides.}$$
$$3 = 2^{x+1} \qquad \textit{Multiply both sides by } -1.$$

We now have a problem! We are going to need a way to solve for a variable that appears in an exponent. Stay tuned for Section 5.5.

Work exercises 13–20 in this textbook or in the _MyMathLab_® Study Plan.

OBJECTIVE 3: SOLVING EXPONENTIAL EQUATIONS BY RELATING THE BASES

One important property of all exponential functions is that they are **one-to-one functions**. You may want to review Section 3.6, which discusses one-to-one functions in detail. The function $f(x) = b^x$ is one-to-one because the graph of f passes the horizontal line test. In Section 3.6, the alternate definition of one-to-one is stated as follows:

A function f is one-to-one if for any two range values $f(u)$ and $f(v)$, $f(u) = f(v)$ implies that $u = v$.

Using this definition and letting $f(x) = b^x$, we can say that if $b^u = b^v$, then $u = v$. In other words, if the bases of an exponential equation of the form $b^u = b^v$ are the same, then the exponents must be the same. Solving exponential equations with this property is known as the **method of relating the bases** for solving exponential equations.

Method of Relating the Bases

If an exponential equation can be written in the form $b^u = b^v$, then $u = v$.

My animation summary

 Example 4 Use the Method of Relating the Bases to Solve Exponential Equations

Solve the following equations:

a. $8 = \dfrac{1}{16^x}$ **b.** $\dfrac{1}{27^x} = \left(\sqrt[4]{3}\right)^{x-2}$

Solution

a. Work through the animation to see how to obtain a solution of $x = -\frac{3}{4}$.

b. Work through the animation to see how to obtain a solution of $x = \frac{2}{13}$.

Work exercises 21–30 in this textbook or in the _MyMathLab_® Study Plan.

OBJECTIVE 4:   SOLVING APPLICATIONS OF EXPONENTIAL FUNCTIONS

Exponential functions are used to describe many real-life situations and natural phenomena. We now look at some examples.

Example 5 Learn to Hit a 3-Wood on a Golf Driving Range

Most golfers find that their golf skills improve dramatically at first and then level off rather quickly. For example, suppose that the distance (in yards) that a typical beginning golfer can hit a 3-wood after t weeks of practice on the driving range is given by the exponential function $d(t) = 225 - 100(2.7)^{-0.7t}$. This function has been developed after many years of gathering data on beginning golfers.

How far can a typical beginning golfer initially hit a 3-wood? How far can a typical beginning golfer hit a 3-wood after 1 week of practice on the driving range? After 5 weeks? After 9 weeks? Round to the nearest hundredth yard.

Solution Initially, when $t = 0$, $d(0) = 225 - 100(2.7)^0 = 225 - 100 = 125$ yards. Therefore, a typical beginning golfer can hit a 3-wood 125 yards.

After 1 week of practice on the driving range, a typical beginning golfer can hit a 3-wood $d(1) = 225 - 100(2.7)^{-0.7(1)} \approx 175.11$ yards. After 5 weeks of practice, $d(5) = 225 - 100(2.7)^{-0.7(5)} \approx 221.91$ yards. After 9 weeks of practice, $d(9) = 225 - 100(2.7)^{-0.7(9)} \approx 224.81$ yards.

Using a graphing utility, we can sketch the graph of $d(t) = 225 - 100(2.7)^{-0.7t}$. You can see from the graph in Figure 5 that the distance increases rather quickly and then tapers off toward a **horizontal asymptote** of 225 yards.

Using Technology

Figure 5 A TI-83 Plus was used to sketch the function $d(t) = 225 - 100(2.7)^{-0.7t}$ and the horizontal asymptote $y = 225$.

Work exercises 31–34 in this textbook or in the ***MyMathLab*®** Study Plan.

Compound Interest

A real-life application of exponential functions is the concept of **compound interest**, that is, interest that is paid on both principal and interest. First, we take a look at how **simple interest** is accrued. If an investment of P dollars is invested at r percent annually (written as a decimal) using simple interest, then the interest earned after 1 year is Pr dollars. Adding this interest to the original investment yields a total amount, A, of

$$A = \underbrace{P}_{\substack{\text{Original} \\ \text{investment}}} + \underbrace{Pr}_{\substack{\text{Interest} \\ \text{earned}}} = P(1 + r)$$

If this amount is reinvested at the same interest rate, then the total amount after 2 years becomes

$$A = \underbrace{P(1 + r)}_{\substack{\text{Total investment} \\ \text{after 1 year}}} + \underbrace{P(1 + r)r}_{\substack{\text{Interest} \\ \text{earned}}} = P(1 + r)(1 + r) = P(1 + r)^2$$

Reinvesting this amount for a third year gives an amount of $P(1 + r)^3$. Continuing this process for k years, we can see that the amount becomes $A = P(1 + r)^k$. This is an exponential function with base $1 + r$.

We can now modify this formula to obtain another formula that will model interest that is compounded periodically throughout the year(s). When interest is compounded periodically, then k no longer represents the number of years but rather the number of pay periods. If interest is paid n times per year for t years, then $k = nt$ pay periods. Thus, in the formula $A = P(1 + r)^k$, we substitute nt for k and get $A = P(1 + r)^{nt}$.

In the earlier simple interest model, the variable r was used to represent annual interest. In the periodically compound interest model being developed here with n pay periods per year, the interest rate per pay period is no longer r but rather $\frac{r}{n}$. Thus, in the formula $A = P(1 + r)^{nt}$, we replace r with $\frac{r}{n}$ and get the periodic compound interest formula $A = P\left(1 + \frac{r}{n}\right)^{nt}$.

Periodic Compound Interest Formula

Periodic compound interest can be calculated using the formula

$$A = P\left(1 + \frac{r}{n}\right)^{nt}$$

where

$A = $ Total amount after t years

$P = $ Principal (original investment)

$r \ = $ Interest rate per year

$n \ = $ Number of times interest is compounded per year

$t \ = $ Number of years

Example 6 Calculate Compound Interest

Which investment will yield the most money after 25 years?

Investment A: $12,000 invested at 3% compounded monthly

Investment B: $10,000 invested at 3.9% compounded quarterly

Solution Investment A: $P = 12{,}000$, $r = .03$, $n = 12$, $t = 25$:

$$A = 12{,}000\left(1 + \frac{.03}{12}\right)^{12(25)} \approx \$25{,}380.23$$

Investment B: $P = 10{,}000$, $r = .039$, $n = 4$, $t = 25$:

$$A = 10{,}000\left(1 + \frac{.039}{4}\right)^{4(25)} \approx \$26{,}386.77$$

Investment B will yield the most money after 25 years.

Work exercises 35–37 in this textbook or in the *MyMathLab*® Study Plan.

Present Value

Sometimes investors want to know how much money to invest now to reach a certain investment goal in the future. This amount of money, P, is known as the

present value of A dollars. To find a formula for present value, start with the formula for periodic compound interest and solve the formula for P:

$$A = P\left(1 + \frac{r}{n}\right)^{nt}$$ *Use the periodic compound interest formula.*

$$\frac{A}{\left(1 + \frac{r}{n}\right)^{nt}} = \frac{P\left(1 + \frac{r}{n}\right)^{nt}}{\left(1 + \frac{r}{n}\right)^{nt}}$$ *Divide both sides by $\left(1 + \frac{r}{n}\right)^{nt}$.*

$$A\left(1 + \frac{r}{n}\right)^{-nt} = P$$ *Rewrite $\frac{1}{\left(1 + \frac{r}{n}\right)^{nt}}$ as $\left(1 + \frac{r}{n}\right)^{-nt}$.*

The formula $P = A\left(1 + \frac{r}{n}\right)^{-nt}$ is known as the **present value formula.**

Present Value Formula

Present value can be calculated using the formula

$$P = A\left(1 + \frac{r}{n}\right)^{-nt}$$

where

P = Principal (original investment)

A = Total amount after t years

r = Interest rate per year

n = Number of times interest is compounded per year

t = Number of years

Example 7 Determine Present Value

Find the present value of $8,000 if interest is paid at a rate of 5.6% compounded quarterly for 7 years. Round to the nearest cent.

Solution Using the present value formula $P = A\left(1 + \frac{r}{n}\right)^{-nt}$ with $A = \$8,000$, $r = .056$, $n = 4$, and $t = 7$, we get

$$P = A\left(1 + \frac{r}{n}\right)^{-nt}$$

$$P = 8,000\left(1 + \frac{.056}{4}\right)^{-(4)(7)} \approx 5,420.35$$

Therefore, the present value of $8,000 in 7 years at 5.6% compounded quarterly is $5,420.35.

Work exercises 38–40 in this textbook or in the **MyMathLab**® Study Plan.

5.1 Exercises

In Exercises 1–5, sketch the graph of each exponential function. Label the y-intercept and at least two other points on the graph using both positive and negative values of x.

1. $f(x) = 4^x$ **2.** $f(x) = \left(\frac{1}{4}\right)^x$ **3.** $f(x) = \left(\frac{3}{2}\right)^x$ **4.** $f(x) = (.4)^x$ **5.** $f(x) = (2.7)^x$

In Exercises 6–12, determine the correct exponential function of the form $f(x) = b^x$ whose graph is given.

6.

$(1, 10)$

7.

$\left(-1, \frac{1}{2}\right)$

8.

$\left(2, \frac{4}{9}\right)$

9.

$\left(-3, \frac{64}{27}\right)$

10.

$\left(\frac{1}{3}, 2\right)$

11.

$\left(-\frac{1}{2}, 4\right)$

12.

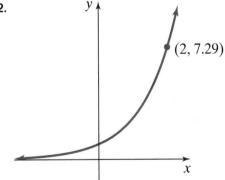

$(2, 7.29)$

In Exercises 13–20, use the graph of $y = 2^x$ or $y = 3^x$ and transformations to sketch each exponential function. Determine the domain and range. Also, determine the y-intercept, and find the equation of the horizontal asymptote.

13. $f(x) = 2^{x-1}$

14. $f(x) = 3^x - 1$

15. $f(x) = -3^{x+2}$

16. $f(x) = -2^{x+1} - 1$

17. $f(x) = \left(\frac{1}{2}\right)^{x+1}$

18. $f(x) = \left(\frac{1}{3}\right)^x - 3$

19. $f(x) = 2^{-x} + 1$

20. $f(x) = 3^{1-x} - 2$

5.1 Exponential Functions   233

In Exercises 21–30, solve each exponential equation using the method of "relating the bases" by first rewriting the equation in the form $b^u = b^v$.

21. $2^x = 16$

22. $3^{x-1} = \dfrac{1}{9}$

23. $\sqrt{5} = 25^x$

24. $\left(\sqrt[3]{3}\right)^x = 9$

25. $\dfrac{1}{\sqrt[5]{8}} = 2^x$

26. $\dfrac{9}{\sqrt[4]{3}} = \left(\dfrac{1}{27}\right)^x$

27. $(49)^x = \left(\dfrac{1}{7}\right)^{x-1}$

28. $\dfrac{125}{\sqrt[3]{5^x}} = \left(\dfrac{1}{25^x}\right)$

29. $\dfrac{3^{x^2}}{9^x} = 27$

30. $2^{x^3} = \dfrac{4^x}{2^{-x^2}}$

31. Typically, weekly sales will drop off rather quickly after the end of an advertising campaign. This drop in sales is known as *sales decay*. Suppose that the gross sales S, in hundreds of dollars, of a certain product is given by the exponential function $S(t) = 3{,}000(1.5^{-0.3t})$, where t is the number of weeks after the end of the advertising campaign.

Answer the following questions, rounding each answer to the nearest whole number:

a. What was the level of sales immediately after the end of the advertising campaign when $t = 0$?

b. What was the level of sales 1 week after the end of the advertising campaign?

c. What was the level of sales 5 weeks after the end of the advertising campaign?

32. Most people who start a serious weight-lifting regimen initially notice a rapid increase in the maximum amount of weight that they can bench press. After a few weeks, this increase starts to level off. The following function models the maximum weight, w, that a particular person can bench press in pounds at the end of t weeks of working out.

$$w(t) = 250 - 120(2.7)^{-0.3t}$$

Answer the following questions, rounding each answer to the nearest whole number:

a. What is the maximum weight that this person can bench press initially?

b. What is the maximum weight that this person can bench press after 3 weeks of weight lifting?

c. What is the maximum weight that this person can bench press after 7 weeks of weight lifting?

33. *Escherichia coli* bacteria reproduce by simple cell division, which is known as binary fission. Under ideal conditions, a population of E. coli bacteria can double every 20 minutes. This behavior can be modeled by the exponential function $N(t) = N_0(2^{.05t})$, where t is in minutes and N_0 is the initial number of E. coli bacteria.

Answer the following questions, rounding each answer to the nearest bacteria:

a. If the initial number of E. coli bacteria is five, how many bacteria will be present in 3 hours?

b. If the initial number of E. coli bacteria is eight, how many bacteria will be present in 3 hours?

c. If the initial number of E. coli bacteria is eight, how many bacteria will be present in 10 hours?

34. A wildlife-management research team noticed that a certain forest had no rabbits, so they decided to introduce a rabbit population into the forest for the first time. The rabbit population will be controlled by wolves and other predators. This rabbit population can be modeled by the function $R(t) = \dfrac{960}{.6 + 23.4(2.7)^{-.045t}}$, where t is the number of weeks after the research team first introduced the rabbits into the forest.

Answer the following questions, rounding each answer to the nearest whole number:

a. How many rabbits did the wildlife management research team bring into the forest?

b. How many rabbits can be expected after 10 weeks?

c. How many rabbits can be expected after the first year?

d. What is the expected rabbit population after 4 years? 5 years? What can the expected rabbit population approach as time goes on?

Use the **periodic compound interest formula** to solve Exercises 35–37.

35. Suppose that $9{,}000 is invested at 3.5% compounded quarterly. Find the total amount of this investment after 10 years. Round to the nearest cent.

36. Suppose that you have $5{,}000 to invest. Which investment yields the greater return over a 10-year period: 7.35% compounded daily or 7.4% compounded quarterly?

37. Which investment yields the greatest return?

Investment A: $4{,}000 invested for 5 years compounded semiannually (twice per year) at 8%

Investment B: $5{,}000 invested for 4 years compounded quarterly at 4.5%

Use the **present value formula** to solve Exercises 38–40.

38. Find the present value of $10{,}000 if interest is paid at a rate of 4.5% compounded semiannually for 12 years. Round to the nearest cent.

39. Find the present value of $1{,}000{,}000 if interest is paid at a rate of 9.5% compounded monthly for 8 years. Round to the nearest cent.

40. How much money would you have to invest at 10% compounded semiannually so that the total investment had a value of $2{,}205 after 1 year? Round to the nearest cent.

5.2 **The Natural Exponential Function**

THINGS TO KNOW

Before working through this section, be sure that you are familiar with the following concepts:

VIDEO   ANIMATION   INTERACTIVE

 You Try It

1. Sketching the Graphs of Exponential Functions Using Transformations (**Section 5.1**)

 You Try It

2. Solving Exponential Equations by Relating the Bases (**Section 5.1**)

 You Try It

3. Solving Periodic Compound Interest Applications (**Section 5.1**)

OBJECTIVES

1 Understanding the Characteristics of the Natural Exponential Function

2 Sketching the Graphs of Natural Exponential Functions Using Transformations

3 Solving Natural Exponential Equations by Relating the Bases

4 Solving Applications of the Natural Exponential Function

OBJECTIVE 1:   UNDERSTANDING THE CHARACTERISTICS OF THE NATURAL EXPONENTIAL FUNCTION

My video summary

We learned in the previous section that any positive number b, where $b \neq 1$, can be used as the base of an **exponential function**. However, there is one number that appears as the base in exponential applications more than any other number. This number is called the **natural base** and is symbolized using the letter e. The number e is an **irrational number** that is defined as the value of the expression $\left(1 + \frac{1}{n}\right)^n$ as n approaches infinity. Table 2 shows the values of the expression $\left(1 + \frac{1}{n}\right)^n$ for increasingly large values of n.

Table 2

n	$\left(1 + \frac{1}{n}\right)^n$
1	2
2	2.25
10	2.5937424601
100	2.7048138294
1,000	2.7169239322
10,000	2.7181459268
100,000	2.7182682372
1,000,000	2.7182804693
10,000,000	2.7182816925
100,000,000	2.7182818149

You can see from Table 2 that as the values of n get large, the value e (rounded to six decimal places) is 2.718282. The function $f(x) = e^x$ is called the **natural exponential function.** Because $2 < e < 3$, it follows that the graph of $f(x) = e^x$ lies between the graph of $y = 2^x$ and $y = 3^x$, as seen in Figure 6.

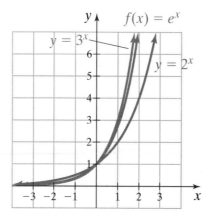

Figure 6 Graph of the natural exponential function $f(x) = e^x$

Characteristics of the Natural Exponential Function

The natural exponential function is the exponential function with base e and is defined as $f(x) = e^x$.

The domain of $f(x) = e^x$ is $(-\infty, \infty)$, and the range is $(0, \infty)$. The graph of $f(x) = e^x$ and some of its characteristics are stated here.

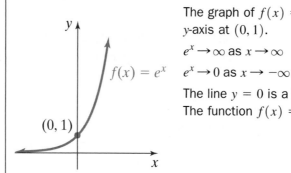

The graph of $f(x) = e^x$ intersects the y-axis at $(0, 1)$.

$e^x \to \infty$ as $x \to \infty$

$e^x \to 0$ as $x \to -\infty$

The line $y = 0$ is a **horizontal asymptote.**
The function $f(x) = e^x$ is **one-to-one.**

It is important that you are able to use your calculator to evaluate various powers of e. Most calculators have an $\boxed{e^x}$ key. Find this special key on your calculator and evaluate the expressions in the following example.

Example 1 Evaluate the Natural Exponential Function

Evaluate each expression correctly to six decimal places.

   a. e^2 **b.** $e^{-.534}$ **c.** $1,000e^{.013}$

Solution Using the $\boxed{e^x}$ key on a calculator, we get

 a. $e^2 \approx 7.389056$

 b. $e^{-.534} \approx .586255$

 c. $1,000e^{0.013} \approx 1,013.084867$

Work exercises 1–5 in this textbook or in the *MyMathLab*® Study Plan.

OBJECTIVE 2: SKETCHING THE GRAPHS OF NATURAL EXPONENTIAL FUNCTIONS USING TRANSFORMATIONS

Again, we can use the transformation techniques that are discussed in Section 3.4 to sketch variations of the natural exponential function.

My video summary

▣ Example 2 Use Transformations to Sketch Natural Exponential Functions

Use transformations to sketch the graph of $f(x) = -e^x + 2$. Determine the domain, range, and y-intercept, and find the equation of any asymptotes.

Solution We can sketch the graph of $f(x) = -e^x + 2$ through a series of the following two transformations.
Start with the graph of $y = e^x$.

1. Reflect the graph of $y = e^x$ about the x-axis, producing the graph of $y = -e^x$.

2. Vertically shift the graph of $y = -e^x$ up two units, producing the graph of $f(x) = -e^x + 2$.

The graph of $f(x) = -e^x + 2$ is shown in Figure 7. Watch the video to see each step. The domain of f is the interval $(-\infty, \infty)$. The range of f is the interval $(-\infty, 2)$. The y-intercept is 1, and the equation of the horizontal asymptote is $y = 2$.

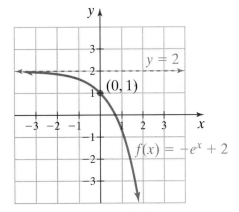

Figure 7 Graph of $f(x) = -e^x + 2$

Work exercises 6–10 in this textbook or in the *MyMathLab*® Study Plan.

OBJECTIVE 3: SOLVING NATURAL EXPONENTIAL EQUATIONS BY RELATING THE BASES

Recall the **method of relating the bases** for solving exponential equations from Section 5.1. If we can write an exponential equation in the form of $b^u = b^v$, then $u = v$. This method for solving exponential equations certainly holds true for the natural base as illustrated in the following example.

My interactive video summary

▣ Example 3 Use the Method of Relating the Bases to Solve Natural Exponential Equations

Use the method of relating the bases to solve each exponential equation:

a. $e^{3x-1} = \dfrac{1}{\sqrt{e}}$　　**b.** $\dfrac{e^{x^2}}{e^{10}} = (e^x)^3$

Solution Work through the interactive video to verify that the solutions are as follows:

a. $x = \dfrac{1}{6}$ **b.** $x = -2$ or $x = 5$

Work exercises 11–18 in this textbook or in the *MyMathLab*®︎ Study Plan.

OBJECTIVE 4:   SOLVING APPLICATIONS OF THE NATURAL EXPONENTIAL FUNCTION

Continuous Compound Interest

Recall the **periodic compound interest formula** that is introduced in Section 5.1. Some banks use **continuous compounding**; that is, they compound the interest every fraction of a second every day! If we start with the formula for periodic compound interest, $A = P\left(1 + \frac{r}{n}\right)^{nt}$, and let n (the number of times the interest is compounded each year) approach infinity, we can derive the formula $A = Pe^{rt}$, which is the formula for continuous compound interest. Work through this animation to see exactly how this formula is derived.

My animation summary

Continuous Compound Interest Formula

Continuous compound interest can be calculated using the formula

$$A = Pe^{rt}$$

where

A = Total amount after t years

P = Principal

r = Interest rate per year

t = Number of years

Example 4 Calculate Continuous Compound Interest

How much money would be in an account after 5 years if an original investment of $6,000 was compounded continuously at 4.5%? Compare this amount to the same investment that was compounded daily. Round to the nearest cent.

Solution First, the amount after 5 years compounded continuously is $A = Pe^{rt} = 6,000e^{.045(5)} \approx \$7,513.94$. The same investment compounded daily yields an amount of $A = P\left(1 + \frac{r}{n}\right)^{nt} = 6,000\left(1 + \frac{.045}{365}\right)^{365(5)} \approx \$7,513.83$. Continuous compound interest yields $.11 more interest after 5 years!

Work exercises 19–21 in this textbook or in the *MyMathLab*®︎ Study Plan.

Present Value

Recall that the present value P is the amount of money to be invested now to obtain A dollars in the future. To find a formula for present value on money that is compounded continuously, we start with the formula for continuous compound interest and solve for P.

$$A = Pe^{rt} \qquad \textit{Write the continuous compound interest formula.}$$

$$\frac{A}{e^{rt}} = P\frac{e^{rt}}{e^{rt}} \qquad \textit{Divide both sides by } e^{rt}.$$

$$Ae^{-rt} = P \qquad \textit{Rewrite } \frac{1}{e^{rt}} \text{ as } e^{-rt}.$$

Present Value Formula

The present value of A dollars after t years of continuous compound interest, with interest rate r, is given by the formula

$$P = Ae^{-rt}$$

Example 5 Calculate Present Value

Find the present value of $18,000 if interest is paid at a rate of 8% compounded continuously for 20 years. Round to the nearest cent.

Solution Using the present value formula $P = Ae^{-rt}$ with $A = \$18,000$, $r = .08$, and $t = 20$, we get:

$$P = Ae^{-rt}$$
$$P = (18,000)e^{-(.08)(20)} \approx \$3,634.14$$

Work exercises 22–23 in this textbook or in the *MyMathLab*® Study Plan.

Exponential Growth Model

You have probably heard that some populations grow exponentially. Most populations grow at a rate proportional to the size of the population. In other words, the larger the population, the faster the population grows. With this in mind, it can be shown in a more advanced math course that the mathematical model that can describe population growth is given by the function $P(t) = P_0e^{kt}$.

Exponential Growth

A model that describes the population, P, after a certain time, t, is

$$P(t) = P_0e^{kt}$$

where $P_0 = P(0)$ is the initial population and $k > 0$ is a constant called the **relative growth rate**. (Note: k may be given as a percent.)

The graph of the exponential growth model is shown in Figure 8. Notice that the graph has a y-intercept of P_0.

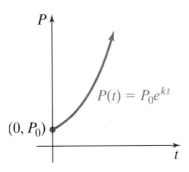

Figure 8 Graph of the exponential growth model $P(t) = P_0e^{kt}$

My video summary

▦ Example 6 Population Growth

The population of a small town follows the exponential growth model $P(t) = 900e^{.015t}$, where t is the number of years after 1900.

Answer the following questions, rounding each answer to the nearest whole number:

a. What was the population of this town in 1900?

b. What was the population of this town in 1950?

c. Use this model to predict the population of this town in 2012.

Solution

a. The initial population was $P(0) = 900e^{.015(0)} = 900$.

b. Because 1950 is 50 years after 1900, we must evaluate $P(50)$.
$P(50) = 900e^{.015(50)} \approx 1,905$.

c. In the year 2012, we can predict that the population will be
$P(112) = 900e^{.015(112)} \approx 4,829$.

My video summary

▣ Example 7 Determine the Initial Population

Twenty years ago, the State of Idaho Fish and Game Department introduced a new breed of wolf into a certain Idaho forest. The current wolf population in this forest is now estimated at 825, with a relative growth rate of 12%.

Answer the following questions, rounding each answer to the nearest whole number:

a. How many wolves did the Idaho Fish and Game Department initially introduce into this forest?

b. How many wolves can be expected after another 20 years?

Solution

a. The relative growth rate is .12, so we use the exponential growth model $P(t) = P_0e^{.12t}$. Because $P(20) = 825$, we get

$$P(20) = P_0e^{.12(20)} \qquad \text{Substitute 20 for } t.$$

$$825 = P_0e^{.12(20)} \qquad \text{Substitute 825 for } P(20).$$

$$P_0 = \frac{825}{e^{.12(20)}} \approx 75 \qquad \text{Solve for } P_0.$$

Therefore, the Idaho Fish and Game Department initially introduced 75 wolves into the forest.

b. Because $P_0 = 75$, we can use the exponential growth model $P(t) = 75e^{.12t}$. In another 20 years, the value of t will be 40. Thus, we must evaluate $P(40)$.

$$P(40) = 75e^{.12(40)} \approx 9,113$$

Therefore, we can expect the wolf population to be approximately 9,113 in another 20 years.

Work exercises 24–27 in this textbook or in the **MyMathLab**® Study Plan.

5.2 **Exercises**

In Exercises 1–5, use a calculator to approximate each exponential expression to six decimal places.

1. e^3 **2.** $e^{-0.2}$ **3.** $e^{1/3}$ **4.** $100e^{-.123}$ **5.** $\sqrt{2}e^{\pi}$

In Exercises 6–10, use transformations to sketch each exponential function. Determine the domain and range. Also, label the *y*-intercept, and find the equation of the horizontal asymptote.

6. $f(x) = e^{x-1}$ **7.** $f(x) = e^x - 1$ **8.** $f(x) = -e^{x+2}$ **9.** $f(x) = -e^{x+1} - 1$ **10.** $f(x) = e^{-x} - 2$

In Exercises 11–18, solve each exponential equation using the method of relating the bases by first rewriting the equation in the form $e^u = e^v$.

11. $e^x = \dfrac{1}{e^2}$

12. $e^{5x+2} = \sqrt[3]{e}$

13. $\dfrac{1}{e^x} = \dfrac{\sqrt{e}}{e^{1-x}}$

14. $(e^{x^2})^2 = e^8$

15. $e^{x^2} = (e^x) \cdot e^{12}$

16. $e^{x^2} = \dfrac{e^3}{(e^x)^5}$

17. $\dfrac{e^{x^3}}{e^x} = \dfrac{e^{2x^2}}{e^2}$

18. $e^{x^3} \cdot e^6 = \dfrac{(e^{2x^2})^2}{e^x}$

19. An original investment of $6,000 earns 6.25% interest compounded continuously. What will the investment be worth in 2 years? 20 years? Round to the nearest cent.

20. How much more will an investment of $10,000 earning 5.5% compounded continuously for 9 years earn compared to the same investment at 5.5% compounded quarterly for 9 years? Round to the nearest cent.

21. Suppose your great-great grandfather invested $500 earning 6.5% interest compounded continuously 100 years ago. How much would his investment be worth today? Round to the nearest cent.

22. Find the present value of $16,000 if interest is paid at a rate of 4.5% compounded continuously for 10 years. Round to the nearest cent.

23. Which has the lower present value: (a) $20,000 if interest is paid at a rate of 5.18% compounded continuously for 2 years or (b) $25,000 if interest is paid at a rate of 3.8% compounded continuously for 30 months?

24. The population of a rural city follows the exponential growth model $P(t) = 2{,}000e^{.035t}$, where t is the number of years after 1995.

  a. What was the population of this city in 1995?

  b. What is the relative growth rate as a percent?

  c. Use this model to approximate the population in 2030, rounding to the nearest whole number.

25. The relative growth rate of a certain bacteria colony is 25%. Suppose there are 10 bacteria initially.

  a. Find a function that describes the population of the bacteria after t hours.

  b. How many bacteria should we expect after 1 day? Round to the nearest whole number.

26. In 2006, the population of a certain American city was 18,221. If the relative growth rate has been 6% since 1986, what was the population of this city in 1986? Round to the nearest whole number.

27. In 1970, a wildlife resource management team introduced a certain rabbit species into a forest for the first time. In 2004, the rabbit population had grown to 7,183. The relative growth rate for this rabbit species is 20%.

  Answer the following questions, rounding each answer to the nearest whole number:

  a. How many rabbits did the wildlife resource management team introduce into the forest in 1970?

  b. How many rabbits can be expected in the year 2025?

5.3 Logarithmic Functions

THINGS TO KNOW

Before working through this section, be sure that you are familiar with the following concepts:

VIDEO   ANIMATION   INTERACTIVE

You Try It

1. Solving Polynomial Inequalities (Section 1.9)

You Try It

2. Solving Rational Inequalities (Section 1.9)

You Try It

3. Determining Whether a Function Is One-to-One Using the Horizontal Line Test (Section 3.6)

You Try It

4. Sketching the Graph of an Inverse Function (Section 3.6)

You Try It

5. Using the Composition Cancellation Properties of Inverse Functions (Section 3.6)

OBJECTIVES

1 Understanding the Definition of a Logarithmic Function

2 Evaluating Logarithmic Expressions

3 Understanding the Properties of Logarithms

4 Using the Common and Natural Logarithms

5 Understanding the Characteristics of Logarithmic Functions

6 Sketching the Graphs of Logarithmic Functions Using Transformations

7 Finding the Domain of Logarithmic Functions

OBJECTIVE 1: UNDERSTANDING THE DEFINITION OF A LOGARITHMIC FUNCTION

My video summary

Every exponential function of the form $f(x) = b^x$, where $b > 0$ and $b \neq 1$, is **one-to-one** and thus has an **inverse function**. (You may want to refer to Section 3.6 to review one-to-one functions and inverse functions.) Remember, given the graph of a one-to-one function f, the graph of the inverse function is a reflection about the line $y = x$. That is, for any point (a, b) that lies on the graph of f, the point (b, a) must lie on the graph of f^{-1}. In other words, the graph of f^{-1} can be obtained by simply interchanging the x and y coordinates of the ordered pairs of $f(x) = b^x$. Watch this video to see how to sketch the graph of $f(x) = b^x$ and its inverse.

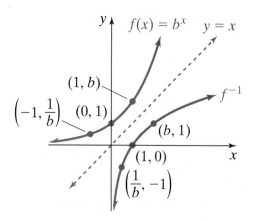

Figure 9 Graph of $f(x) = b^x$, $b > 1$, and its inverse function

The graphs of f and f^{-1} are sketched in Figure 9, but what is the equation of the inverse of $f(x) = b^x$? To find the equation of f^{-1}, we follow the **four-step process** for finding inverse functions that is discussed in Section 3.6.

Step 1 Change $f(x)$ to y: $y = b^x$

Step 2 Interchange x and y: $x = b^y$

Step 3 Solve for y: ??

Before we can solve for y we must introduce the following definition:

DEFINITION Logarithmic Function

For $x > 0$, $b > 0$, and $b \neq 1$, the **logarithmic function** with base b is defined by

$$y = \log_b x \text{ if and only if } x = b^y$$

The equation $y = \log_b x$ is said to be in **logarithmic form**, whereas the equation $x = b^y$ is in **exponential form**. We can now continue to find the inverse of $f(x) = b^x$ by completing steps 3 and 4.

Step 3 Solve for y: $x = b^y$ can be written as $y = \log_b x$

Step 4 Change y to $f^{-1}(x)$: $f^{-1}(x) = \log_b x$

In general, if $f(x) = b^x$ for $b > 0$ and $b \neq 1$, then the inverse function is $f^{-1}(x) = \log_b x$. For example, the inverse of $f(x) = 2^x$ is $f^{-1}(x) = \log_2 x$, which is read as "the log base 2 of x." We revisit the graphs of logarithmic functions later on in this section, but first it is very important to understand the definition of the logarithmic function and practice how to go back and forth writing exponential equations as logarithmic equations and vice versa.

My video summary

▣ Example 1 Change from Exponential Form to Logarithmic Form

Write each exponential equation as an equation involving a logarithm.

a. $2^3 = 8$ **b.** $5^{-2} = \dfrac{1}{25}$ **c.** $1.1^M = z$

Solution We use the fact that the equation $x = b^y$ is equivalent to the equation $y = \log_b x$.

a. $2^3 = 8$ is equivalent to $\log_2 8 = 3$.

b. $5^{-2} = \frac{1}{25}$ is equivalent to $\log_5 \frac{1}{25} = -2$.

c. $1.1^M = z$ is equivalent to $\log_{1.1} z = M$.

Note that the exponent of the original (exponential) equation ends up by itself on the right side of the second (logarithmic) equation. Therefore, a logarithmic expression can be thought of as describing the exponent of a certain exponential equation.

Watch the video to see this example worked out in more detail.

Work exercises 1–6 in this textbook or in the *MyMathLab*® Study Plan.

My video summary

Example 2 Change from Logarithmic Form to Exponential Form

Write each logarithmic equation as an equation involving an exponent.

   a. $\log_3 81 = 4$ **b.** $\log_4 16 = y$ **c.** $\log_{3/5} x = 2$

Solution We use the fact that the equation $y = \log_b x$ is equivalent to the equation $x = b^y$.

   a. $\log_3 81 = 4$ is equivalent to $3^4 = 81$.

   b. $\log_4 16 = y$ is equivalent to $4^y = 16$.

   c. $\log_{3/5} x = 2$ is equivalent to $\left(\frac{3}{5}\right)^2 = x$.

Watch the video to see this example worked out in more detail.

Work exercises 7–11 in this textbook or in the *MyMathLab*® study plan.

OBJECTIVE 2:   EVALUATING LOGARITHMIC EXPRESSIONS

Because a logarithmic expression represents the exponent of an exponential equation, it is possible to evaluate many logarithms by inspection or by creating the corresponding exponential equation. Remember that the expression $\log_b x$ is the exponent to which b must be raised to in order to get x. For example, suppose we are to evaluate the expression $\log_4 64$. To evaluate this expression, we must ask ourselves, "4 raised to what power is 64?" Because $4^3 = 64$, we conclude that $\log_4 64 = 3$. For some logarithmic expressions, it is often convenient to create an exponential equation and use the **method of relating the bases** for solving exponential equations. For more complex logarithmic expressions, additional techniques are required. These techniques are discussed in Sections 5.4 and 5.5.

My interactive video summary

Example 3 Evaluate Logarithmic Expressions

Evaluate each logarithm:

   a. $\log_5 25$ **b.** $\log_3 \dfrac{1}{27}$ **c.** $\log_{\sqrt{2}} \dfrac{1}{4}$

Solution

   a. To evaluate $\log_5 25$, we must ask, "5 raised to what exponent is 25?" Because $5^2 = 25$, $\log_5 25 = 2$.

   b. The expression $\log_3 \frac{1}{27}$ requires more analysis. In this case, we ask, "3 raised to what exponent is $\frac{1}{27}$?" Suppose we let y equal this exponent. Then $3^y = \frac{1}{27}$. To solve for y, we can use the **method of relating the bases** for solving exponential equations.

$$3^y = \frac{1}{27} \qquad \text{\textit{Write the exponential equation.}}$$

$$3^y = \frac{1}{3^3} \qquad \text{\textit{Rewrite 27 as } } 3^3.$$

$$3^y = 3^{-3} \qquad \text{\textit{Use }} \frac{1}{b^n} = b^{-n}.$$

$$y = -3 \qquad \text{\textit{Use the method of relating the bases.}}$$
$$\text{\textit{(If } } b^u = b^v, \text{\textit{ then } } u = v.)$$

Thus, $\log_3 \frac{1}{27} = -3$.

c. Watch the interactive video to verify that $\log_{\sqrt{2}} \frac{1}{4} = -4$ and to see each solution worked out in detail.

Work exercises 12–18 in this textbook or in the *MyMathLab*® Study Plan.

OBJECTIVE 3:   UNDERSTANDING THE PROPERTIES OF LOGARITHMS

Because $b^1 = b$ for any real number b, we can use the definition of the logarithmic function ($y = \log_b x$ if and only if $x = b^y$) to rewrite this expression as $\log_b b = 1$. Similarly, because $b^0 = 1$ for any real number b, we can rewrite this expression as $\log_b 1 = 0$. These two general properties are summarized as follows.

General Properties of Logarithms

For $b > 0$ and $b \neq 1$,

1. $\log_b b = 1$ and
2. $\log_b 1 = 0$.

In Section 3.6, we see that a function f and its inverse function f^{-1} satisfy the following two composition cancellation equations:

$$f(f^{-1}(x)) = x \text{ for all } x \text{ in the domain of } f^{-1} \text{ and}$$
$$f^{-1}(f(x)) = x \text{ for all } x \text{ in the domain of } f.$$

If $f(x) = b^x$, then $f^{-1}(x) = \log_b x$. Applying the two composition cancellation equations we get

$$f(f^{-1}(x)) = b^{\log_b x} = x \text{ and}$$
$$f^{-1}(f(x)) = \log_b b^x = x.$$

Cancellation Properties of Exponentials and Logarithms

For $b > 0$ and $b \neq 1$,

1. $b^{\log_b x} = x$ and
2. $\log_b b^x = x$.

Example 4 Use the Properties of Logarithms

Use the properties of logarithms to evaluate each expression.

   a. $\log_3 3^4$ **b.** $\log_{12} 12$ **c.** $7^{\log_7 13}$ **d.** $\log_8 1$

Solution

   a. By the second cancellation property, $\log_3 3^4 = 4$.
   b. Because $\log_b b = 1$ for all $b > 0$ and $b \neq 1$, $\log_{12} 12 = 1$.
   c. By the first cancellation property, $7^{\log_7 13} = 13$.
   d. Because $\log_b 1 = 0$ for all $b > 0$ and $b \neq 1$, $\log_8 1 = 0$.

Work exercises 19–26 in this textbook or in the *MyMathLab*® Study Plan.

OBJECTIVE 4: USING THE COMMON AND NATURAL LOGARITHMS

There are two bases that are used more frequently than any other base. They are base 10 and base e. (Refer to Section 5.2 to review the natural base e). Because our counting system is based on the number 10, the base 10 logarithm is known as the **common logarithm**. Instead of using the notation $\log_{10} x$ to denote the common logarithm, it is usually abbreviated without the subscript 10 as simply $\log x$. The base e logarithm is called the **natural logarithm** and is abbreviated as $\ln x$ instead of $\log_e x$. Most scientific calculators are equipped with a $\boxed{\log}$ key and a $\boxed{\ln}$ key. We can apply the definition of the logarithmic function for the base 10 and for the base e logarithm as follows.

DEFINITION Common Logarithmic Function

For $x > 0$, the **common logarithmic function** is defined by
$$y = \log x \text{ if and only if } x = 10^y.$$

DEFINITION Natural Logarithmic Function

For $x > 0$, the **natural logarithmic function** is defined by
$$y = \ln x \text{ if and only if } x = e^y.$$

My video summary

Example 5 Change from Exponential Form to Logarithmic Form

Write each exponential equation as an equation involving a common logarithm or natural logarithm.

a. $e^0 = 1$ **b.** $10^{-2} = \dfrac{1}{100}$ **c.** $e^K = w$

Solution

a. $e^0 = 1$ is equivalent to $\ln 1 = 0$.
b. $10^{-2} = \frac{1}{100}$ is equivalent to $\log\left(\frac{1}{100}\right) = -2$.
c. $e^K = w$ is equivalent to $\ln w = K$.

Watch the video to see this example worked out in more detail.

Work exercises 27–31 in this textbook or in the *MyMathLab*® Study Plan.

My video summary

Example 6 Change from Logarithmic Form to Exponential Form

Write each logarithmic equation as an equation involving an exponent.

a. $\log 10 = 1$ **b.** $\ln 20 = Z$ **c.** $\log(x - 1) = T$

Solution

a. $\log 10 = 1$ is equivalent to $10^1 = 10$.
b. $\ln 20 = Z$ is equivalent to $e^Z = 20$.
c. $\log(x - 1) = T$ is equivalent to $10^T = x - 1$.

Watch the video to see this example worked out in more detail.

Work exercises 32–35 in this textbook or in the *MyMathLab*® Study Plan.

My video summary

Example 7 Evaluate Common and Natural Logarithmic Expressions

Evaluate each expression without the use of a calculator.

a. $\log 100$ **b.** $\ln \sqrt{e}$ **c.** $e^{\ln 51}$ **d.** $\log 1$

Solution

a. $\log 100 = 2$ because $10^2 = 100$ by the definition of the logarithmic function or $\log 100 = \log 10^2 = 2$ by cancellation property **(2)**.

b. $\ln \sqrt{e} = \ln e^{1/2} = \frac{1}{2}$ by cancellation property **(2)**.

c. $e^{\ln 51} = 51$ by cancellation property **(1)**.

d. $\log 1 = 0$ by general property **(1)**.

Watch the video to see this example worked out in more detail.

Work exercises 36–43 in this textbook or in the *MyMathLab*® Study Plan.

OBJECTIVE 5: UNDERSTANDING THE CHARACTERISTICS OF LOGARITHMIC FUNCTIONS

To sketch the graph of a logarithmic function of the form $f(x) = \log_b x$, where $b > 0$ and $b \neq 1$, follow these three steps:

Step 1 Start with the graph of the exponential function $y = b^x$, labeling several ordered pairs.

Step 2 Because $f(x) = \log_b x$ is the inverse of $y = b^x$, we can find several points on the graph of $f(x) = \log_b x$ by reversing the coordinates of the ordered pairs of $y = b^x$.

Step 3 Plot the ordered pairs from step 2, and complete the graph of $f(x) = \log_b x$ by connecting the ordered pairs with a smooth curve. The graph of $f(x) = \log_b x$ is a reflection of the graph of $y = b^x$ about the line $y = x$.

My video summary

Example 8 Sketch the Graph of a Logarithmic Function

Sketch the graph of $f(x) = \log_3 x$.

Solution

Step 1 The graph of $y = 3^x$ passes through the points $\left(-1, \frac{1}{3}\right)$, $(0, 1)$, and $(1, 3)$. See Figure 10.

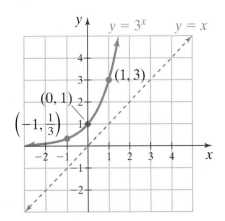

Figure 10 Graph of $y = 3^x$

Step 2 We reverse the three ordered pairs from step 1 to obtain the following three points: $\left(\frac{1}{3}, -1\right)$, $(1, 0)$, and $(3, 1)$.

Step 3 Plot the points $\left(\frac{1}{3}, -1\right)$, $(1, 0)$, and $(3, 1)$, and connect them with a smooth curve to obtain the graph of $f(x) = \log_3 x$. See Figure 11.

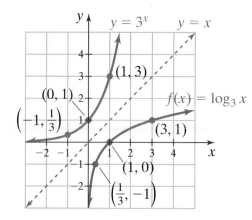

Figure 11 Graph of $y = 3^x$ and $f(x) = \log_3 x$

Notice in Figure 11 that the y-axis is a **vertical asymptote** of the graph of $f(x) = \log_3 x$. Every logarithmic function of the form $y = \log_b x$, where $b > 0$ and $b \neq 1$ has a vertical asymptote at the y-axis. The graphs and the characteristics of logarithmic functions are outlined as follows.

Characteristics of Logarithmic Functions

For $b > 0$, $b \neq 1$, the logarithmic function with base b is defined by $y = \log_b x$. The domain of $f(x) = \log_b x$ is $(0, \infty)$, and the range is $(-\infty, \infty)$. The graph of $f(x) = \log_b x$ has one of the following two shapes.

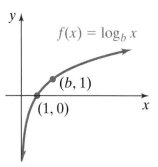

$$f(x) = \log_b x, \, b > 1$$

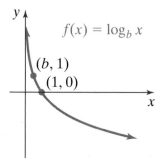

$$f(x) = \log_b x, \, 0 < b < 1$$

The graph intersects the x-axis at $(1, 0)$.
The graph contains the point $(b, 1)$.
The graph is **increasing** on the interval $(0, \infty)$.
The y-axis ($x = 0$) is a **vertical asymptote**.
The function is **one-to-one**.

The graph intersects the x-axis at $(1, 0)$.
The graph contains the point $(b, 1)$.
The graph is **decreasing** on the interval $(0, \infty)$.
The y-axis ($x = 0$) is a **vertical asymptote**.
The function is **one-to-one**.

Work exercises 44 and 45 in this textbook or in the *MyMathLab*® Study Plan.

OBJECTIVE 6:   SKETCHING THE GRAPHS OF LOGARITHMIC FUNCTIONS
USING TRANSFORMATIONS

Often we can use the **transformation techniques** that are discussed in Section 3.4 to sketch the graph of logarithmic functions.

My video summary

⊞ **Example 9** Use Transformations to Sketch the Graph of a
Logarithmic Function

Sketch the graph of $f(x) = -\ln(x + 2) - 1$.

Solution Recall that the function $y = \ln x$ has a base of e, where $2 < e < 3$. This means that the graph of $y = \ln x$ is increasing on the interval $(0, \infty)$ and contains the points $(1, 0)$ and $(e, 1)$. Starting with the graph of $y = \ln x$, we can obtain the graph of $f(x) = -\ln(x + 2) - 1$ through the following series of transformations:

1. Shift the graph of $y = \ln x$ horizontally to the left two units to obtain the graph of $y = \ln(x + 2)$.

2. Reflect the graph of $y = \ln(x + 2)$ about the x-axis to obtain the graph of $y = -\ln(x + 2)$.

3. Shift the graph of $y = -\ln(x + 2)$ vertically down one unit to obtain the final graph of $f(x) = -\ln(x + 2) - 1$.

The graph of $f(x) = -\ln(x + 2) - 1$ is sketched in Figure 12. You can see from the graph that the domain of $f(x) = -\ln(x + 2) - 1$ is $(-2, \infty)$. The vertical asymptote is $x = -2$, and the x-intercept is $\left(\frac{1}{e} - 2, 0\right)$. Watch the video to see each step worked out in detail.

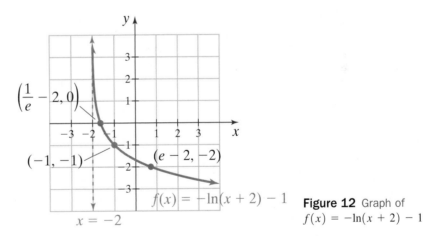

Figure 12 Graph of
$f(x) = -\ln(x + 2) - 1$

Work exercises **46–51** in this textbook or in the **MyMathLab**® Study Plan.

OBJECTIVE 7:   FINDING THE DOMAIN OF LOGARITHMIC FUNCTIONS

In Example 9, we sketch the function $f(x) = -\ln(x + 2) - 1$ and observe that the domain was $(-2, \infty)$. We do not have to sketch the graph of a logarithmic function to determine the domain. The domain of a logarithmic function consists of all values of x for which the **argument** of the logarithm is greater than zero. In other words, if $f(x) = \log_b[g(x)]$, then the domain of f can be found by solving the inequality $g(x) > 0$. For example, given the function $f(x) = -\ln(x + 2) - 1$ from Example 9, we can determine the domain by solving the linear inequality

$x + 2 > 0$. Solving this inequality for x, we obtain $x > -2$. Thus, the domain of $f(x) = -\ln(x + 2) - 1$ is $(-2, \infty)$.

Example 10 is a bit more challenging because the argument of the logarithm is a **rational expression**.

My interactive video summary

🎥 **Example 10 Find the Domain of a Logarithmic Function with a Rational Argument**

Find the domain of $f(x) = \log_5\left(\frac{2x - 1}{x + 3}\right)$.

Solution To find the domain of f, we must find all values of x for which the argument $\frac{2x - 1}{x + 3}$ is greater than zero. That is, you must solve the rational inequality $\left(\frac{2x - 1}{x + 3}\right) > 0$. See Section 1.9 if you need help remembering how to solve this inequality. By the techniques discussed in Section 1.9, we find that the solution to $\left(\frac{2x - 1}{x + 3}\right) > 0$ is $x < -3$ or $x > \frac{1}{2}$. Therefore, the domain of $f(x) = \log_5\left(\frac{2x - 1}{x + 3}\right)$ in set notation is $\left\{x \mid x < -3 \text{ or } x > \frac{1}{2}\right\}$. In interval notation, the domain is $(-\infty, -3) \cup \left(\frac{1}{2}, \infty\right)$. Watch the interactive video to see this problem worked out in detail.

Work exercises 52–58 in this textbook or in the *MyMathLab*® Study Plan.

5.3 Exercises

In Exercises 1–6, write each exponential equation as an equation involving a logarithm.

1. $3^2 = 9$

2. $16^{1/2} = 4$

3. $2^{-3} = \dfrac{1}{8}$

4. $\sqrt{2}^{\pi} = W$

5. $\left(\dfrac{1}{3}\right)^t = 27$

6. $7^{5k} = L$

In Exercises 7–11, write each logarithmic equation as an exponential equation.

7. $\log_5 1 = 0$

8. $\log_7 343 = 3$

9. $\log_{\sqrt{2}} 8 = 6$

10. $\log_4 K = L$

11. $\log_a(x - 1) = 3$

In Exercises 12–18, evaluate each logarithm without the use of a calculator.

12. $\log_2 8$

13. $\log_6 \sqrt{6}$

14. $\log_3 \dfrac{1}{9}$

15. $\log_{\sqrt{5}} 25$

16. $\log_4\left(\dfrac{1}{\sqrt[5]{64}}\right)$

17. $\log_{1/7} \sqrt[3]{7}$

18. $\log_{0.1} 100$

In Exercises 19–26, use the properties of logarithms to evaluate each expression without the use of a calculator.

19. $2^{\log_2 11}$

20. $\log_4 4$

21. $\log_9 1$

22. $\log_7 7^{-3}$

23. $\log_a a, a > 1$

24. $5^{\log_5 M}, M > 0$

25. $\log_y 1, y > 0$

26. $\log_x x^{20}, x > 1$

In Exercises 27–31, write each exponential equation as an equation involving a common logarithm or a natural logarithm.

27. $10^3 = 1,000$

28. $e^{-1} = \dfrac{1}{e}$

29. $e^k = 2$

30. $10^e = M$

31. $e^{10} = Z$

In Exercises 32–35, write each logarithmic equation as an exponential equation.

32. $\ln 1 = 0$

33. $\log(1,000,000) = 6$

34. $\log K = L$

35. $\ln Z = 4$

In Exercises 36–43, evaluate each expression without the use of a calculator, and then verify your answer using a calculator.

36. $\log 10,000$

37. $\log\left(\dfrac{1}{1,000}\right)$

38. $\ln 1$

39. $\ln \sqrt[3]{e^2}$

40. $10^{\log e}$

41. $e^{\ln 49}$

42. $\log 10^6$

43. $\ln e + \ln e^3$

In Exercises 44–51, sketch each logarithmic function. Label at least two points on the graph, and determine the domain and the equation of any vertical asymptotes.

44. $h(x) = \log_4 x$

45. $g(x) = \log_{\frac{1}{3}} x$

46. $f(x) = \log_2(x) - 1$

47. $f(x) = \log_5(x - 1)$

48. $f(x) = -\ln(x)$

49. $y = \log_{1/2}(x + 1) + 2$

50. $y = \log_3(1 - x)$

51. $h(x) = -\dfrac{1}{2}\log_3(x + 3) + 1$

In Exercises 52–58, find the domain of each logarithmic function.

52. $f(x) = \log(-x)$

53. $f(x) = \log_{1/4}(2x + 6)$

54. $f(x) = \ln(1 - 3x)$

55. $f(x) = \log_2(x^2 - 9)$

56. $f(x) = \log_7(x^2 - x - 20)$

57. $f(x) = \ln\left(\dfrac{x + 5}{x - 8}\right)$

58. $f(x) = \log\left(\dfrac{x^2 - x - 6}{x + 10}\right)$

5.4 Properties of Logarithms

THINGS TO KNOW

Before working through this section, be sure that you are familiar with the following concepts:

VIDEO   ANIMATION   INTERACTIVE

 1. Solving Exponential Equations by Relating the Bases (Section 5.1)
You Try It

 2. Change from Exponential Form to Logarithmic Form (Section 5.3)
You Try It

 3. Change from Logarithmic Form to Exponential Form (Section 5.3)
You Try It

 4. Evaluating Logarithmic Expressions (Section 5.3)
You Try It

 5. Using the Common and Natural Logarithms (Section 5.3)
You Try It

 6. Finding the Domain of Logarithmic Functions (Section 5.3)
You Try It

OBJECTIVES

1 Using the Product Rule, Quotient Rule, and Power Rule for Logarithms

2 Expanding and Condensing Logarithmic Expressions

3 Solving Logarithmic Equations Using the Logarithm Property of Equality

4 Using the Change of Base Formula

OBJECTIVE 1:   USING THE PRODUCT RULE, QUOTIENT RULE, AND POWER RULE FOR LOGARITHMS

In this section, we learn how to manipulate logarithmic expressions using properties of logarithms. Understanding how to use these properties will help us solve exponential and logarithmic equations that are encountered in the next section. Recall from Section 5.3 the **general properties** and **cancellation properties** of logarithms. We now look at three additional properties of logarithms.

Properties of Logarithms

If $b > 0$, $b \neq 1$, u and v represent positive numbers and r is any real number, then

$$\log_b uv = \log_b u + \log_b v \qquad \text{product rule for logarithms}$$

$$\log_b \frac{u}{v} = \log_b u - \log_b v \qquad \text{quotient rule for logarithms}$$

$$\log_b u^r = r \log_b u \qquad \text{power rule for logarithms}$$

To prove the product rule and quotient rule for logarithms, we use properties of exponents and the **method of relating the bases** to solve exponential equations. The power rule for logarithms is a direct result of the product rule. Click on a video proof link above to see a proof of one or more of these properties.

My video summary

 Example 1 Use the Product Rule

Use the product rule for logarithms to expand each expression. Assume $x > 0$.

  a. $\ln(5x)$ **b.** $\log_2(8x)$

Solution

  a. $\ln(5x) = \ln 5 + \ln x$ *Use the product rule for logarithms.*

  b. $\log_2(8x) = \log_2 8 + \log_2 x$ *Use the product rule for logarithms.*

         $= \ \ \ 3 \ \ \ + \log_2 x$ *Use the definition of the logarithmic function to rewrite $\log_2 8$ as 3 because $2^3 = 8$.*

CAUTION $\log_b(u + v)$ is *not* equivalent to $\log_b u + \log_b v$.

Example 2 Use the Quotient Rule

Use the quotient rule for logarithms to expand each expression. Assume $x > 0$.

  a. $\log_5\left(\dfrac{12}{x}\right)$ **b.** $\ln\left(\dfrac{x}{e^5}\right)$

Solution

  a. $\log_5\left(\dfrac{12}{x}\right) = \log_5 12 - \log_5 x$ *Use the quotient rule for logarithms.*

  b. $\ln\left(\dfrac{x}{e^5}\right) = \ln x - \ln e^5$ *Use the quotient rule for logarithms.*

         $= \ln x - 5$ *Use cancellation property (2) to rewrite $\ln e^5$ as 5.*

◇ CAUTION
$\log_b(u - v)$ is *not* equivalent to $\log_b u - \log_b v$, and $\dfrac{\log_b u}{\log_b v}$ is *not* equivalent to $\log_b u - \log_b v$.

My video summary
✎

▣ Example 3 Use the Power Rule

Use the power rule for logarithms to rewrite each expression. Assume $x > 0$.

a. $\log 6^3$ **b.** $\log_{1/2} \sqrt[4]{x}$

Solution

a. $\log 6^3 = 3 \log 6$ *Use the power rule for logarithms.*

b. $\log_{1/2} \sqrt[4]{x} = \log_{1/2} x^{1/4}$ *Rewrite the fourth root of x using a rational exponent.*

$= \dfrac{1}{4} \log_{1/2} x$ *Use the power rule for logarithms.*

The process of using the power rule to simplify a logarithmic expression is often casually referred to as "bringing down the exponent."

◇ CAUTION
$(\log_b u)^r$ is *not* equivalent to $r \log_b u$.

Work exercises 1–10 in this textbook or in the *MyMathLab*® Study Plan.

OBJECTIVE 2:   EXPANDING AND CONDENSING LOGARITHMIC EXPRESSIONS

Sometimes it is necessary to combine several properties of logarithms to expand a logarithmic expression into the sum and/or difference of logarithms or to condense several logarithms into a single logarithm.

My interactive video summary
✎

Example 4 Expand a Logarithmic Expression

Use properties of logarithms to expand each logarithmic expression as much as possible.

a. $\log_7(49x^3\sqrt{y^2})$ **b.** $\ln\!\left(\dfrac{(x^2 - 4)}{9e^{x^3}}\right)$

Solution

a. $\log_7(49x^3\sqrt{y^2})$ *Write the original expression.*

$= \log_7 49 + \log_7 x^3\sqrt{y^2}$ *Use the product rule.*

$= \log_7 49 + \log_7 x^3 + \log_7 \sqrt{y^2}$ *Use the product rule again.*

$= \log_7 49 + \log_7 x^3 + \log_7 y^{2/5}$ *Rewrite $\sqrt[5]{y^2}$ using a rational exponent.*

$= 2 + 3 \log_7 x + \dfrac{2}{5} \log_7 y$ *Rewrite $\log_7 49$ as 2 and use the power rule.*

b. $\ln\!\left(\dfrac{(x^2 - 4)}{9e^{x^3}}\right) = \ln\!\left(\dfrac{(x - 2)(x + 2)}{9e^{x^3}}\right)$ *Factor the expression in the numerator.*

$= \ln(x - 2)(x + 2) - \ln 9e^{x^3}$ *Use the quotient rule.*

$= \ln(x - 2) + \ln(x + 2) - \left[\ln 9 + \ln e^{x^3}\right]$ *Use the product rule twice.*

$$= \ln(x - 2) + \ln(x + 2) - \left[\ln 9 + x^3\right] \qquad \textit{Use cancellation property (2)}$$
$$\textit{to rewrite } \ln e^{x^3} \textit{ as } x^3.$$

$$= \ln(x - 2) + \ln(x + 2) - \ln 9 - x^3 \qquad \textit{Simplify.}$$

Watch the interactive video to see this example worked out in detail.

Work exercises 11–20 in this textbook or in the *MyMathLab*® Study Plan.

My interactive video summary

▶ Example 5 Condense a Logarithmic Expression

Use properties of logarithms to rewrite each expression as a single logarithm.

a. $\dfrac{1}{2}\log(x - 1) - 3\log z + \log 5$ b. $\dfrac{1}{3}(\log_3 x - 2\log_3 y) + \log_3 10$

Solution

a. $\dfrac{1}{2}\log(x - 1) - 3\log z + \log 5$ *Write the original expression.*

$= \log(x - 1)^{1/2} - \log z^3 + \log 5$ *Use the power rule twice.*

$= \log \dfrac{(x - 1)^{1/2}}{z^3} + \log 5$ *Use the quotient rule.*

$= \log \dfrac{5(x - 1)^{1/2}}{z^3}$ or $\log \dfrac{5\sqrt{x - 1}}{z^3}$ *Use the product rule.*

b. $\dfrac{1}{3}(\log_3 x - 2\log_3 y) + \log_3 10$ *Write the original expression.*

$= \dfrac{1}{3}(\log_3 x - \log_3 y^2) + \log_3 10$ *Use the power rule.*

$= \dfrac{1}{3}\log_3 \dfrac{x}{y^2} + \log_3 10$ *Use the quotient rule.*

$= \log_3\left(\dfrac{x}{y^2}\right)^{1/3} + \log_3 10$ *Use the power rule.*

$= \log_3\left[10\left(\dfrac{x}{y^2}\right)^{1/3}\right]$ or $\log_3\left[10\sqrt[3]{\dfrac{x}{y^2}}\right]$ *Use the product rule.*

Watch the interactive video to see this example worked out in detail.

Work exercises 21–30 in this textbook or in the *MyMathLab*® Study Plan.

OBJECTIVE 3:   SOLVING LOGARITHMIC EQUATIONS USING THE LOGARITHM PROPERTY OF EQUALITY

Remember that all logarithmic functions of the form $f(x) = \log_b x$ for $b > 0$ and $b \neq 1$ are **one-to-one**. In Section 3.6, the alternate definition of **one-to-one** stated that:

A function f is one-to-one if for any two range values $f(u)$ and $f(v)$, $f(u) = f(v)$ implies that $u = v$.

Using this definition and letting $f(x) = \log_b x$, we can say that if $\log_b u = \log_b v$, then $u = v$. In other words, if the bases of a logarithmic equation of the form $\log_b u = \log_b v$ are equal, then the **arguments** must be equal. This is known as the **logarithm property of equality**.

Logarithm Property of Equality

If a logarithmic equation can be written in the form $\log_b u = \log_b v$, then $u = v$.
Furthermore, if $u = v$, then $\log_b u = \log_b v$.

The second statement of the logarithm property of equality says that if we start with the equation $u = v$, then we can rewrite the equation as $\log_b u = \log_b v$. This process is often casually referred to as "taking the log of both sides."

My interactive video summary

▣ Example 6 Use the Logarithm Property of Equality to Solve Logarithmic Equations

Solve the following equations:

   a. $\log_7(x - 1) = \log_7 12$
   b. $2 \ln x = \ln 16$

Solution

   a. Because the base of each logarithm is 7, we can use the logarithm property of equality to eliminate the logarithms.

$$\log_7(x - 1) = \log_7 12 \quad \textit{Write the original equation.}$$
$$(x - 1) = 12 \quad \textit{If } \log_b u = \log_b v, \textit{ then } u = v.$$
$$x = 13 \quad \textit{Solve for } x.$$

   b. $2 \ln x = \ln 16$ *Write the original expression.*
   $\ln x^2 = \ln 16$ *Use the power rule.*
   $x^2 = 16$ *If $\log_b u = \log_b v$, then $u = v$.*
   $x = \pm 4$ *Use the square root property.*

The domain of $\ln x$ is $x > 0$; this implies that $x = -4$ is an **extraneous solution**, and hence, we must discard it. Therefore, this equation has only one solution, $x = 4$.

Work exercises 31–36 in this textbook or in the *MyMathLab*® Study Plan.

OBJECTIVE 4: USING THE CHANGE OF BASE FORMULA

Most scientific calculators are equipped with a $\boxed{\log}$ key and a $\boxed{\ln}$ key to evaluate common logarithms and natural logarithms. But how do we use a calculator to evaluate logarithmic expressions having a base other than 10 or e? The answer is to use the following **change of base formula**.

Change of Base Formula

For any positive base $b \neq 1$ and for any positive real number u, then

$$\log_b u = \frac{\log_a u}{\log_a b}$$

where a is any positive number such that $a \neq 1$.

My video summary

Click here to watch the video proof of the change of base formula.

The change of base formula allows us to change the base of a logarithmic expression into a ratio of two logarithms using any base we choose. For example, suppose we are given the logarithmic expression $\log_3 10$. We can use the change of base formula to write this logarithm as a logarithm involving any base we choose:

$$\log_3 10 = \frac{\log_7 10}{\log_7 3} \quad \text{or} \quad \log_3 10 = \frac{\log_2 10}{\log_2 3} \quad \text{or} \quad \log_3 10 = \frac{\log 10}{\log 3} \quad \text{or} \quad \log_3 10 = \frac{\ln 10}{\ln 3}$$

In each of the previous four cases, we introduced a new base (7, 2, 10, and e, respectively). However, if we want to use a calculator to get a numerical approximation of $\log_3 10$, then it really only makes sense to change $\log_3 10$ into an expression involving base 10 or base e because these are the only two bases most calculators can handle.

Note: $\log_3 10 = \dfrac{\log 10}{\log 3} \approx 2.0959$ or $\log_3 10 = \dfrac{\ln 10}{\ln 3} \approx 2.0959$

Example 7 Use the Change of Base Formula

Approximate the following expressions. Round each to four decimal places.

   a. $\log_9 200$ **b.** $\log_{\sqrt{3}} \pi$

Solution

   a. $\log_9 200 = \dfrac{\log 200}{\log 9} \approx 2.4114$ or $\log_9 200 = \dfrac{\ln 200}{\ln 9} \approx 2.4114$

   b. $\log_{\sqrt{3}} \pi = \dfrac{\log \pi}{\log \sqrt{3}} \approx 2.0840$ or $\log_{\sqrt{3}} \pi = \dfrac{\ln \pi}{\ln \sqrt{3}} \approx 2.0840$

Work exercises 37–40 in this textbook or in the *MyMathLab*®️ Study Plan.

My video summary

Example 8 Use the Change of Base Formula and Properties of Logarithms

Use the change of base formula and the properties of logarithms to rewrite as a single logarithm involving base 2.

$$\log_4 x + 3 \log_2 y$$

Solution To use properties of logarithms, the base of each logarithmic expression must be the same. We use the change of base formula to rewrite $\log_4 x$ as a logarithmic expression involving base 2:

$$\log_4 x + 3 \log_2 y = \frac{\log_2 x}{\log_2 4} + 3 \log_2 y \qquad \text{Use the change of base formula } \log_4 x = \frac{\log_2 x}{\log_2 4}.$$

$$= \frac{\log_2 x}{2} + 3 \log_2 y \qquad \text{Rewrite } \log_2 4 \text{ as 2 because } 2^2 = 4.$$

$$= \frac{1}{2} \log_2 x + 3 \log_2 y \qquad \text{Rewrite } \frac{\log_2 x}{2} \text{ as } \frac{1}{2} \log_2 x.$$

$$= \log_2 x^{1/2} + \log_2 y^3 \qquad \text{Use the power rule.}$$

$$= \log_2 x^{1/2} y^3 \text{ or } \log_2 \sqrt{x} y^3 \qquad \text{Use the product rule.}$$

Therefore, the expression $\log_4 x + 3\log_2 y$ is equivalent to $\log_2 \sqrt{xy^3}$. Note that we could have chosen to rewrite the original expression as a single logarithm involving base 4. Watch the video to see that the expression $\log_4 x + 3\log_2 y$ is also equivalent to $\log_4 xy^6$.

Work exercises 41–44 in this textbook or in the MyMathLab® Study Plan.

My video summary

■ **Example 9 Use the Change of Base Formula to Solve Logarithmic Equations**

Use the change of base formula and the properties of logarithms to solve the equation

$$2\log_3 x = \log_9 16$$

Solution Watch the video to see how the change of base formula and the power rule for logarithms can be used to solve this equation.

Work exercises 45–48 in this textbook or in the MyMathLab® Study Plan.

5.4 Exercises

In Exercises 1–10, use the product rule, quotient rule, or power rule to expand each logarithmic expression. Wherever possible, evaluate logarithmic expressions.

1. $\log_4 (xy)$ **2.** $\log\left(\dfrac{9}{t}\right)$ **3.** $\log_5 y^3$ **4.** $\log_3(27w)$ **5.** $\ln 5e^2$

6. $\log_9 \sqrt[4]{k}$ **7.** $\log 100P$ **8.** $\ln\left(\dfrac{e^5}{r}\right)$ **9.** $\log_{\sqrt{2}} 8x$ **10.** $\log_2\left(\dfrac{M}{32}\right)$

In Exercises 11–20, use the properties of logarithms to expand each logarithmic expression. Wherever possible, evaluate logarithmic expressions.

11. $\log_7 x^2 y^3$ **12.** $\ln \dfrac{a^2 b^3}{c^4}$ **13.** $\log \dfrac{\sqrt{x}}{10y^3}$ **14.** $\log_3 9(x^2 - 25)$

15. $\log_2 \sqrt{4xy}$ **16.** $\log_5 \dfrac{\sqrt{5x^5}}{\sqrt[3]{25y^4}}$ **17.** $\ln \dfrac{\sqrt[5]{ez}}{\sqrt{x-1}}$ **18.** $\log_3 \sqrt[4]{\dfrac{x^3 y^5}{9}}$

19. $\ln\left[\dfrac{x+1}{(x^2-1)^3}\right]^{2/3}$ **20.** $\log \dfrac{(10x)^3 \sqrt{x-4}}{(x^2-16)^5}$

In Exercises 21–30, use properties of logarithms to rewrite each expression as a single logarithm. Wherever possible, evaluate logarithmic expressions.

21. $\log_b A + \log_b C$ **22.** $\log_4 M - \log_4 N$ **23.** $2\log_8 x + \dfrac{1}{3}\log_8 y$

24. $\log 20 + \log 5$ **25.** $\log_2 80 - \log_2 5$ **26.** $\ln\sqrt{x} - \dfrac{1}{3}\ln x + \ln\sqrt[4]{x}$

27. $\log_5(x-2) + \log_5(x+2)$ **28.** $\log_9(x^2 - 5x + 6) - \log_9(x^2 - 4) + \log_9(x+2)$

29. $\log(x-3) + 2\log(x+3) - \log(x^3 + x^2 - 9x - 9)$ **30.** $\dfrac{1}{2}[\ln(x-1)^2 - \ln(2x^2 - x - 1)^4] + 2\ln(2x+1)$

In Exercises 31–36, use the properties of logarithms and the logarithm property of equality to solve each logarithmic equation.

31. $\log_3(2x+1) = \log_3 11$ **32.** $\log_{11}\sqrt{x} = \log_{11} 6$

33. $2\log(x+5) = \log 12 + \log 3$ **34.** $\ln 5 + \ln x = \ln 7 + \ln(3x - 2)$

35. $\log_7(x+6) - \log_7(x+2) = \log_7 x$ **36.** $\log_2(x+3) + \log_2(x-4) = \log_2(x+12)$

In Exercises 37–40, use the change of base formula and a calculator to approximate each logarithmic expression. Round your answers to four decimal places.

37. $\log_4 51$ **38.** $\log_7 0.8$ **39.** $\log_{1/5} 72$ **40.** $\log_{\sqrt{7}} 100$

In Exercises 41–44, use the change of base formula and the properties of logarithms to rewrite each expression as a single logarithm in the indicated base.

41. $\log_3 x + 4 \log_9 w$, base 3 **42.** $\log_5 x + \log_{1/5} x^3$, base 5

43. $\log_{16} x^4 + \log_8 y^3 + \log_4 w^2$, base 2 **44.** $\log_{e^2} x^5 + \log_{e^3} x^6 + \log_{e^4} x^{12}$, base e

In Exercises 45–48, solve each logarithmic equation.

45. $\log_2 x = \log_4 25$ **46.** $\log_{1/3} x = \log_3 20$ **47.** $\log_5 x = \log_{\sqrt{5}} 6$ **48.** $2 \ln x = \log_{e^3} 125$

5.5 Exponential and Logarithmic Equations

THINGS TO KNOW

Before working through this section, be sure that you are familiar with the following concepts:

VIDEO   ANIMATION   INTERACTIVE

 1. Solving Exponential Equations by Relating the Bases (Section 5.1)

 2. Changing from Exponential to Logarithmic Form (Section 5.3)

 3. Changing from Logarithmic to Exponential Form (Section 5.3)

 4. Using the Cancellation Properties of Exponentials and Logarithms (Section 5.3)

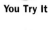 **5.** Expanding and Condensing Logarithmic Expressions (Section 5.4)

 6. Solving Logarithmic Equations Using the Logarithm Property of Equality (Section 5.4)

OBJECTIVES

1 Solving Exponential Equations

2 Solving Logarithmic Equations

In this section, we learn how to solve exponential and logarithmic equations. The techniques and strategies learned in this section help us solve applied problems that are discussed in Section 5.6. We start by developing a strategy to solve exponential equations.

OBJECTIVE 1:   SOLVING EXPONENTIAL EQUATIONS

We have already solved exponential equations using the **method of relating the bases**. For example, we can solve the equation $4^{x+3} = \frac{1}{2}$ by converting the base on both sides of the equation to base 2. Click here to see how to solve this equation.

But suppose we are given an exponential equation in which the bases cannot be related, such as $2^{x+1} = 3$. Remember in Section 5.1, Example 3, we wanted to find the one x-intercept of the graph of $f(x) = -2^{x+1} + 3$ (Figure 13).

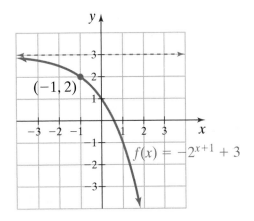

Figure 13 Graph of $f(x) = -2^{x+1} + 3$

To find the x-intercept of $f(x) = -2^{x+1} + 3$, we need to set $f(x) = 0$ and solve for x.

$$f(x) = -2^{x+1} + 3$$
$$0 = -2^{x+1} + 3$$
$$2^{x+1} = 3$$

In Section 5.1, we could not solve this equation for x because we had not yet defined the logarithm. We can now use some properties of logarithms to solve this equation. Recall the following logarithmic properties.

If $u = v$, then $\log_b u = \log_b v$.   *logarithm property of equality*

$\log_b u^r = r \log_b u$ *power rule for logarithms*

We can use these two properties to solve the equation $2^{x+1} = 3$ and thus determine the x-intercept of $f(x) = -2^{x+1} + 3$. We solve the equation $2^{x+1} = 3$ in Example 1.

My video summary

▦ Example 1 Solve an Exponential Equation

Solve $2^{x+1} = 3$.

Solution

$2^{x+1} = 3$	*Write the original equation.*
$\ln 2^{x+1} = \ln 3$	*Use the logarithm property of equality.*
$(x + 1)\ln 2 = \ln 3$	*Use the power rule for logarithms.*
$x \ln 2 + \ln 2 = \ln 3$	*Use the distributive property.*
$x \ln 2 = \ln 3 - \ln 2$	*Subtract $\ln 2$ from both sides.*
$x = \dfrac{\ln 3 - \ln 2}{\ln 2}$	*Divide both sides by $\ln 2$.*

5.5 Exponential and Logarithmic Equations   259

The solution to Example 1 verifies that the x-intercept of $f(x) = -2^{x+1} + 3$ is $x = \frac{\ln 3 - \ln 2}{\ln 2} \approx .5850$. See Figure 14.

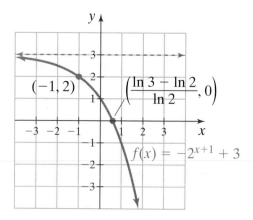

Figure 14 Graph of $f(x) = -2^{x+1} + 3$

When we cannot easily relate the bases of an exponential equation, as in Example 1, we use logarithms and their properties to solve them. The methods used to solve exponential equations are outlined as follows.

Solving Exponential Equations

- If the equation can be written in the form $b^u = b^v$, then solve the equation $u = v$.

- If the equation cannot easily be written in the form $b^u = b^v$,
  1. Use the logarithm property of equality to "take the log of both sides" (typically using base 10 or base e).
  2. Use the product rule of logarithms to "bring down" any exponents.
  3. Solve for the given variable.

My interactive video summary

▶ Example 2 Solve Exponential Equations

Solve each equation. For part b, round to four decimal places.

a. $3^{x-1} = \left(\dfrac{1}{27}\right)^{2x+1}$ b. $7^{x+3} = 4^{2-x}$

Solution

a. Watch the interactive video, or click here to see that the solution is $x = -\frac{2}{7}$.

b. We cannot easily use the method of relating the bases because we cannot easily write both 7 and 4 using a common base. Therefore, we use logarithms to solve.

$$7^{x+3} = 4^{2-x}$$ *Write the original equation.*

$$\ln 7^{x+3} = \ln 4^{2-x}$$ *Use the logarithm property of equality.*

$$(x + 3)\ln 7 = (2 - x)\ln 4$$ *Use the power rule for logarithms.*

$$x \ln 7 + 3 \ln 7 = 2 \ln 4 - x \ln 4$$ *Use the distributive property.*

$$x \ln 7 + x \ln 4 = 2 \ln 4 - 3 \ln 7$$ *Add $x \ln 4$ to both sides, and subtract 3 ln 7 from both sides.*

$$x(\ln 7 + \ln 4) = 2 \ln 4 - 3 \ln 7 \qquad \textit{Factor out an } x \textit{ from the left-hand side.}$$

$$x = \frac{2 \ln 4 - 3 \ln 7}{\ln 7 + \ln 4} \qquad \textit{Divide both sides by } \ln 7 + \ln 4.$$

$$= \frac{\ln 16 - \ln 343}{\ln 28} \qquad \textit{Use the power rule for logarithms in the numerator, and use the product rule for logarithms in the denominator.}$$

$$= \frac{\ln\left(\dfrac{16}{343}\right)}{\ln 28} \qquad \textit{Use the quotient rule for logarithms to rewrite } \ln 16 - \ln 343 \textit{ as } \ln\left(\dfrac{16}{343}\right).$$

$$\approx -.9199 \qquad \textit{Use a calculator to round to four decimal places.}$$

Work exercises 1–10 in this textbook or in the MyMathLab® Study Plan.

My interactive video summary

▶ Example 3 Solve Exponential Equations Involving the Natural Exponential Function

Solve each equation. Round to four decimal places.

a. $25e^{x-5} = 17$　　　**b.** $e^{2x-1} \cdot e^{x+4} = 11$

Solution

a. Isolate the exponential term on the left by dividing both sides of the equation by 25.

$$25e^{x-5} = 17 \qquad \textit{Write the original equation.}$$

$$e^{x-5} = \frac{17}{25} \qquad \textit{Divide both sides by 25.}$$

$$\ln e^{x-5} = \ln \frac{17}{25} \qquad \textit{Use the logarithm property of equality.}$$

$$x - 5 = \ln \frac{17}{25} \qquad \textit{Use cancellation property (2) to rewrite } \ln e^{x-5} \textit{ as } x - 5.$$

$$x = \ln \frac{17}{25} + 5 \qquad \textit{Add 5 to both sides.}$$

$$\approx 4.6143 \qquad \textit{Use a calculator to round to four decimal places.}$$

b.
$$e^{2x-1} \cdot e^{x+4} = 11 \qquad \textit{Write the original equation.}$$

$$e^{(2x-1)+(x+4)} = 11 \qquad \textit{Use } b^m \cdot b^n = b^{m+n}.$$

$$e^{3x+3} = 11 \qquad \textit{Combine like terms in the exponent.}$$

$$\ln e^{3x+3} = \ln 11 \qquad \textit{Use the logarithm property of equality.}$$

$$3x + 3 = \ln 11 \qquad \textit{Use cancellation property (2) to rewrite } \ln e^{3x+3} \textit{ as } 3x + 3.$$

$$3x = \ln 11 - 3 \qquad \textit{Subtract 3 from both sides.}$$

$$x = \frac{\ln 11 - 3}{3} \qquad \textit{Divide both sides by 3.}$$

$$\approx -.2007 \qquad \textit{Use a calculator to round to four decimal places.}$$

5.5 Exponential and Logarithmic Equations　**261**

Watch the interactive video to see the solutions to this example worked out in detail.

Work exercises 11–15 in this textbook or in the *MyMathLab*® Study Plan.

OBJECTIVE 2:   SOLVING LOGARITHMIC EQUATIONS

We now turn our attention to solving logarithmic equations. In Section 5.4, we learn how to solve certain logarithmic equations by using the **logarithm property of equality**. That is, if we can write a logarithmic equation in the form $\log_b u = \log_b v$, then $u = v$. Before we look at an example, let's review three of the properties of logarithms.

Properties of Logarithms

If $b > 0$, $b \neq 1$, u and v represent positive numbers and r is any real number, then

$$\log_b uv = \log_b u + \log_b v \qquad \textit{product rule for logarithms}$$

$$\log_b \frac{u}{v} = \log_b u - \log_b v \qquad \textit{quotient rule for logarithms}$$

$$\log_b u^r = r \log_b u \qquad \textit{power rule for logarithms}$$

My video summary

🎞 **Example 4   Solve a Logarithmic Equation Using the Logarithm Property of Equality**

Solve $2 \log_5(x - 1) = \log_5 64$.

Solution We can use the power rule for logarithms and the logarithmic property of equality to solve.

$$
\begin{aligned}
2 \log_5(x - 1) &= \log_5 64 & &\textit{Write the original equation.} \\
\log_5(x - 1)^2 &= \log_5 64 & &\textit{Use the power rule.} \\
(x - 1)^2 &= 64 & &\textit{Use the logarithm property of equality.} \\
x - 1 &= \pm 8 & &\textit{Use the square root property.} \\
x &= 1 \pm 8 & &\textit{Solve for } x. \\
x = 9 \quad \text{or} \quad x &= -7 & &\textit{Simplify.}
\end{aligned}
$$

Recall that the domain of a logarithmic function must contain only positive numbers; thus, $x - 1$ must be positive. Therefore, the solution of $x = -7$ must be discarded. The only solution is $x = 9$. You may want to review how to determine the domain of a logarithmic function, which is discussed in Section 5.3.

⬦ CAUTION   When solving logarithmic equations, it is important to always verify the solutions. Logarithmic equations often lead to **extraneous solutions**, as in Example 4.

When a logarithmic equation cannot be written in the form $\log_b u = \log_b v$, as in Example 4, we adhere to the steps outlined as follows:

Solving Logarithmic Equations

1. Determine the domain of the variable.

2. Use properties of logarithms to combine all logarithms, and write as a single logarithm, if needed.

---

3. Eliminate the logarithm by rewriting the equation in exponential form. Click here to review how to change from logarithmic form to exponential form, which is discussed in Section 5.3.
4. Solve for the given variable.
5. Check for any extraneous solutions. Verify that each solution is in the domain of the variable.

Work exercises 16–19 in this textbook or in the *MyMathLab*® Study Plan.

My video summary

▦ Example 5 Solve a Logarithmic Equation

Solve $\log_4(2x - 1) = 2$.

Solution The domain of the variable in this equation is the solution to the inequality $2x - 1 > 0$ or $x > \frac{1}{2}$. Thus, our solution must be greater than $\frac{1}{2}$. Because the equation involves a single logarithm, we can proceed to the third step.

$$\log_4(2x - 1) = 2 \qquad \text{Write the original equation.}$$
$$4^2 = 2x - 1 \qquad \text{Rewrite in exponential form.}$$
$$16 = 2x - 1 \qquad \text{Simplify.}$$
$$17 = 2x \qquad \text{Add 1 to both sides.}$$
$$x = \frac{17}{2} \qquad \text{Divide by 2.}$$

Because the solution satisfies the inequality, there are no extraneous solutions. We can verify the solution by substituting $x = \frac{17}{2}$ into the original equation.

Check:
$$\log_4(2x - 1) = 2 \qquad \text{Write the original equation.}$$
$$\log_4\left(2\left(\frac{17}{2}\right) - 1\right) \overset{?}{=} 2 \qquad \text{Substitute } x = \frac{17}{2}.$$
$$\log_4(17 - 1) \overset{?}{=} 2 \qquad \text{Simplify.}$$
$$\log_4(16) = 2 \qquad \text{This is a true statement because } 4^2 = 16.$$

Work exercises 20–24 in this textbook or in the *MyMathLab*® Study Plan.

My interactive video summary

🎦 Example 6 Solve a Logarithmic Equation

Solve $\log_2(x + 10) + \log_2(x + 6) = 5$.

Solution The domain of the variable in this equation is the solution to the compound inequality $x + 10 > 0$ and $x + 6 > 0$. The solution to this compound inequality is $x > -6$. (You may want to review compound inequalities from Section 1.7.)

$$\log_2(x + 10) + \log_2(x + 6) = 5 \qquad \text{Write the original equation.}$$
$$\log_2(x + 10)(x + 6) = 5 \qquad \text{Use the product rule.}$$
$$(x + 10)(x + 6) = 2^5 \qquad \text{Rewrite in exponential form.}$$
$$x^2 + 16x + 60 = 32 \qquad \text{Simplify.}$$
$$x^2 + 16x + 28 = 0 \qquad \text{Subtract 32 from both sides.}$$
$$(x + 14)(x + 2) = 0 \qquad \text{Factor.}$$
$$x = -14 \quad \text{or} \quad x = -2 \qquad \text{Use the zero product property to solve.}$$

Because the domain of the variable is $x > -6$, we must *exclude* the solution $x = -14$. Therefore, the only solution to this logarithmic equation is $x = -2$. Work through the interactive video to see this solution worked out in detail.

Work exercises 25–30 in this textbook or in the *MyMathLab*® Study Plan.

Example 7 Solve a Logarithmic Equation

Solve $\ln(x - 4) - \ln(x - 5) = 2$. Round to four decimal places.

Solution The domain of the variable is the solution to the compound inequality $x - 4 > 0$ and $x - 5 > 0$. The solution to this compound inequality is $x > 5$. (You may want to review compound inequalities from Section 1.7.)

$\ln(x - 4) - \ln(x - 5) = 2$	*Write the original equation.*
$\ln\left(\dfrac{x - 4}{x - 5}\right) = 2$	*Use the quotient rule.*
$e^2 = \dfrac{x - 4}{x - 5}$	*Rewrite in exponential form.*
$e^2(x - 5) = x - 4$	*Multiply both sides by $x - 5$.*
$e^2 x - 5e^2 = x - 4$	*Use the distributive property.*
$e^2 x - x = 5e^2 - 4$	*Add $5e^2$ to both sides and subtract x from both sides.*
$x(e^2 - 1) = 5e^2 - 4$	*Factor out an x from the left-hand side.*
$x = \dfrac{5e^2 - 4}{e^2 - 1} \approx 5.1565$	*Solve for x. Use a calculator to round to four decimal places.*

We approximate the exact answer $x = \frac{5e^2 - 4}{e^2 - 1}$ in order to verify that the solution is in the domain of the variable. In this example, we see that 5.1565 is clearly greater than 5. In some cases, we may need to use the exact answer to verify a solution to a logarithmic equation. Click here to see this verification for Example 7.

Work exercises 31 and 32 in this textbook or in the *MyMathLab*® Study Plan.

5.5 Exercises

In Exercises 1–15, solve each exponential equation. For irrational solutions, round to four decimal places.

1. $3^x = 5$

2. $2^{x/3} = 19$

3. $4^{x^2 - 2x} = 64$

4. $3^{x+7} = -20$

5. $(1.52)^{-3x/7} = 11$

6. $\left(\dfrac{1}{5}\right)^{x-1} = 25^x$

7. $8^{4x-7} = 11^{5+x}$

8. $3(9)^{x-1} = (81)^{2x+1}$

9. $(3.14)^x = \pi^{1-2x}$

10. $7(2 - 10^{4x-2}) = 8$

11. $e^x = 2$

12. $150e^{x-4} = 5$

13. $e^{x-3} \cdot e^{3x+7} = 24$

14. $2(e^{x-1})^2 \cdot e^{3-x} = 80$

15. $8e^{-x/3} \cdot e^x = 1$

In Exercises 16–30, solve each logarithmic equation.

16. $\log_4(x + 1) = \log_4(6x - 5)$

17. $\log_3(x^2 - 21) = \log_3 4x$

18. $2\log_5(3 - x) - \log_5 2 = \log_5 18$

19. $2\ln x - \ln(2x - 3) = \ln 2x - \ln(x - 1)$

20. $\log_2(4x - 7) = 3$

21. $\log(1 - 5x) = 2$

22. $\log_3(2x - 5) = -2$

23. $\log_x 3 = -1$

24. $\log_{x/2} 16 = 2$

25. $\log_2(x - 2) + \log_2(x + 2) = 5$

26. $\log_7(x + 9) + \log_7(x + 15) = 1$

27. $\log_7(3x + 1) - \log_2(x - 2) = 2$

28. $\log_4(x - 7) + \log_4 x = \dfrac{3}{2}$

29. $\ln 3 + \ln\left(x^2 + \dfrac{2x}{3}\right) = 0$

30. $\log_2(x - 4) + \log_2(x + 6) = 2 + \log_2 x$

In Exercises 31 and 32, solve each logarithmic equation. Round to four decimal places.

31. $\ln x - \ln(x + 6) = 1$

32. $\ln(x + 3) - \ln(x - 2) = 4$

5.6 Applications of Exponential and Logarithmic Functions

THINGS TO KNOW

Before working through this section, be sure that you are familiar with the following concepts:

VIDEO   ANIMATION   INTERACTIVE

You Try It
1. Solving Applications of Exponential Functions (Section 5.1)

You Try It
2. Solving Applications of the Natural Exponential Function (Section 5.2)

You Try It
3. Solving Applications of the Natural Exponential Function (Section 5.2)

You Try It
4. Understanding the Properties of Logarithms (Section 5.3)

OBJECTIVES

1 Solving Compound Interest Applications

2 Exponential Growth and Decay

3 Solving Logistic Growth Applications

4 Using Newton's Law of Cooling

We have seen that exponential functions appear in a wide variety of settings, including biology, chemistry, physics, and business. In this section, we revisit some applications that are discussed previously in this chapter and then introduce several new applications. The difference between the applications presented earlier and the applications presented in this section is that we are now equipped with the tools necessary to solve for variables that appear as exponents. We start with applications involving **compound interest**. You may want to review periodic compound interest from Section 5.1 and continuous compound interest from Section 5.2 before proceeding.

OBJECTIVE 1:   SOLVING COMPOUND INTEREST APPLICATIONS

In Sections 5.1 and 5.2, the formula for compound interest and continuous compound interest are defined as follows.

Compound Interest Formulas

Periodic Compound Interest Formula

$$A = P\left(1 + \frac{r}{n}\right)^{nt}$$

Continuous Compound Interest Formula

$$A = Pe^{rt}$$

where

A = Total amount after t years

P = Principal (original investment)

r = Interest rate per year

n = Number of times interest is compounded per year

t = Number of years

My video summary

▦ Example 1 Find the Doubling Time

How long will it take (in years and months) for an investment to double if it earns 7.5% compounded monthly?

Solution We use the periodic compound interest formula $A = P\left(1 + \frac{r}{n}\right)^{nt}$ with $r = .075$ and $n = 12$ and solve for t. Notice that the principal is not given. As it turns out, any value of P will suffice. If the principal is P, then the amount needed to double the investment is $A = 2P$. We now have all of the information necessary to solve for t:

$$A = P\left(1 + \frac{r}{n}\right)^{nt}$$ *Use the periodic compound interest formula.*

$$2P = P\left(1 + \frac{.075}{12}\right)^{12t}$$ *Substitute the appropriate values.*

$$2 = \left(1 + \frac{.075}{12}\right)^{12t}$$ *Divide both sides by P.*

$$2 = (1.00625)^{12t}$$ *Simplify within the parentheses.*

$$\ln 2 = \ln(1.00625)^{12t}$$ *Use the logarithm property of equality.*

$$\ln 2 = 12t \ln(1.00625)$$ *Use the power rule: $\log_b u^r = r \log_b u$.*

$$t = \frac{\ln 2}{12 \ln(1.00625)}$$ *Divide both sides by $12 \ln(1.00625)$.*

$$t \approx 9.27 \text{ years}$$ *Round to two decimal places.*

Note that .27 years = .27 years $\times \frac{12 \text{ months}}{1 \text{ year}}$ = 3.24 months. Because the interest is compounded at the end of each month, the investment will not double until 9 years and 4 months.

My video summary

▦ Example 2 Continuous Compound Interest

Suppose an investment of $5,000 compounded continuously grew to an amount of $5,130.50 in 6 months. Find the interest rate, and then determine how long it will take for the investment to grow to $6,000. Round the interest rate to the nearest hundredth of a percent and the time to the nearest hundredth of a year.

Solution Because the investment is compounded continuously, we use the formula $A = Pe^{rt}$.

We are given that $P = 5,000$, so $A = 5,000e^{rt}$. In 6 months, or when $t = .5$ years, we know that $A = 5,130.50$. Substituting these values into the compound interest formula will enable us to solve for r:

$$5,130.50 = 5,000e^{r(0.5)} \qquad \text{\textit{Substitute the appropriate values.}}$$

$$\frac{5,130.50}{5,000} = e^{.5r} \qquad \text{\textit{Divide by 5,000.}}$$

$$\ln\left(\frac{5,130.50}{5,000}\right) = \ln e^{.5r} \qquad \text{\textit{Use the logarithm property of equality.}}$$

$$\ln\left(\frac{5,130.50}{5,000}\right) = .5r \qquad \text{\textit{Use cancellation property (2) to rewrite} } \ln e^{.5r} \text{ \textit{as} } .5r.$$

$$r = \frac{\ln\left(\dfrac{5,130.50}{5,000}\right)}{.5} \approx .051530 \qquad \text{\textit{Divide by .5.}}$$

Therefore, the interest rate is 5.15%. To find the time that it takes for the investment to grow to $6,000, we use the formula $A = Pe^{rt}$, with $A = 6,000$, $P = 5,000$, and $r = 0.0515$, and solve for t. Watch the video to verify that it will take approximately 3.54 years.

Work exercises 1–5 in this textbook or in the *MyMathLab*® Study Plan.

OBJECTIVE 2: EXPONENTIAL GROWTH AND DECAY

In Section 5.2, the exponential growth model is introduced. This model is used when a population grows at a rate proportional to the size of its current population. This model is often called the uninhibited growth model. We review this exponential growth model and sketch the graph in Figure 15.

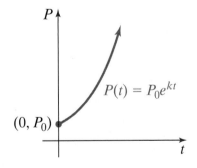

Figure 15 Graph of $P(t) = P_0e^{kt}$ for $k > 0$

Exponential Growth

A model that describes the exponential uninhibited growth of a population, P, after a certain time, t, is

$$P(t) = P_0e^{kt}$$

where $P_0 = P(0)$ is the initial population and $k > 0$ is a constant called the **relative growth rate**. (*Note:* k is sometimes given as a percent.)

My video summary

▣ Example 3 Population Growth

The population of a small town grows at a rate proportional to its current size. In 1900, the population was 900. In 1920, the population had grown to 1,600. What was the population of this town in 1950? Round to the nearest whole number.

Solution Using the model $P(t) = P_0 e^{kt}$, we must first determine the constants P_0 and k. The initial population was 900 in 1900 so $P_0 = 900$. Therefore, $P(t) = 900e^{kt}$. To find k, we use the fact that in 1920, or when $t = 20$, the population was 1,600; thus,

$$P(20) = 900e^{k(20)} = 1,600 \qquad \text{\textit{Substitute }} P(20) = 1,600.$$

$$900e^{20k} = 1,600$$

$$e^{20k} = \frac{16}{9} \qquad \text{\textit{Divide by 900 and simplify.}}$$

$$\ln e^{20k} = \ln \frac{16}{9} \qquad \text{\textit{Use the logarithm property of equality.}}$$

$$20k = \ln \frac{16}{9} \qquad \text{\textit{Use cancellation property (2) to rewrite }} \ln e^{20k} \text{\textit{ as }} 20k.$$

$$k = \frac{\ln\left(\frac{16}{9}\right)}{20} \qquad \text{\textit{Divide by 20.}}$$

The function that models the population of this town at any time t is given by $P(t) = 900e^{\frac{\ln(16/9)}{20}t}$. To determine the population in 1950, or when $t = 50$, we evaluate $P(50)$:

$$P(50) = 900e^{\frac{\ln(16/9)}{20}(50)} \approx 3,793$$

Exponential Decay

Some populations exhibit *negative exponential growth*. In other words, the population, quantity, or amount *decreases* over time. Such models are called **exponential decay** models. The only difference between an exponential growth model and an exponential decay model is that the constant, k, is less than zero. See Figure 16.

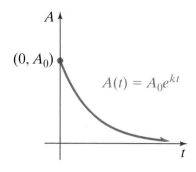

Figure 16 Graph of $A(t) = A_0 e^{kt}$ for $k < 0$

Exponential Decay

A model that describes the exponential decay of a population, quantity, or amount A, after a certain time, t, is

$$A(t) = A_0 e^{kt}$$

where $A_0 = A(0)$ is the initial quantity and $k < 0$ is a constant called the **relative decay constant**. (*Note:* k is sometimes given as a percent.)

Half-Life

Every radioactive element has a half-life, which is the required time for a given quantity of that element to decay to half of its original mass. For example, the half-life of Cesium-137 is 30 years. Thus, it takes 30 years for any amount of Cesium-137 to decay to $\frac{1}{2}$ of its original mass. It takes an additional 30 years to decay to $\frac{1}{4}$ of its original mass and so on. See Figure 17, and view the animation that illustrates the half-life of Cesium-137.

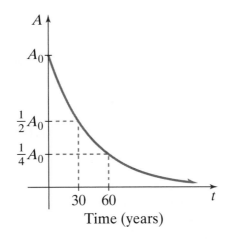

Figure 17 Half-life of Cesium-137.

Example 4 Radioactive Decay

Suppose that a meteorite is found containing 4% of its original Krypton-99. If the half-life of Krypton-99 is 80 years, how old is the meteorite? Round to the nearest year.

Solution We use the formula $A(t) = A_0 e^{kt}$, where A_0 is the original amount of Krypton-99. We first must find the constant k. To find k, we use the fact that the half-life of Krypton-99 is 80 years. Therefore, $A(80) = \frac{1}{2}A_0$. Because $A(80) = A_0 e^{k(80)}$, we can set $\frac{1}{2}A_0 = A_0 e^{k(80)}$ and solve for k.

$$\frac{1}{2}A_0 = A_0 e^{k(80)}$$

$$\frac{1}{2} = e^{80k} \qquad \textit{Divide both sides by } A_0.$$

$$\ln \frac{1}{2} = \ln e^{80k} \qquad \textit{Use the logarithm property of equality.}$$

$$\ln \frac{1}{2} = 80k \qquad \textit{Use cancellation property (2) to rewrite } \ln e^{80k} \textit{ as } 80k.$$

$$\frac{\ln \frac{1}{2}}{80} = k \qquad \textit{Divide both sides by } 80.$$

$$\frac{-\ln 2}{80} = k \qquad \ln \frac{1}{2} = \ln 1 - \ln 2 = 0 - \ln 2 = -\ln 2$$

Now that we know $k = \frac{-\ln 2}{80}$, our function becomes $A(t) = A_0 e^{\frac{-\ln 2}{80}t}$. To find out the age of the meteorite, we set $A(t) = .04A_0$ because the meteorite now contains 4% of the original amount of Krypton-99.

$$.04A_0 = A_0 e^{\frac{-\ln 2}{80}t}$$ *Substitute .04A_0 for $A(t)$.*

$$.04 = e^{\frac{-\ln 2}{80}t}$$ *Divide both sides by A_0.*

$$\ln .04 = \ln e^{\frac{-\ln 2}{80}t}$$ *Use the logarithm property of equality.*

$$\ln .04 = \frac{-\ln 2}{80}t$$ *Use cancellation property (2) to rewrite*

$\ln e^{\frac{-\ln 2}{80}t}$ as $\dfrac{-\ln 2}{80}t$.

$$\frac{\ln .04}{\frac{-\ln 2}{80}} = t \approx 372 \text{ years}$$ *Divide both sides by $\dfrac{-\ln 2}{80}$.*

The meteorite is about 372 years old.

Work exercises 6–12 in this textbook or in the *MyMathLab*® Study Plan.

OBJECTIVE 3:   SOLVING LOGISTIC GROWTH APPLICATIONS

The uninhibited exponential growth model $P(t) = P_0 e^{kt}$ for $k > 0$ is used when there are no outside limiting factors such as predators or disease that affect the population growth. When such outside factors exist, scientists often use a **logistic model** to describe the population growth. One such logistic model is described and sketched in Figure 18.

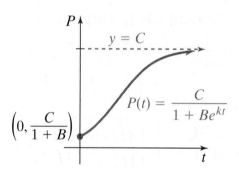

Figure 18 Graph of $P(t) = \frac{C}{1 + Be^{kt}}$

Logistic Growth

A model that describes the logistic growth of a population P at any time t is given by the function

$$P(t) = \frac{C}{1 + Be^{kt}}$$

where B, C, and k are constants with $C > 0$ and $k < 0$.

The number C is called the **carrying capacity**. In the logistic model, the population will approach the value of the carrying capacity over time but never exceed it. You can see in the graph sketched in Figure 18 that the graph of the logistic growth model approaches the **horizontal asymptote** $y = C$.

My interactive video summary

▶ Example 5 Logistic Growth

Ten goldfish were introduced into a small pond. Because of limited food, space, and oxygen, the carrying capacity of the pond is 400 goldfish. The goldfish population at any time t, in days, is modeled by the logistic growth function $F(t) = \frac{C}{1 + Be^{kt}}$. If 30 goldfish are in the pond after 20 days,

a. Find B.

b. Find k.

c. When will the pond contain 250 goldfish? Round to the nearest whole number.

Solution

a. The carrying capacity is 400; thus, $C = 400$. Also, initially (at $t = 0$) there were 10 goldfish, so $F(0) = 10$. Therefore,

$$10 = \frac{400}{1 + Be^{k(0)}} \qquad \text{Substitute } C = 400 \text{ and } F(0) = 10.$$

$$10 = \frac{400}{1 + B} \qquad \text{Evaluate } e^0 = 1.$$

$$10 + 10B = 400 \qquad \text{Multiply both sides by } 1 + B.$$

$$B = 39 \qquad \text{Solve for B.}$$

b. Use the function $F(t) = \frac{400}{1 + 39e^{kt}}$ and the fact that $F(20) = 30$ (there are 30 goldfish after 20 days) to solve for k.

$$30 = \frac{400}{1 + 39e^{k(20)}} \qquad \text{Substitute } F(20) = 30.$$

$$30(1 + 39e^{20k}) = 400 \qquad \text{Multiply both sides by } 1 + 39e^{20k}.$$

$$30 + 1{,}170e^{20k} = 400 \qquad \text{Use the distributive property.}$$

$$1{,}170e^{20k} = 370 \qquad \text{Subtract 30 from both sides.}$$

$$e^{20k} = \frac{370}{1{,}170} \qquad \text{Divide both sides by 1,170.}$$

$$e^{20k} = \frac{37}{117} \qquad \text{Simplify.}$$

$$\ln e^{20k} = \ln \frac{37}{117} \qquad \text{Use the logarithm property of equality.}$$

$$20k = \ln \frac{370}{117} \qquad \text{Use cancellation property (2) to rewrite } \ln e^{20k} \text{ as } 20k.$$

$$k = \frac{\ln \frac{37}{117}}{20} \qquad \text{Divide both sides by 20.}$$

c. Use the function $F(t) = \dfrac{400}{1 + 39e^{\frac{\ln \frac{37}{117}}{20} t}}$, and then find t when $F(t) = 250$.

By repeating the exact same process as in part b, we find that it will take approximately 73 days until there are 250 goldfish in the pond. Watch the interactive video to verify the solution.

Work exercises 13–15 in this textbook or in the *MyMathLab*® Study Plan.

OBJECTIVE 4: USING NEWTON'S LAW OF COOLING

Newton's law of cooling states that the temperature of an object changes at a rate proportional to the difference between its temperature and that of its surroundings. It can be shown in a more advanced course that the function describing Newton's law of cooling is given by the following.

My animation summary

Newton's Law of Cooling
The temperature T of an object at any time t is given by $$T(t) = S + (T_0 - S)e^{kt}$$ where T_0 is the original temperature of the object, S is the constant temperature of the surroundings, and k is the cooling constant.

View the animation to see how this function behaves.

My video summary

▣ Example 6 Newton's Law of Cooling

Suppose that the temperature of a cup of hot tea obeys Newton's law of cooling. If the tea has a temperature of 200°F when it is initially poured and 1 minute later has cooled to 189°F in a room that maintains a constant temperature of 69°F, determine when the tea reaches a temperature of 146°F. Round to the nearest minute.

Solution We start using the formula for Newton's law of cooling with $T_0 = 200$ and $S = 69$.

$$T(t) = S + (T_0 - S)e^{kt} \qquad \textit{Use Newton's law of cooling formula.}$$
$$T(t) = 69 + (200 - 69)e^{kt} \qquad \textit{Substitute } T_0 = 200 \textit{ and } S = 69.$$
$$T(t) = 69 + 131e^{kt} \qquad \textit{Simplify.}$$

We now proceed to find k. The object cools to 189°F in 1 minute; thus, $T(1) = 189$. Therefore,

$$189 = 69 + 131e^{k(1)} \qquad \textit{Substitute } T(1) = 189.$$
$$120 = 131e^{k} \qquad \textit{Subtract 69 from both sides.}$$
$$\frac{120}{131} = e^{k} \qquad \textit{Divide both sides by 131.}$$
$$\ln \frac{120}{131} = \ln e^{k} \qquad \textit{Use the logarithm property of equality.}$$
$$\ln \frac{120}{131} = k \qquad \textit{Use cancellation property (2) to rewrite } \ln e^{k} \textit{ as } k.$$

Now that we know the cooling constant $k = \ln \frac{120}{131}$, we can use the function $T(t) = 69 + 131e^{\ln \frac{120}{131}t}$ and determine the value of t when $T(t) = 146$.

$$146 = 69 + 131e^{\ln \frac{120}{131}t} \qquad \textit{Set } T(t) = 146.$$
$$77 = 131e^{\ln \frac{120}{131}t} \qquad \textit{Subtract 69 from both sides.}$$
$$\frac{77}{131} = e^{\ln \frac{120}{131}t} \qquad \textit{Divide both sides by 131.}$$

$$\ln \frac{77}{131} = \ln e^{\ln \frac{120}{131}t}$$ Use the logarithm property of equality.

$$\ln \frac{77}{131} = \ln \frac{120}{131}t$$ Use cancellation property (2) to rewrite $\ln e^{\ln \frac{120}{131}t}$ as $\ln \frac{120}{131}t$.

$$\frac{\ln \frac{77}{131}}{\ln \frac{120}{131}} = t$$ Divide both sides by $\ln \frac{120}{131}$.

$$t \approx 6 \text{ minutes}$$ Use a calculator to approximate the time rounded to the nearest minute.

So, it takes approximately 6 minutes for the tea to cool to 146°F.

Using Technology

The graph of $T(t) = 69 + 131e^{\ln \frac{120}{131}t}$, which describes the temperature of the tea t minutes after being poured, was created using a graphing utility. Note that the line $y = 69$, which represents the temperature of the surroundings, is a **horizontal asymptote**.

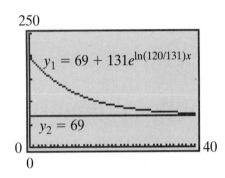

Work exercises 16–18 in this textbook or in the *MyMathLab*® Study Plan.

5.6 Exercises

1. Jimmy invests $15,000 in an account that pays 6.25% compounded quarterly. How long (in years and months) will it take for his investment to reach $20,000?

2. How long (in years and months) will it take for an investment to double at 9% compounded monthly?

3. How long will it take for an investment to triple if it is compounded continuously at 8%? Round to 2 decimal places.

4. What is the interest rate necessary for an investment to quadruple after 8 years of continuous compound interest? (Round to the nearest hundredth of a percent.)

5. Marsha and Jan both invested money on March 1, 2005. Marsha invested $5,000 at Bank A, where the interest was compounded quarterly. Jan invested $3,000 at Bank B, where the interest was compounded continuously. On March 1, 2007, Marsha had a balance of $5,468.12, whereas Jan had a balance of $3,289.09. What was the interest rate at each bank? (Round to the nearest tenth of a percent.)

6. The population of Adamsville grew from 9,000 to 15,000 in 6 years. Assuming uninhibited exponential growth, what is the expected population in an additional 4 years? Round to the nearest whole number.

7. During a research experiment, it was found that the number of bacteria in a culture grew at a rate proportional to its size. At 8:00 AM, there were 2,000 bacteria present in the culture. At noon, the number of bacteria grew to 2,400. How many bacteria will there be at midnight? Round to the nearest whole number.

8. A skull-cleaning factory cleans animal skulls such as deer, buffalo, and other types of animal skulls using flesh-eating beetles to clean the skulls. The factory owner started with only 10 adult beetles. After 40 days, the beetle population grew to 30 adult beetles. How long did it take before the beetle population reached 10,000 beetles? Round to the nearest whole number.

9. The population of a Midwest industrial town decreased from 210,000 to 205,000 in just 3 years. Assuming that this trend continues, what will the population be after an additional 3 years? Round to the nearest whole number.

10. A certain radioactive isotope is leaked into a small stream. Three hundred days after the leak, 2% of the original amount of the substance remained. Determine the half-life of this radioactive isotope. Round to the nearest whole number.

11. Radioactive Iodine-131 is a by-product of certain nuclear reactors. On April 26, 1986, one of the nuclear reactors in Chernobyl, Ukraine, a republic of the former Soviet Union, experienced a massive release of radioactive iodine. Fortunately, Iodine-131 has a very short half-life of 8 days. Estimate the percentage of the original amount of Iodine-131 released by the Chernobyl explosion on May 1, 1986, 5 days after the explosion. Round to 2 decimal places.

12. Superman is rendered powerless when exposed to 50 or more grams of kryptonite. A 500-year-old rock that originally contained 300 grams of kryptonite was recently stolen from a rock museum by Superman's enemies. The half-life of kryptonite is known to be 200 years.

  a. How many grams of kryptonite are still contained in the stolen rock? Round to two decimal places.

  b. For how many years can this rock be used by Superman's enemies to render him powerless? Round to the nearest whole number.

13. The logistic growth model $H(t) = \frac{6,000}{1 + 2e^{-.65t}}$ represents the number of families that own a home in a certain small (but growing) Idaho city t years after 1980.

  a. What is the maximum number of families that will own a home in this city?

  b. How many families owned a home in 1980?

  c. In what year did 5,920 families own a home?

14. The number of students that hear a rumor on a small college campus t days after the rumor starts is modeled by the logistic function $R(t) = \frac{3,000}{1 + Be^{kt}}$. Determine the following if 8 students initially heard the rumor and 100 students heard the rumor after 1 day.

  a. What is the carrying capacity for the number of students who will hear the rumor?

  b. Find B.

  c. Find k.

  d. How long will it take 2,900 students to hear the rumor?

15. In 1999, 1,500 runners entered the inaugural Run-for-Your-Life marathon in Joppetown, USA. In 2005, 21,500 runners entered the race. Because of the limited number of hotels, restaurants, and portable toilets in the area, the carrying capacity for the number of racers is 61,500. The number of racers at any time, t, in years, can be modeled by the logistic function $P(t) = \frac{C}{1 + Be^{kt}}$.

  a. What is the value of C?

  b. Find B.

  c. Find k.

  d. In what year should at least 49,500 runners be expected to run in the race? Round to the nearest year.

16. Estabon poured himself a hot beverage that had a temperature of 198°F and then set it on the kitchen table to cool. The temperature of the kitchen was a constant 75°F. If the drink cooled to 180°F in 2 minutes, how long will it take for the drink to cool to 100°F?

17. Police arrive at a murder scene at 1:00 AM and immediately record the body's temperature, which was 92°F. At 2:30 AM, after thoroughly inspecting and fingerprinting the area, they again took the temperature of the body, which had dropped to 85°F. The temperature of the crime scene has remained at a constant 60°F. Determine when the person was murdered. (Assume that the victim was healthy at the time of death. That is, assume that the temperature of the body at the time of death was 98.6°F.)

18. Jodi poured herself a cold soda that had an initial temperature of 40°F and immediately went outside to sunbathe where the temperature was a steady 99°F. After 5 minutes, the temperature of the soda was 47°F. Jodi had to run back into the house to answer the phone. What is the expected temperature of the soda after an additional 10 minutes?

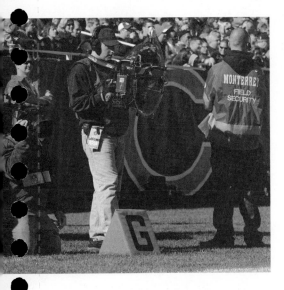

Conic Sections

CHAPTER SIX CONTENTS

Introduction to Conic Sections

Introduction to Conic Sections

In this chapter, we focus on the geometric study of conic sections. Conic sections (or conics) are formed when a plane intersects a pair of **right circular cones**. The surface of the cones comprises the set of all line segments that intersect the outer edges of the circular bases of the cones and pass through a fixed point. The fixed point is called the **vertex** of the cone, and the line segments are called the **elements**. See Figure 1.

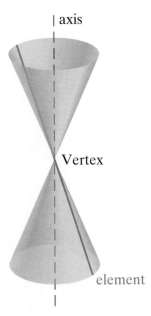

| axis

Vertex

element

|

Figure 1 Pair of right circular cones

When a plane intersects a right circular cone, a conic section is formed. The four conic sections that can be formed are circles, ellipses, parabolas, and hyperbolas. Click on one of the following four animations to see how each conic section is formed.

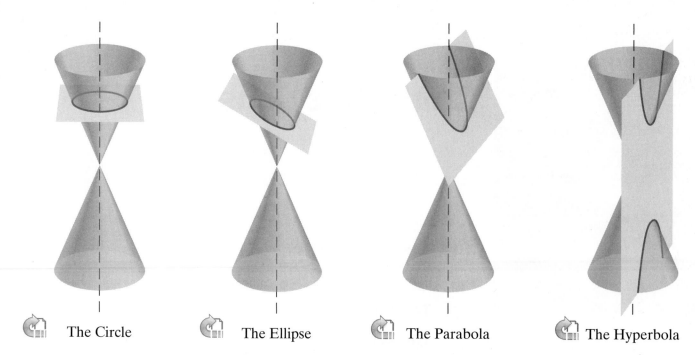

The Circle   The Ellipse   The Parabola   The Hyperbola

Because we studied circles in Section 2.2, they are not covered again in this chapter. It may however, be useful to review circles before going on. Watch the video to review how to write the equation of a circle in standard form by completing the square. We also need to be able to complete the square to write each of the other conic sections in standard form. These conic sections are introduced in the following order:

Section 6.1 The Parabola
Section 6.2 The Ellipse
Section 6.3 The Hyperbola

DEGENERATE CONIC SECTIONS

It is worth noting that when the circle, ellipse, parabola, or hyperbola is formed, the intersecting plane does not pass through the vertex of the cones. When a plane does intersect the vertex of the cones, a degenerate conic section is formed. The three degenerate conic sections are a point, a line, and a pair of intersecting lines. See Figure 2. We do not concern ourselves with degenerate conic sections in this chapter.

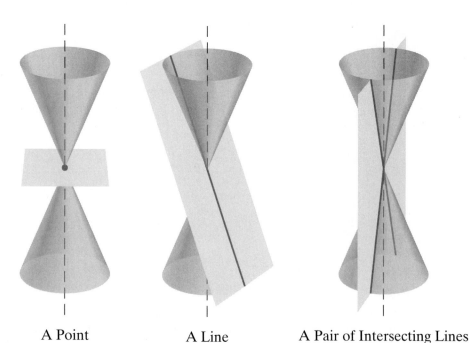

A Point   A Line   A Pair of Intersecting Lines

Figure 2 Degenerate conic sections

6.1 **The Parabola**

THINGS TO KNOW

Before working through this section, be sure that you are familiar with the following concepts:

VIDEO   ANIMATION   INTERACTIVE

 You Try It   1.   Finding the Distance between Two Points Using the Distance Formula (**Section 2.1**)

 You Try It   2.   Converting the General Form of a Circle into Standard Form (**Section 2.2**)

 You Try It   3.   Finding the Equations of Horizontal and Vertical Lines (**Section 2.3**)

OBJECTIVES

1 Determining the Equation of a Parabola with a Vertical Axis of Symmetry

2 Determining the Equation of a Parabola with a Horizontal Axis of Symmetry

3 Determining the Equation of a Parabola Given Information about the Graph

4 Completing the Square to Find the Equation of a Parabola in Standard Form

5 Solving Applied Problems Involving Parabolas

OBJECTIVE 1:   DETERMINING THE EQUATION OF A PARABOLA WITH A VERTICAL AXIS OF SYMMETRY

In Section 4.1, we studied quadratic functions of the form $f(x) = ax^2 + bx + c$, $a \neq 0$. We learned that every quadratic function has a u-shaped graph called a *parabola*. You may want to review the different characteristics of a parabola. Work through the following animation, and click on each characteristic to get a detailed description.

Characteristics of a Parabola

My animation summary

1.   Vertex
2.   Axis of symmetry
3.   *y*-Intercept
4.   *x*-Intercept(s) or real zeros
5.   Domain and range

My video summary

 In Section 4.1, we studied quadratic functions and parabolas from an algebraic point of view. We now look at parabolas from a geometric perspective. We see in the introduction to this chapter that when a plane is parallel to an **element** of the cone, the plane will intersect the cone in a parabola. (Click on the animation.)

My animation summary

The Parabola

The set of points that define the parabola formed by the intersection described previously is stated in the following geometric definition of the parabola.

Geometric Definition of the Parabola

A **parabola** is the set of all points in a plane equidistant from a fixed point F and a fixed line D. The fixed point is called the **focus**, and the fixed line is called the **directrix**.

Watch the video to see how to sketch the parabola seen in Figure 3 using the geometric definition of the parabola.

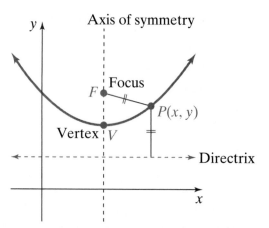

Figure 3 The distance from any point P to the focus is the same as the distance from point P to the directrix.

In Figure 3, we can see that for any point $P(x, y)$ that lies on the graph of the parabola, the distance from point P to the focus is exactly the same as the distance from point P to the directrix. Similarly, because the vertex, V, lies on the graph of the parabola, the distance from V to the focus must also be the same as the distance from V to the directrix. Therefore, if the distance from V to F is p units, then the distance from V to the directrix is also p units. If the coordinates of the vertex in Figure 3 are (h, k), then the coordinates of the focus must be $(h, k + p)$ and the equation of the directrix is $y = k - p$. We can use this information and the fact that the distance from $P(x, y)$ to the focus is equal to the distance from $P(x, y)$ to the directrix to derive the equation of a parabola. The equation of the parabola with a vertical axis of symmetry is derived in Appendix A. Click here to see this derivation.

Equation of a Parabola in Standard Form with a Vertical Axis of Symmetry

The equation of a parabola with a vertical axis of symmetry is $(x - h)^2 = 4p(y - k)$,

where

the vertex is $V(h, k)$,

$|p|$ = distance from the vertex to the focus = distance from the vertex to the directrix,

the focus is $F(h, k + p)$, and

the equation of the directrix is $y = k - p$.

The parabola opens *upward* if $p > 0$ or *downward* if $p < 0$.

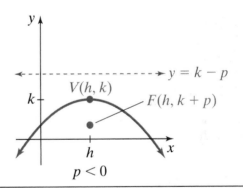

Example 1 Find the Vertex, Focus, and Directrix of a Parabola and Sketch Its Graph

Find the vertex, focus, and directrix of the parabola $x^2 = 8y$ and sketch its graph.

Solution Notice that we can rewrite the equation $x^2 = 8y$ as $(x - 0)^2 = 8(y - 0)$. We can now compare the equation $(x - 0)^2 = 8(y - 0)$ to the standard form equation $(x - h)^2 = 4p(y - k)$ to see that $h = 0$ and $k = 0$; hence, the vertex is at the origin, $(0, 0)$. To find the focus and directrix, we need to find p.

$$4p = 8$$
$$p = 2 \qquad \text{Divide both sides by 4.}$$

Because the value of p is positive, the parabola opens upward and the focus is located two units vertically *above* the vertex, whereas the directrix is the horizontal line located two units vertically *below* the vertex. The focus has coordinates $(0, 2)$, and the equation of the directrix is $y = -2$. The graph is shown in Figure 4.

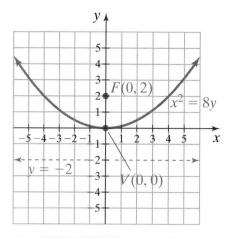

Figure 4

My video summary

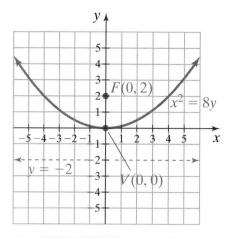

Example 2 Find the Vertex, Focus, and Directrix
of a Parabola and Sketch Its Graph

Find the vertex, focus, and directrix of the parabola $-(x + 1)^2 = 4(y - 3)$ and sketch its graph.

Solution Watch the video to verify that the vertex has coordinates $(-1, 3)$, the focus has coordinates $(-1, 2)$, and the equation of the directrix is $y = 4$. The graph of this parabola is sketched in Figure 5.

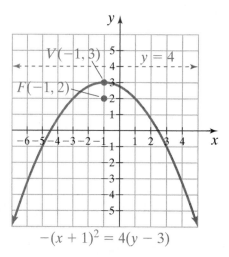

$$-(x + 1)^2 = 4(y - 3)$$ **Figure 5**

Work exercises 1–5 in this textbook or in the *MyMathLab*® Study Plan.

OBJECTIVE 2:   DETERMINING THE EQUATION OF A PARABOLA
WITH A HORIZONTAL AXIS OF SYMMETRY

In Examples 1 and 2, the graphs of both parabolas had vertical axes of symmetry. The graph of a parabola could also have a horizontal axis of symmetry and open "sideways." We derive the standard form of the parabola with a horizontal axis of symmetry in much the same way as we did with the parabola with a vertical axis of symmetry. This equation is derived in Appendix A. Click here to see this derivation.

Equation of a Parabola in Standard Form with a Horizontal Axis of Symmetry

The equation of a parabola with a horizontal axis of symmetry is $(y - k)^2 = 4p(x - h)$,

where

  the vertex is $V(h, k)$,

  $|p|$ = distance from the vertex to the focus = distance from the vertex to the directrix,

  the focus is $F(h + p, k)$, and

  the equation of the directrix is $x = h - p$.

The parabola opens *right* if $p > 0$ or *left* if $p < 0$.

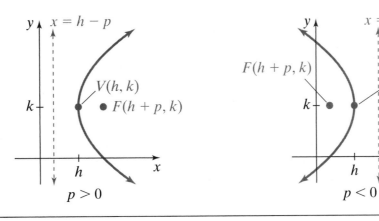

My video summary

◫ Example 3 **Find the Vertex, Focus, and Directrix**
             of a Parabola and Sketch Its Graph

Find the vertex, focus, and directrix of the parabola $(y - 3)^2 = 8(x + 2)$ and sketch
its graph.

Solution Watch the video to verify that the vertex has coordinates $(-2, 3)$, the
focus has coordinates $(0, 3)$, and the equation of the directrix is $x = -4$. The
graph of this parabola is sketched in Figure 6.

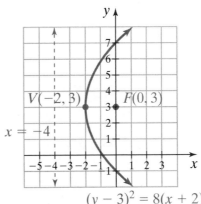

$(y - 3)^2 = 8(x + 2)$ **Figure 6**

Work exercises 6–10 in this textbook or in the *MyMathLab*® Study Plan.

OBJECTIVE 3: DETERMINING THE EQUATION OF A PARABOLA GIVEN INFORMATION ABOUT THE GRAPH

It is often necessary to determine the equation of a parabola given certain information. It is always useful to first determine whether the parabola has a vertical axis of symmetry or a horizontal axis of symmetry. Try to work through Examples 4 and 5. Then watch the solutions to the corresponding videos to determine whether you are correct.

My video summary

▤ Example 4 Find the Equation of a Parabola

Find the standard form of the equation of the parabola with focus $\left(-3, \frac{5}{2}\right)$ and directrix $y = \frac{11}{2}$.

Solution Watch the video to see that the equation of this parabola is $(x + 3)^2 = -6(y - 4)$. The graph is shown in Figure 7.

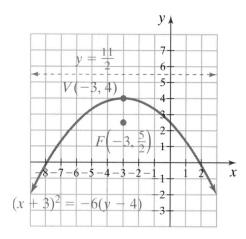

Figure 7

My video summary

▤ Example 5 Find the Vertex, Focus, and Directrix of a Parabola and Sketch Its Graph

Find the standard form of the equation of the parabola with focus $(4, -2)$ and vertex $\left(\frac{13}{2}, -2\right)$.

Solution Watch the video to see that the equation of this parabola is $(y + 2)^2 = -10\left(x - \frac{13}{2}\right)$. The graph is shown in Figure 8.

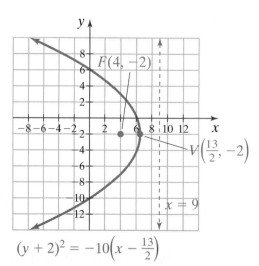

$$(y + 2)^2 = -10\left(x - \frac{13}{2}\right)$$

Figure 8

Using Technology

We can use a graphing utility to graph the parabola
from Example 5 by solving the equation for y.

$$(y + 2)^2 = -10\left(x - \frac{13}{2}\right)$$

$$y + 2 = \pm\sqrt{-10\left(x - \frac{13}{2}\right)}$$

$$y = -2 \pm \sqrt{-10\left(x - \frac{13}{2}\right)}$$

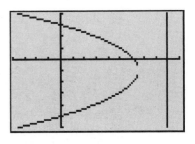

Figure 9

Using $y_1 = -2 + \sqrt{-10\left(x - \frac{13}{2}\right)}$, and
$y_2 = -2 - \sqrt{-10\left(x - \frac{13}{2}\right)}$, we obtain the graph
seen in Figure 9. *Note:* The directrix was created
using the calculator's DRAW feature.

Work exercises 11–19 in this textbook or in the *MyMathLab*® Study Plan.

OBJECTIVE 4:   COMPLETING THE SQUARE TO FIND THE EQUATION
OF A PARABOLA IN STANDARD FORM

If the equation of a parabola is in the standard form of $(x - h)^2 = 4p(y - k)$ or
$(y - k)^2 = 4p(x - h)$, it is not too difficult to determine the vertex, focus, and
directrix and sketch its graph. However, the equation might not be given in standard form. If this is the case, we complete the square on the variable that is
squared to rewrite the equation in standard form as in Example 6.

My video summary

■ **Example 6 Rewrite the Equation of a Parabola
in Standard Form by Completing
the Square**

Find the vertex, focus, and directrix and sketch the graph of the parabola
$x^2 - 8x + 12y = -52$.

Solution Because x is squared, we will complete the square on the variable x.

$$\begin{aligned}
x^2 - 8x + 12y &= -52 \\
x^2 - 8x &= -12y - 52 \\
x^2 - 8x + 16 &= -12y - 52 + 16 \\
(x - 4)^2 &= -12y - 36 \\
(x - 4)^2 &= -12(y + 3)
\end{aligned}$$

The equation is now in standard form with vertex $(4, -3)$ and $4p = -12$
so $p = -3$. The parabola must open down because the variable x is
squared and $p < 0$. The focus is located three units below the vertex, whereas
the directrix is three units above the vertex. Thus, the focus has coordinates $(4, -6)$, and the equation of the directrix is $y = 0$ or the x-axis. You can
watch the video to see this solution worked out in detail. The graph is shown in
Figure 10.

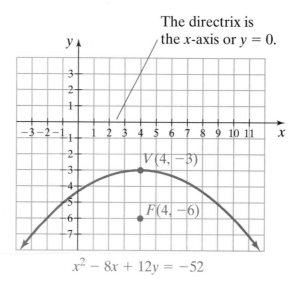

The directrix is the *x*-axis or *y* = 0.

$$V(4, -3)$$

$$F(4, -6)$$

$$x^2 - 8x + 12y = -52$$

Figure 10

Work exercises 20–23 in this textbook or in the *MyMathLab*® Study Plan.

OBJECTIVE 5:   **SOLVING APPLIED PROBLEMS INVOLVING PARABOLAS**

The Romans were one of the first civilizations to use the engineering properties of parabolic structures in their creation of **arch bridges**. The cables of many suspension bridges, such as the **Golden Gate Bridge** in San Francisco, span from tower to tower in the shape of a parabola.

Parabolic surfaces are used in the manufacture of many satellite dishes, search lights, car headlights, telescopes, lamps, heaters, and other objects. This is because parabolic surfaces have the property that incoming rays of light or radio waves traveling parallel to the axis of symmetry of a parabolic reflector or receiver will reflect off the parabolic surface and travel directly toward the antenna that is placed at the focus. See Figure 11. When a light source such as the headlight of a car is placed at the focus of a parabolic reflector, the light reflects off the surface outward, producing a narrow beam of light and thus maximizing the output of illumination.

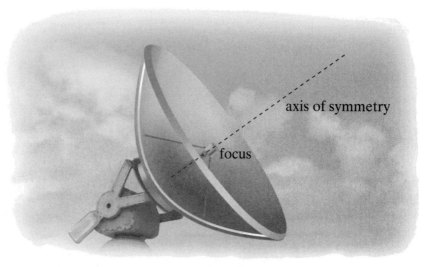

axis of symmetry

focus

Figure 11 Incoming rays reflect off the parabolic surface toward the antenna placed at the focus.

Example 7 Find the Focus of a Parabolic Microphone

Parabolic microphones can be seen on the sidelines of professional sporting events so that television networks can capture audio sounds from the players on the field. If the surface of a parabolic microphone is 27 centimeters deep and has a diameter of 72 centimeters at the top, where should the microphone be placed relative to the vertex of the parabola?

Solution We can draw a parabola with the vertex at the origin representing the center cross section of the parabolic microphone. The equation of this parabola in standard form is $x^2 = 4py$. Substitute the point $(36, 27)$ into the equation to get

$$x^2 = 4py$$
$$(36)^2 = 4p(27)$$
$$1{,}296 = 108p$$
$$p = 12$$

The microphone must be placed 12 centimeters from the vertex.

Work exercises 24–28 in this textbook or in the MyMathLab® Study Plan.

6.1 Exercises

In Exercises 1–10, determine the vertex, focus, and directrix of the parabola and sketch its graph.

1. $x^2 = 16y$

2. $x^2 = -8y$

3. $(x - 1)^2 = -12(y - 4)$

4. $(x + 3)^2 = 6(y - 1)$

5. $(x + 2)^2 = 5(y + 6)$

6. $y^2 = 4x$

7. $y^2 = -8x$

8. $(y - 5)^2 = -4(x - 2)$

9. $(y + 3)^2 = 20(x - 4)$

10. $(y + 4)^2 = 9(x + 3)$

In Exercises 11–19, find the equation in standard form of the parabola described.

11. The focus has coordinates $(2, 0)$, and the equation of the directrix is $x = -2$.

12. The focus has coordinates $(0, -\frac{1}{2})$, and the equation of the directrix is $y = \frac{1}{2}$.

13. The focus has coordinates $(3, -5)$, and the equation of the directrix is $y = -1$.

14. The focus has coordinates $(2, 4)$, and the equation of the directrix is $x = -4$.

15. The vertex has coordinates $(-\frac{11}{4}, -2)$, and the focus has coordinates $(-3, -2)$.

16. The vertex has coordinates $(4, -\frac{1}{4})$, and the focus has coordinates $(4, \frac{1}{4})$.

17. The vertex has coordinates $(-3, 4)$, and the equation of the directrix is $x = -7$.

18. Find the equations of the two parabolas in standard form that have a horizontal axis of symmetry and a focus at the point $(0, 4)$, and that pass through the origin.

19. Find the equations of the two parabolas in standard form that have a vertical axis of symmetry and a focus at the point $(2, \frac{3}{2})$, and that pass through the origin.

In Exercises 20–23, determine the vertex, focus, and directrix of the parabola and sketch its graph.

20. $x^2 + 10x = 5y - 10$

21. $y^2 - 12y + 6x + 30 = 0$

22. $x^2 - 2x = -y + 1$

23. $y^2 + x + 6y = -10$

24. A parabolic eavesdropping device is used by a CIA agent to record terrorist conversations. The parabolic surface measures 120 centimeters in diameter at the top and is 90 centimeters deep at its center. How far from the vertex should the microphone be located?

25. A parabolic space heater is 18 inches in diameter and 8 inches deep. How far from the vertex should the heat source be located to maximize the heating output?

26. A large NASA parabolic satellite dish is 22 feet across and has a receiver located at the focus 4 feet from its base. The satellite dish should be how deep?

27. A parabolic arch bridge spans 160 feet at the base and is 40 feet above the water at the center. Find the equation of the parabola if the vertex is placed at the point $(0, 40)$. Can a sailboat that is 35 feet tall fit under the bridge 30 feet from the center?

28. The cable between two 40-meter towers of a suspension bridge is in the shape of a parabola that just touches the bridge halfway between the towers. The two towers are 100 meters apart. Vertical cables are spaced every 10 meters along the bridge. What are the lengths of the vertical cables located 30 meters from the center of the bridge?

6.2 The Ellipse

THINGS TO KNOW

Before working through this section, be sure that you are familiar with the following concepts:

VIDEO   ANIMATION   INTERACTIVE

You Try It

1. Finding the Distance between Two Points Using the Distance Formula (**Section 2.1**)

You Try It

2. Completing the Square to Find the Equation of a Parabola in Standard Form (**Section 6.1**)

OBJECTIVES

1 Sketching the Graph of an Ellipse

2 Determining the Equation of an Ellipse Given Information about the Graph

3 Completing the Square to Find the Equation of an Ellipse in Standard Form

4 Solving Applied Problems Involving Ellipses

....

My animation summary

When a plane intersects a **right circular cone** at an angle between 0 and 90 degrees to the axis of the cone, the conic section formed is an ellipse. (Click on the animation.)

The set of points in the plane that defines an ellipse formed by the intersection of a plane and a cone described previously is stated in the following geometric definition.

The Ellipse

> **Geometric Definition of the Ellipse**
>
> An **ellipse** is the set of all points in a plane, the sum of whose distances from two fixed points is a positive constant. The two fixed points, F_1 and F_2, are called the *foci*.

The previous geometric definition implies that for any two points P and Q that lie on the graph of the ellipse, the sum of the distance between P and F_1 plus the distance between P and F_2 is equal to the sum of the distance between Q and F_1 plus the distance between Q and F_2. In symbols, we write $d(P, F_1) + d(P, F_2) = d(Q, F_1) + d(Q, F_2)$. See Figure 12.

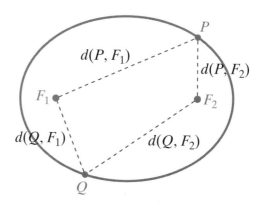

Figure 12 For any points P and Q on an ellipse,
$d(P, F_1) + d(P, F_2) = d(Q, F_1) + d(Q, F_2)$.

OBJECTIVE 1:   SKETCHING THE GRAPH OF AN ELLIPSE

My video summary

An ellipse has two **axes of symmetry**. The longer axis, which is the line segment that connects the two vertices, is called the *major axis*. The foci are always located along the major axis. The shorter axis is called the *minor axis*. It is the line segment perpendicular to the major axis that passes through the center having endpoints that lie on the ellipse. Watch the video to see how to sketch the two ellipses seen in Figure 13. In Figure 13a, the ellipse has a **horizontal major axis**. The ellipse in Figure 13b has a **vertical major axis**.

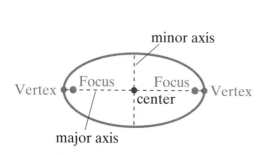

Figure 13 (a) Ellipse with Horizontal Major Axis

(b) Ellipse with Vertical Major Axis

Consider the ellipse with a **horizontal major axis** centered at (h, k) as shown in Figure 14, where $c > 0$ is the distance between the center and a focus, $a > 0$ is the distance between the center and one of the vertices, and $b > 0$ is the distance from the center to an endpoint of the **minor axis**.

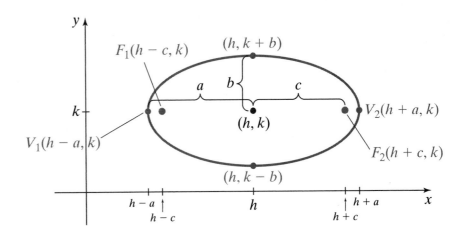

Figure 14

Before we can derive the equation of this ellipse, we must establish the following two facts:

Fact 1: The sum of the distances from any point on the ellipse to the two foci is $2a$.

Fact 2: $b^2 = a^2 - c^2$, or equivalently, $c^2 = a^2 - b^2$.

Click on Fact 1 or Fact 2 to see how these two facts are established in Appendix A.

Once we have established these two facts, we can derive the equation of the ellipse. The derivation of the equation of an ellipse can be seen in Appendix A. Click here to see this derivation. We now state the two standard equations of an ellipse.

Equation of an Ellipse in Standard Form with Center (h, k)

Horizontal Major Axis

$$\frac{(x - h)^2}{a^2} + \frac{(y - k)^2}{b^2} = 1$$

- $a > b > 0$
- Foci: $F_1(h - c, k)$ and $F_2(h + c, k)$
- Vertices: $V_1(h - a, k)$ and $V_2(h + a, k)$
- Endpoints of minor axis: $(h, k - b)$ and $(h, k + b)$
- $c^2 = a^2 - b^2$

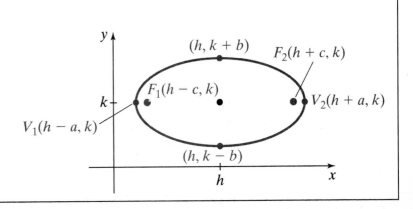

Equation of an Ellipse in Standard Form with Center (h, k)

Vertical Major Axis

$$\frac{(x - h)^2}{b^2} + \frac{(y - k)^2}{a^2} = 1$$

- $a > b > 0$
- Foci:
  $F_1(h, k - c)$ and $F_2(h, k + c)$
- Vertices:
  $V_1(h, k - a)$ and $V_2(h, k + a)$
- Endpoints of minor axis:
  $(h - b, k)$ and $(h + b, k)$
- $c^2 = a^2 - b^2$

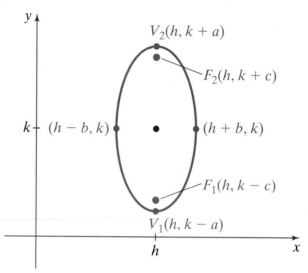

Notice that the two equations are identical except for the placement of a^2. If the equation of an ellipse is in standard form, we can quickly determine whether the ellipse has a horizontal major axis or a vertical major axis by looking at the denominator. If the larger denominator, a^2, appears under the x-term, then the ellipse has a horizontal major axis. If the larger denominator appears under the y-term, then the ellipse has a vertical major axis. If the denominators are equal $(a^2 = b^2)$, then the ellipse is a circle!

If $h = 0$ and $k = 0$, then the ellipse is centered at the **origin**. Ellipses centered at the origin have the following equations.

Equation of an Ellipse in Standard Form Centered at the Origin

Horizontal Major Axis	**Vertical Major Axis**

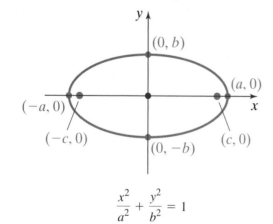

$$\frac{x^2}{a^2} + \frac{y^2}{b^2} = 1$$

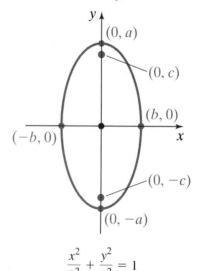

$$\frac{x^2}{b^2} + \frac{y^2}{a^2} = 1$$

Example 1 Sketch the Graph of an Ellipse
Centered at the Origin

Sketch the graph of the ellipse $\frac{x^2}{25} + \frac{y^2}{4} = 1$, and label the center, foci, and vertices.

My video summary

✎ Solution Watch the video to see how to sketch the ellipse shown in Figure 15.

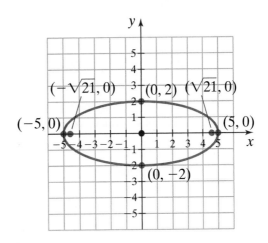

Figure 15

Work exercises 1–4 in this textbook or in the *MyMathLab*® Study Plan.

My video summary

■ **Example 2** Sketch the Graph of an Ellipse

Sketch the graph of the ellipse $\frac{(x + 2)^2}{20} + \frac{(y - 3)^2}{36} = 1$, and label the center, foci, and vertices.

Solution Note that the larger denominator appears under the *y*-term. This indicates that the ellipse has a vertical major axis. Watch the video to verify that the center of the ellipse is $(-2, 3)$. The foci have coordinates $(-2, -1)$ and $(-2, 7)$, the vertices have coordinates $(-2, -3)$ and $(-2, 9)$, and the coordinates of the minor axis are $(-2 - 2\sqrt{5}, 3)$ and $(-2 + 2\sqrt{5}, 3)$. The graph is shown in Figure 16.

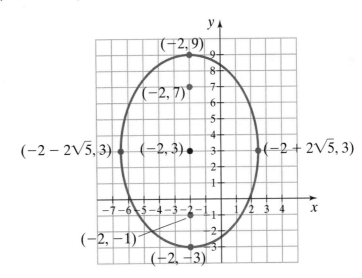

Figure 16

Work exercises 5–8 in this textbook or in the *MyMathLab*® Study Plan.

OBJECTIVE 2:   DETERMINING THE EQUATION OF AN ELLIPSE GIVEN INFORMATION ABOUT THE GRAPH

It is often necessary to determine the equation of an ellipse given certain information. It is always useful to first determine whether the ellipse has a horizontal major axis or a vertical major axis. Try to work through Examples 3 and 4. Then watch the solutions to the corresponding videos to determine whether you are correct.

My video summary

▦ Example 3 Determine the Equation of an Ellipse

Find the standard form of the equation of the ellipse with foci at $(-6, 1)$ and $(-2, 1)$ such that the length of the major axis is eight units.

Solution Watch the video to verify that this is an ellipse centered at $(-4, 1)$ with a horizontal major axis such that $a = 4$ and $b = \sqrt{12}$. The equation in standard form is $\frac{(x + 4)^2}{4^2} + \frac{(y - 1)^2}{(\sqrt{12})^2} = 1$ or $\frac{(x + 4)^2}{16} + \frac{(y - 1)^2}{12} = 1$. The graph of this ellipse is shown in Figure 17.

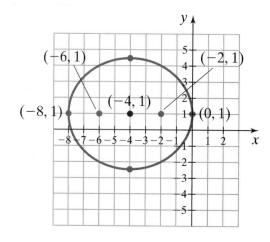

Figure 17 $\frac{(x + 4)^2}{16} + \frac{(y - 1)^2}{12} = 1$

Using Technology

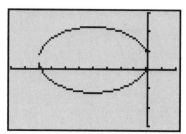

Figure 18 We can use a graphing utility to graph the ellipse from Example 3 by solving the equation for y:

$$y = 1 \pm \sqrt{12\left(1 - \frac{(x + 4)^2}{16}\right)}$$

(Click here to see how to solve for y.)

Using $y_1 = 1 + \sqrt{12\left(1 - \frac{(x + 4)^2}{16}\right)}$ and $y_2 = 1 - \sqrt{12\left(1 - \frac{(x + 4)^2}{16}\right)}$, we obtain the graph seen in Figure 18.

Work exercises 9–18 in this textbook or in the *MyMathLab*® Study Plan.

My video summary

▦ Example 4 Determine the Equation of an Ellipse

Determine the equation of the ellipse with foci located at $(0, 6)$ and $(0, -6)$ that passes through the point $(-5, 6)$.

Solution To find the equation, we can use **Fact 1** and **Fact 2** of ellipses. Watch the video to see how we can use these two facts to determine that the equation of this ellipse is $\frac{x^2}{45} + \frac{y^2}{81} = 1$.

Work exercises 19 and 20 in this textbook or in the *MyMathLab*® Study Plan.

OBJECTIVE 3:   COMPLETING THE SQUARE TO FIND THE EQUATION
OF AN ELLIPSE IN STANDARD FORM

If the equation of an ellipse is in the standard form of $\frac{(x-h)^2}{a^2} + \frac{(y-k)^2}{b^2} = 1$ or $\frac{(x-h)^2}{b^2} + \frac{(y-k)^2}{a^2} = 1$, it is not too difficult to determine the center and foci and sketch its graph. However, the equation might not be given in standard form. If this is the case, we complete the square on both variables to rewrite the equation in standard form as in Example 5.

My video summary

▦ Example 5 Rewrite the Equation of an Ellipse in Standard
Form by Completing the Square

Find the center and foci and sketch the ellipse

$$36x^2 + 20y^2 + 144x - 120y - 396 = 0.$$

Solution Rearrange the terms, leaving some room to complete the square and move any constants to the right-hand side:

$$36x^2 + 144x \quad + 20y^2 - 120y \quad = 396 \qquad \text{Rearrange the terms.}$$

Then factor and complete the square.

$$36(x^2 + 4x \quad) + 20(y^2 - 6y \quad) \quad = 396 \qquad \text{Factor out 36 and 20.}$$

$$36(x^2 + 4x + 4) + 20(y^2 - 6y + 9) = 396 + 144 + 180 \qquad \text{Complete the square on } x \text{ and } y. \text{ Remember to add } 36 \cdot 4 = 144 \text{ and } 20 \cdot 9 = 180 \text{ to the right side.}$$

$$36(x+2)^2 + 20(y-3)^2 = 720 \qquad \text{Factor the left side, and simplify the right side.}$$

$$\frac{36(x+2)^2}{720} + \frac{20(y-3)^2}{720} = \frac{720}{720} \qquad \text{Divide both sides by 720.}$$

$$\frac{(x+2)^2}{20} + \frac{(y-3)^2}{36} = 1 \qquad \text{Simplify.}$$

The equation is now in standard form. Watch the video to see the solution to this example worked out in detail. Notice that this is the exact same ellipse that we sketched in Example 2. To see how to sketch this ellipse, refer to Example 2 or watch the last part of this video.

Work exercises 21–24 in this textbook or in the **MyMathLab**® Study Plan.

OBJECTIVE 4:   SOLVING APPLIED PROBLEMS INVOLVING ELLIPSES

Ellipses have many applications. The planets of our solar system travel around the Sun in an elliptical orbit with the Sun at one focus. Some comets, such as Halley's comet, also travel in elliptical orbits. See Figure 19. (Some comets are only seen in our solar system once because they travel in hyperbolic orbits.) We saw in Section 6.1 that the parabola had a special reflecting property where incoming rays of light or radio waves traveling parallel to the axis of symmetry of a parabolic reflector or receiver will reflect off the parabolic surface and travel directly toward the antenna that is placed at the focus.

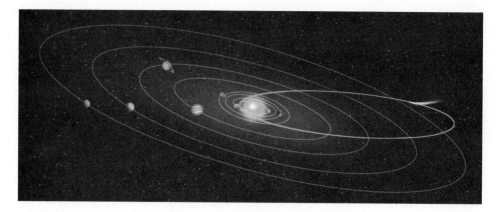

Figure 19 Planets travel around the Sun in an elliptical orbit.

Ellipses have a similar reflecting property. When light or sound waves originate from one focus of an ellipse, the waves will reflect off the surface of the ellipse and travel directly toward the other focus. See Figure 20. This reflecting property is used in a medical procedure called *sound wave lithotripsy* in which the patient having a kidney stone is placed in an elliptical tank with the kidney stone placed at one focus. An ultrasound wave emitter is positioned at the other focus. The sound waves reflect off the walls of the tank directly to the kidney stone, thus obliterating the stone into fragments that are easily passed naturally through the patient's body. See Example 5.

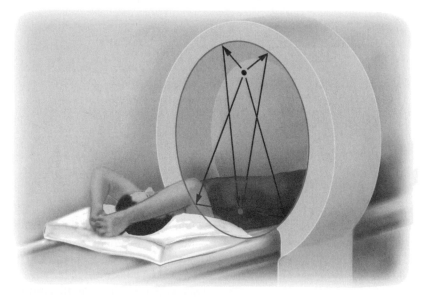

Figure 20 Sound waves or light rays emitted from one focus of an ellipse reflect off the surface directly to the other focus.

Example 6 Position a Patient During Kidney Stone Treatment

A patient is placed in an elliptical tank that is 200 centimeters long and 80 centimeters wide to undergo sound wave lithotripsy treatment for kidney stones. Determine where the sound emitter and the stone should be positioned relative to the center of the ellipse.

Solution The kidney stone and the sound emitter must be placed at the foci of the ellipse. We know that the major axis is 200 centimeters. Therefore, the vertices

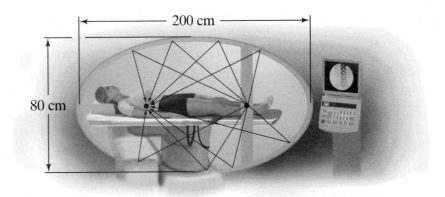

must be 100 centimeters from the center. Similarly, the endpoints of the minor axis are located 40 centimeters from the center.

To find c, we use the fact that $c^2 = a^2 - b^2$.

$$c^2 = 100^2 - 40^2$$
$$c^2 = 8,400$$
$$c = \sqrt{8,400} \approx 91.65$$

The stone and the sound emitter should be positioned approximately 91.65 centimeters from the center of the tank on the major axis.

Work exercises 25–31 in this textbook or in the *MyMathLab*® Study Plan.

6.2 **Exercises**

In Exercises 1–10, determine the center, foci, and vertices of the ellipse and sketch its graph.

1. $\dfrac{x^2}{16} + \dfrac{y^2}{4} = 1$

2. $\dfrac{x^2}{12} + \dfrac{y^2}{25} = 1$

3. $4x^2 + 9y^2 = 36$

4. $20x^2 + 5y^2 = 100$

5. $\dfrac{(x-2)^2}{36} + \dfrac{(y-4)^2}{25} = 1$

6. $\dfrac{(x-1)^2}{9} + \dfrac{(y+4)^2}{49} = 1$

7. $9(x-5)^2 + 25(y+1)^2 = 225$

8. $(x+7)^2 + 9(y-2)^2 = 9$

In Exercises 9–18, determine the standard equation of each ellipse using the given graph or the stated information.

9.

10.

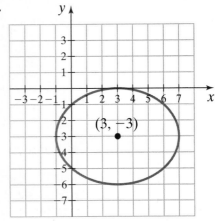

11. Foci at $(-5, 0)$ and $(5, 0)$; length of the major axis is fourteen units.

12. Foci at $(8, -1)$ and $(-2, -1)$; length of the major axis is twelve units.

13. Vertices at $(-6, 10)$ and $(-6, 0)$; length of the minor axis is eight units.

14. Center at $(4, 5)$; vertical minor axis with length sixteen units; $c = 6$.

15. Center at $(-1, 4)$; vertex at $(-1, 8)$; focus at $(-1, 7)$.

16. Vertices at $(2, -7)$ and $(2, 5)$; focus at $(2, 3)$.

17. Center at $(4, 1)$; focus at $(4, 8)$; ellipse passes through the point $(6, 1)$.

18. Center at $(1, 3)$; focus at $(1, 6)$; ellipse passes through the point $(2, 3)$.

In Exercises 19 and 20, determine the standard equation of each ellipse by first using **Fact 1** to find a and then by using **Fact 2** to find b.

19. The foci have coordinates $(-2, 0)$ and $(2, 0)$. The ellipse contains the point $(2, 3)$.

20. The foci have coordinates $(0, -8)$ and $(0, 8)$. The ellipse contains the point $(-2, 6)$.

In Exercises 21–24, complete the square to write each equation in the form $\dfrac{(x - h)^2}{a^2} + \dfrac{(y - k)^2}{b^2} = 1$ or $\dfrac{(x - h)^2}{b^2} + \dfrac{(y - k)^2}{a^2} = 1$. Determine the center, foci, and vertices of the ellipse and sketch its graph.

21. $16x^2 + 20y^2 + 64x - 40y - 236 = 0$

22. $9y^2 + 16x^2 + 224x + 54y + 721 = 0$

23. $x^2 + 4x + 16y^2 - 32y + 4 = 0$

24. $50x^2 + y^2 + 20y = 0$

25. An elliptical arch railroad tunnel 16 feet high at the center and 30 feet wide is cut through the side of a mountain. Find the equation of the ellipse if a vertex is represented by the point $(0, 16)$.

16 ft

30 ft

26. A 78-inch by 36-inch door contains a decorative glass elliptical pattern. One vertex of the ellipse is located 9 inches from the top of the door. The other vertex is located 27 inches from the bottom of the door. The endpoints of the minor axis are located 9 inches from each side. Find the equation of the elliptical pattern. (Assume that the center of the elliptical pattern is at the origin.)

36 in.

9 in.

9 in.   9 in.

78 in.

27 in.

27. A rectangular playing field lies in the interior of an elliptical track that is 50 yards wide and 140 yards long. What is the width of the rectangular playing field if the width is located 10 yards from either vertex?

28. A patient is placed in an elliptical tank that is 180 centimeters long and 100 centimeters wide to undergo sound wave lithotripsy treatment for kidney stones. Determine where the sound emitter and the stone should be positioned relative to the center of the tank. (Round to the nearest hundredth of a centimeter.)

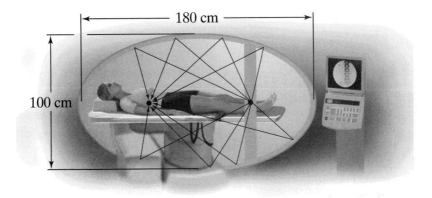

29. A window is constructed with the top half of an ellipse on top of a square. The square portion of the window has a 36-inch base. If the window is 50 inches tall at its highest point, find the height, h, of the window 14 inches from the center of the base. (Round the height to the nearest hundredth of an inch.)

30. A government spy satellite is in an elliptical orbit around the Earth with the center of the Earth at a focus. If the satellite is 150 miles from the surface of the Earth at one vertex of the orbit and 500 miles from the surface of the Earth at the other vertex, find the equation of the elliptical orbit. (Assume that the Earth is a sphere with a diameter of 8,000 miles.)

31. An elliptical arch bridge spans 160 feet. The elliptical arch has a maximum height of 40 feet. What is the height of the arch at a distance of 10 feet from the center?

6.3 **The Hyperbola**

THINGS TO KNOW

Before working through this section, be sure that you are familiar with the following concepts:

VIDEO   ANIMATION   INTERACTIVE

You Try It

1. Finding the Distance between Two Points Using the Distance Formula (**Section 2.1**)

You Try It

2. Finding the Equation of a Line Using the Point–Slope Form (**Section 2.3**)

You Try It

3. Completing the Square to Find the Equation of an Ellipse in Standard Form (**Section 6.2**)

OBJECTIVES

1 Sketching the Graph of a Hyperbola

2 Determining the Equation of a Hyperbola in Standard Form

3 Completing the Square to Find the Equation of a Hyperbola in Standard Form

4 Solving Applied Problems Involving Hyperbolas

....

When a plane intersects two **right circular cones** at the same time, the conic section formed is a hyperbola. (Click on the animation.)

    The set of points in the plane that defines a hyperbola formed by the intersection of a plane and the cones described previously is stated in the following geometric definition.

The Hyperbola

Geometric Definition of the Hyperbola

A **hyperbola** is the set of all points in a plane, the difference of whose distances from two fixed points is a positive constant. The two fixed points, F_1 and F_2, are called the foci.

Notice that the previous geometric definition is very similar to the **geometric definition of the ellipse**. Recall that for a point to lie on the graph of an ellipse, the **sum** of the distances from the point to the two foci is constant. For a point to lie on the graph of a hyperbola, the **difference** between the distances from the point to the two foci is constant. Because subtraction is not **commutative**, we consider the absolute value of the difference in the distances between a point on the hyperbola and the foci to ensure that the constant is positive. Thus, for any two points P and Q that lie on the graph of a hyperbola, $\left|d(P, F_1) - d(P, F_2)\right| = \left|d(Q, F_1) - d(Q, F_2)\right|$. See Figure 21.

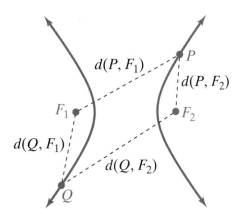

Figure 21 For any points P and Q that lie on the graph of a hyperbola, $\left|d(P, F_1) - d(P, F_2)\right| = \left|d(Q, F_1) - d(Q, F_2)\right|$.

OBJECTIVE 1:   SKETCHING THE GRAPH OF A HYPERBOLA

My video summary

The graph of a hyperbola has two branches. These branches look somewhat like parabolas, but they are certainly not because the branches do not satisfy the geometric definition of the parabola. Every hyperbola has a center, two vertices, and two foci. The vertices are located at the endpoints of an invisible line segment called the **transverse axis**. The transverse axis is either parallel to the x-axis (horizontal transverse axis) or parallel to the y-axis (vertical transverse axis). The center of a hyperbola is located midway between the two vertices (or two foci). The hyperbola has another invisible line segment called the **conjugate axis** that passes through the center and lies perpendicular to the transverse axis. Each branch of the hyperbola approaches (but never intersects) a pair of lines called *asymptotes*. A **reference rectangle** is typically used as a guide to help sketch the asymptotes. The reference rectangle is a rectangle whose midpoints of each side are the vertices of the hyperbola or the endpoints of the conjugate axis. The asymptotes pass diagonally through opposite corners of the reference rectangle. Watch the video to see how to sketch the graphs of the two hyperbolas shown in Figure 22.

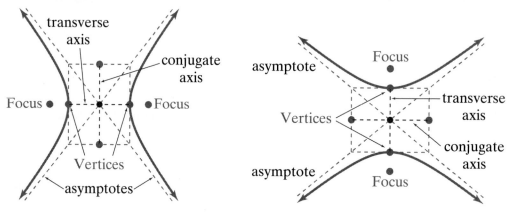

Figure 22 (a) Hyperbola with a Horizontal Transverse Axis (b) Hyperbola with a Vertical Transverse Axis

To derive the equation of a hyperbola, consider a hyperbola with a horizontal transverse axis centered at (h, k). If the distance between the center and either vertex is $a > 0$, then the coordinates of the vertices are $V_1(h - a, k)$ and $V_2(h + a, k)$, and the length of the transverse axis is equal to $2a$. If $c > 0$ is the distance between the center and either foci, then the foci have coordinates $F_1(h - c, k)$ and $F_2(h + c, k)$. See Figure 23.

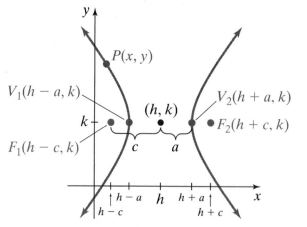

Figure 23

By the geometric definition of a hyperbola, we know that for any point $P(x, y)$ that lies on the hyperbola, the difference of the distances from P to the two foci is a constant. It can be shown that this constant is equal to $2a$, the length of the transverse axis. We now state this fact and denote it as Fact 1 for Hyperbolas.

Fact 1 for Hyperbolas

Given the foci of a hyperbola F_1 and F_2 and any point P that lies on the graph of the hyperbola, the difference of the distance between P and the foci is equal to $2a$. In other words, $\left| d(P, F_1) - d(P, F_2) \right| = 2a$. The constant $2a$ represents the length of the transverse axis.

We prove Fact 1 for Hyperbolas in Appendix A. Click here to see this proof.

Once we know that the constant stated in the geometric definition is equal to $2a$, we can derive the equation of a hyperbola. The derivation of the equation of a hyperbola can be seen in Appendix A. Click here to see this derivation. We now state the two standard equations of hyperbolas.

Equation of a Hyperbola in Standard Form with Center (h, k)

Horizontal Transverse Axis

$$\frac{(x - h)^2}{a^2} - \frac{(y - k)^2}{b^2} = 1$$

- Foci: $F_1(h - c, k)$ and $F_2(h + c, k)$
- Vertices: $V_1(h - a, k)$ and $V_2(h + a, k)$
- Endpoints of conjugate axis:
  $(h, k - b)$ and $(h, k + b)$
- $b^2 = c^2 - a^2$
- Asymptotes: $y - k = \pm\frac{b}{a}(x - h)$

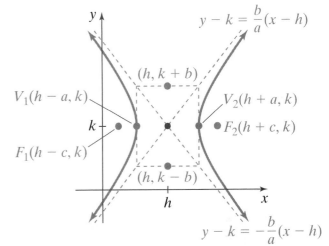

Vertical Transverse Axis

$$\frac{(y - k)^2}{a^2} - \frac{(x - h)^2}{b^2} = 1$$

- Foci: $F_1(h, k - c)$ and $F_2(h, k + c)$
- Vertices: $V_1(h, k - a)$ and $V_2(h, k + a)$
- Endpoints of conjugate axis:
  $(h - b, k)$ and $(h + b, k)$
- $b^2 = c^2 - a^2$
- Asymptotes: $y - k = \pm\frac{a}{b}(x - h)$

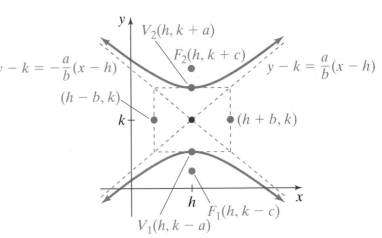

The two hyperbola equations are nearly identical except for two differences. The x-terms are positive and the y-terms are negative in the equation with a horizontal transverse axis. The signs are reversed in the equation with a vertical transverse axis. Also note that a^2 appears in the denominator of the positive squared term in each equation. We can derive the equations of the asymptotes by using the **point–slope form of the equation of a line.**

Note: If $h = 0$ and $k = 0$, then the hyperbola is centered at the **origin**. Hyperbolas centered at the origin have the following equations.

Standard Equations of a Hyperbola with the Center at the Origin

$$\frac{x^2}{a^2} - \frac{y^2}{b^2} = 1 \qquad\qquad \frac{y^2}{a^2} - \frac{x^2}{b^2} = 1$$

(a)

(b)

My interactive video summary

📹 **Example 1 Sketch the Graph of a Hyperbola in Standard Form**

Sketch the following hyperbolas. Determine the center, transverse axis, vertices, and foci and find the equations of the asymptotes.

a. $\dfrac{(y-4)^2}{36} - \dfrac{(x+5)^2}{9} = 1$ **b.** $25x^2 - 16y^2 = 400$

Solution Watch the interactive video to see how to sketch each hyperbola shown in Figure 24(a) and Figure 24(b).

Figure 24 (a)

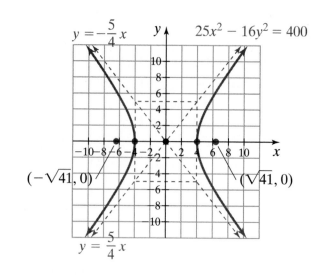

Figure 24 (b)

Work exercises 1–6 in this textbook or in the ***MyMathLab*** ® Study Plan.

OBJECTIVE 2:   DETERMINING THE EQUATION OF A HYPERBOLA
IN STANDARD FORM

My video summary

▤ Example 2 Find the Equation of a Hyperbola

Find the equation of the hyperbola with the center at $(-1, 0)$, a focus at $(-11, 0)$, and a vertex at $(5, 0)$.

Solution A focus and a vertex lie along the x-axis. This indicates that the hyperbola has a horizontal transverse axis. Because the center is at $(-1, 0)$, we know that the equation of the hyperbola is $\frac{(x + 1)^2}{a^2} - \frac{y^2}{b^2} = 1$. Watch the video to verify that the equation in standard form is $\frac{(x + 1)^2}{36} - \frac{y^2}{64} = 1$. The equations of the asymptotes are $y = -\frac{4}{3}(x + 1)$ and $y = \frac{4}{3}(x + 1)$. The graph of this hyperbola is shown in Figure 25.

Figure 25

Using Technology

Figure 26 We can use a graphing utility to graph the parabola from Example 2 by solving the equation for y:

$$y = \pm \frac{4}{3}\sqrt{(x + 1)^2 - 36}$$

(Click here to see how to solve for y.)
Using $y_1 = \frac{4}{3}\sqrt{(x + 1)^2 - 36}$,

$y_2 = -\frac{4}{3}\sqrt{(x + 1)^2 - 36}$, $y_3 = -\frac{4}{3}(x + 1)$,

and $y_4 = -\frac{4}{3}(x + 1)$, we obtain the graph seen in Figure 26.

Work exercises 7–14 in this textbook or in the *MyMathLab*® Study Plan.

OBJECTIVE 3: COMPLETING THE SQUARE TO FIND THE EQUATION OF A HYPERBOLA IN STANDARD FORM

If the equation of a hyperbola is in the standard form of $\frac{(x - h)^2}{a^2} - \frac{(y - k)^2}{b^2} = 1$ or $\frac{(y - k)^2}{a^2} - \frac{(x - h)^2}{b^2} = 1$, then it is not too difficult to determine the center, vertices, foci, and asymptotes and sketch its graph. However, the equation might not be given in standard form. If this is the case, we may need to complete the square on both variables as in Example 3.

My video summary

Example 3 Rewrite the Equation of a Hyperbola in Standard Form by Completing the Square

Find the center, vertices, foci, and equations of asymptotes and sketch the hyperbola $12x^2 - 4y^2 - 72x - 16y + 140 = 0$.

Solution Rearrange the terms leaving some room to complete the square, and move any constants to the right-hand side:

$$12x^2 - 72x \qquad - 4y^2 - 16y \qquad = -140 \qquad \text{Rearrange the terms.}$$

Then factor and complete the square.

$$12(x^2 - 6x \quad) - 4(y^2 + 4y \quad) = -140 \qquad \text{Factor out 12 and } -4.$$

$$12(x^2 - 6x + 9) - 4(y^2 + 4y + 4) = -140 + 108 - 16 \qquad \begin{array}{l}\textit{Complete the} \\ \textit{square on x and y.} \\ \textit{Remember to add} \\ 12 \cdot 9 = 108 \textit{ and} \\ -4 \cdot 4 = -16 \textit{ to the} \\ \textit{right side.}\end{array}$$

$$12(x - 3)^2 - 4(y + 2)^2 = -48 \qquad \begin{array}{l}\textit{Factor the left side,} \\ \textit{and simplify the right} \\ \textit{side.}\end{array}$$

$$\frac{12(x - 3)^2}{-48} - \frac{4(y + 2)^2}{-48} = \frac{-48}{-48} \qquad \begin{array}{l}\textit{Divide both sides} \\ \textit{by } -48.\end{array}$$

$$-\frac{(x - 3)^2}{4} + \frac{(y + 2)^2}{12} = 1 \qquad \textit{Simplify.}$$

$$\frac{(y + 2)^2}{12} - \frac{(x - 3)^2}{4} = 1 \qquad \textit{Rewrite the equation.}$$

The equation is now in standard form. Watch the video to see this example worked out in detail. You should verify that this is the equation of a hyperbola with a vertical transverse axis with center $(3, -2)$. The vertices have coordinates $(3, -2 - 2\sqrt{3})$ and $(3, -2 + 2\sqrt{3})$. The foci have coordinates $(3, -6)$ and $(3, 2)$. The equations of the asymptotes are $y + 2 = -\sqrt{3}(x - 3)$ and $y + 2 = \sqrt{3}(x - 3)$.

Work exercises 15–20 in this textbook or in the *MyMathLab*® Study Plan.

OBJECTIVE 4: SOLVING APPLIED PROBLEMS INVOLVING HYPERBOLAS

Hyperbolas have many applications. We saw in Section 6.2 that the planets in our solar system and some comets, such as Halley's comet, travel through our solar system in elliptical orbits. However, some comets are only seen once in our solar system because they travel through the solar system on the **path of a** hyperbola with the Sun at a focus. On October 14, 1947, Chuck Yeager became the first person to break the sound barrier. As an airplane moves faster than the speed of sound, a **cone-shaped shock wave** is produced. The cone intersects the ground in the shape of a hyperbola. When two rocks are simultaneously tossed into a calm pool of water, ripples move outward in the form of **concentric circles**. These circles intersect in points that form a hyperbola. Hyperbolas can be used to locate ships by sending radio signals simultaneously from radio transmitters placed at some fixed distance apart. A device measures the difference in the time it takes the radio signals to reach the ship. The equation of a hyperbola can then be determined to describe the current path of the ship. If three transmitters are used, two hyperbolic equations can be determined. The precise location of the ship can be determined by finding the intersection of the two hyperbolas. This system of locating ships is known as long-range navigation or LORAN. See Example 4.

Example 4 Use a Hyperbola to Locate a Ship

One transmitting station is located 100 miles due east from another transmitting station. Each station simultaneously sends out a radio signal. The signal from the west tower is received by a ship $\frac{1,600}{3}$ microseconds after the signal from the east tower. If the radio signal travels at 0.18 miles per microsecond, find the equation of the hyperbola on which the ship is presently located.

Solution Start by plotting the two foci of the hyperbola at points $F_1(-50, 0)$ and $F_2(50, 0)$. These two points represent the position of the two transmitting towers. Note that $c = 50$. Because the hyperbola is centered at the origin with a horizontal transverse axis, the equation must be of the form $\frac{x^2}{a^2} - \frac{y^2}{b^2} = 1$. The difference in the distances from the two transmitters to the ship is $\left(\frac{1,600}{3}\text{ microseconds}\right) \cdot \left(0.18\frac{\text{miles}}{\text{microsecond}}\right) = 96$ miles. This distance represents the constant stated in the **Fact 1 for Hyperbolas**. Therefore, $2a = 96$, so $a = 48$ or $a^2 = 2,304$. To find b^2, we use the fact that $b^2 = c^2 - a^2$.

$$b^2 = c^2 - a^2$$
$$b^2 = 50^2 - 48^2$$
$$b^2 = 196$$

We now substitute the values of $a^2 = 2,304$ and $b^2 = 196$ into the previous equation to obtain the equation $\frac{x^2}{2,304} - \frac{y^2}{196} = 1$.

Work exercises 21–25 or in this textbook in the _MyMathLab_® **Study Plan.**

6.3 Exercises

In Exercises 1–6, determine the center, transverse axis, vertices, foci, and the equations of the asymptotes and sketch the hyperbola.

1. $\dfrac{x^2}{16} - \dfrac{y^2}{9} = 1$ **2.** $\dfrac{y^2}{9} - \dfrac{x^2}{16} = 1$ **3.** $\dfrac{(y-4)^2}{25} - \dfrac{(x-2)^2}{36} = 1$ **4.** $\dfrac{(x+1)^2}{9} - \dfrac{(y+3)^2}{49} = 1$

5. $20x^2 - 5y^2 = 100$ **6.** $20(x-1)^2 - 16(y-3)^2 = -320$

In Exercises 7–12, determine the standard equation of the hyperbola with the given characteristics and sketch the graph.

7. The center is at $(0, 0)$, a focus is at $(5, 0)$, and a vertex is at $(3, 0)$.

8. The center is at $(0, 0)$, a focus is at $(0, 10)$, and a vertex is at $(0, -6)$.

9. The center is at $(4, -4)$, a focus is at $(6, -4)$, and a vertex is at $(5, -4)$.

10. The center is at $(-6, -1)$, a focus is at $(-6, -9)$, and a vertex is at $(-6, -5)$.

11. The foci are at $(9, 3)$ and $(9, 9)$; the vertex is at $(9, 8)$.

12. The vertices are at $(-2, -3)$ and $(10, -3)$; an asymptote has equation $y + 3 = \dfrac{7}{6}(x - 4)$.

In Exercises 13 and 14, determine the equation of the hyperbola with the given characteristics and sketch the graph. *Hint:* $\left| d(P, F_1) - d(P, F_2) \right| = 2a$.

13. The hyperbola has foci with coordinates $F_1(0, -6)$ and $F_2(0, 6)$ and passes through the point $P(8, 10)$.

14. The hyperbola has foci with coordinates $F_1(-6, 1)$ and $F_2(10, 1)$ and passes through the point $P(10, 13)$.

In Exercises 15–20, complete the square to write each equation in the form $\dfrac{(x-h)^2}{a^2} - \dfrac{(y-k)^2}{b^2} = 1$ or $\dfrac{(y-k)^2}{a^2} - \dfrac{(x-h)^2}{b^2} = 1$. Determine the center, vertices, foci, endpoints of the conjugate axis, and the equations of the asymptotes of the hyperbola, and sketch its graph.

15. $x^2 - y^2 - 4x + 2y - 1 = 0$ **16.** $x^2 - y^2 + 8x - 6y + 9 = 0$

17. $y^2 - 9x^2 - 12y - 36x - 9 = 0$ **18.** $x^2 - 16y^2 + 10x + 64y - 55 = 0$

19. $25y^2 - 144x^2 + 1{,}728x + 400y + 16 = 0$ **20.** $49y^2 - 576x^2 - 98y - 1{,}152x - 28{,}751 = 0$

21. A light on a wall produces a shadow in the shape of a hyperbola. If the distance between the two vertices of the hyperbola is 14 inches and the distance between the two foci is 16 inches, find the equation of the hyperbola.

22. This figure shows the hyperbolic orbit of a comet with the center of the Sun positioned at a focus, 100 million miles from the origin. (The units are in millions of miles.) The comet will be 50 million miles from the center of the Sun at its nearest point during the orbit. Find the equation of the hyperbola describing the comet's orbit.

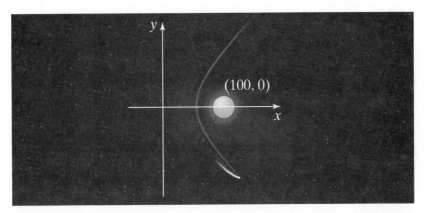

23. A nuclear power plant has a large cooling tower with sides curved in the shape of a hyperbola. The radius of the base of the tower is 60 meters. The radius of the top of the tower is 50 meters. The sides of the tower are 60 meters apart at the closest point located 90 meters above the ground.

    a. Find the equation of the hyperbola that describes the sides of the cooling tower. (Assume that the center is at the origin.)

    b. Determine the height of the tower. (Round your answer to the nearest meter.)

24. One transmitting station is located 80 miles north of another transmitting station. Each station simultaneously sends out a radio signal. The signal from the north station is received 200 microseconds after the signal from the south station. If the radio signal travels at 0.18 miles per microsecond, find the equation of the hyperbola on which the ship is presently located.

CHAPTER SEVEN

Systems
of Equations
and Matrices

CHAPTER SEVEN CONTENTS

7.1 Systems of Linear Equations in Two Variables

THINGS TO KNOW

Before working through this section, be sure that you are familiar with the following concepts:

VIDEO   ANIMATION   INTERACTIVE

 You Try   1. Applications of Linear Equations (Section 1.2)

You Try   2. Writing the Equation of a Line in Standard Form (Section 2.3)

You Try   3. Sketching Lines by Plotting Intercepts (Section 2.3)

 You Try   4. Understanding the Definition of Parallel Lines (Section 2.4)

OBJECTIVES

1 Verifying Solutions to a System of Linear Equations in Two Variables

2 Solving a System of Linear Equations Using the Substitution Method

3 Solving a System of Linear Equations Using the Elimination Method

4 Solving Applied Problems Using a System of Linear Equations

---

OBJECTIVE 1:   VERIFYING SOLUTIONS TO A SYSTEM OF LINEAR EQUATIONS IN TWO VARIABLES

My video summary

In Section 1.1, a linear equation in one variable was defined as an equation that can be written in the form $ax + b = c$, where a, b, and c are real numbers with $a \neq 0$. This definition can be extended to more variables as follows.

DEFINITION Linear Equation in n Variables

A linear equation in n variables is an equation that can be written in the form $a_1x_1 + a_2x_2 + \cdots + a_nx_n = b$ for variables $x_1, x_2, \ldots x_n$ and real numbers $a_1, a_2, \ldots a_n, b$, where at least one of $a_1, a_2, \ldots a_n$ is nonzero.

For example, $2x - 5y = -7$ is a linear equation in two variables, whereas $4x_1 + 9x_2 - 17x_3 = 11$ is a linear equation in three variables. The important thing to remember is that all variables of a linear equation have an exponent of 1. For much of this chapter, we learn how to solve **systems of linear equations**. We start our discussion of linear systems by considering linear systems in two variables.

DEFINITION System of Linear Equations in Two Variables

A system of linear equations in two variables is the collection of two linear equations in two variables considered simultaneously. The solution to a system of equations in two variables is the set of all ordered pairs for which *both* equations are true.

Consider the following three systems of linear equations in two variables.

$$3x - 2y = -9 \qquad -3x_1 + 2x_2 = 11 \qquad \sqrt{2}a + b = \pi$$
$$x + y = 2 \qquad 8x_1 - 9x_2 = -2 \qquad -4a + \sqrt{3}b = 1$$

In each system, different variables were used. It does not matter what we call the two variables as long as both equations are linear. According to the definition of a system of linear equations in two variables, the solution is the set of all ordered pairs for which both equations are true. We verify that the ordered pair $(-1, 3)$ is a solution to the system $\begin{array}{l} 3x - 2y = -9 \\ x + y = 2 \end{array}$ in Example 1.

Example 1 Verify the Solution to a System of Linear Equations in Two Variables

Show that the ordered pair $(-1, 3)$ is a solution to the system

$$3x - 2y = -9$$
$$x + y = 2.$$

Solution We can verify that the ordered pair $(-1, 3)$ is a solution to this system by substituting -1 for x and 3 for y in both equations.

First Equation

$$3x \quad - \quad 2y \quad = -9$$
$$3(-1) - 2(3) \overset{?}{=} -9 \qquad \text{\textit{Substitute} } x = -1$$
$$\text{\textit{and} } y = 3.$$
$$-9 = -9 \,\checkmark \qquad \text{\textit{Both equations are true.}}$$

Second Equation

$$x \quad + \quad y = 2$$
$$(-1) + (3) \overset{?}{=} 2$$
$$2 = 2 \,\checkmark$$

Both equations yielding true statements indicate that the ordered pair $(-1, 3)$ is a solution to this system.

Work exercises 1–4 in this textbook or in the *MyMathLab*® Study Plan.

If a linear system has at least one solution, like the system shown in Example 1, it is said to be **consistent**, and the solution is the set of all ordered pairs that satisfy both equations. If the system does not have a solution, it is said to be **inconsistent**. The solution to a system of two linear equations involving two variables can be viewed geometrically. Because the graph of each equation of the system is a line, we can sketch the two lines and geometrically view the solution. There are only three possibilities for the graph of a linear system of equations involving two variables:

1. The lines can have different slopes and thus intersect at a single point (Figure 1a). The solution is the ordered pair representing the point of intersection. This system is said to be **consistent**. Consistent systems have at least one solution. The equations are said to be **independent**.

2. The two lines can be parallel (have the same slope) and thus have no intersecting point (Figure 1b). This system is said to be **inconsistent** and has no solution. The equations are **independent**.

3. The two equations can be different representations of the same line and have the same graph (Figure 1c). In this case, there are infinitely many solutions. Because this system has at least one solution, it is **consistent**. The equations are said to be **dependent**.

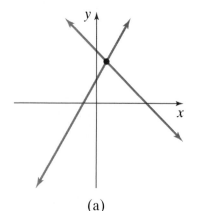

(a)
Consistent, Independent
One solution
Two lines are different, having one common point. The lines have different slopes.

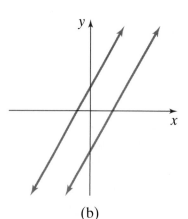

(b)
Inconsistent, Independent
No solution
Two lines are different, having no common points. The lines have the same slope but different y-intercepts.

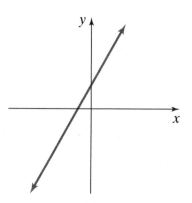

(c)
Consistent, Dependent
Infinitely many solutions
Two lines are the same, having infinitely many common points. The lines have the same slope and same y-intercepts.

Figure 1

We now learn two algebraic methods used for solving linear systems involving two equations and two variables. These methods are known as the *substitution method* and the *elimination method.*

OBJECTIVE 2:   SOLVING A SYSTEM OF LINEAR EQUATIONS USING THE SUBSTITUTION METHOD

My video summary

 The substitution method involves solving one of the equations for one variable in terms of the other, and then substituting that expression into the other equation. The substitution method can be summarized in the following four steps.

Solving a System of Equations by the Method of Substitution

Step 1   Choose an equation and solve for one variable in terms of the other variable.

Step 2   Substitute the expression from step 1 into the other equation.

Step 3   Solve the equation in one variable.

Step 4   Substitute the value found in step 3 into one of the original equations to find the value of the other variable.

The substitution method is used to solve the system in Example 2.

Example 2 Solve a System of Equations Using the Substitution Method

Solve the following system using the method of substitution: $\begin{array}{l} 2x - 3y = -5 \\ x + y = 5 \end{array}$.

Solution

Step 1   If possible, choose an equation in which at least one **coefficient** is 1. In this example, both coefficients of the second equation are 1. We can choose to solve the second equation for either variable. In this case, we solve the second equation for y:

$$x + y = 5 \qquad \textit{Write the second equation.}$$
$$y = 5 - x \qquad \textit{Subtract } x \textit{ from both sides.}$$

Step 2   Substitute $5 - x$ for y in the first equation:

$$2x - 3(5 - x) = -5$$

Step 3   Solve for x:

$$2x - 3(5 - x) = -5 \qquad \textit{Rewrite the equation from step 2.}$$
$$2x - 15 + 3x = -5 \qquad \textit{Use the distributive property.}$$
$$5x - 15 = -5 \qquad \textit{Combine like terms.}$$
$$5x = 10 \qquad \textit{Add 15 to both sides.}$$
$$x = 2 \qquad \textit{Divide both sides by 5.}$$

Step 4   Because $x = 2$, we can solve for y by substituting $x = 2$ into one of the original equations. Here we use the simplified equation found in step 1.

$$y = 5 - (2) = 3$$

The solution of this system is the ordered pair $(2, 3)$. You can see in Figure 2 that this is the intersection of the two lines of this system.

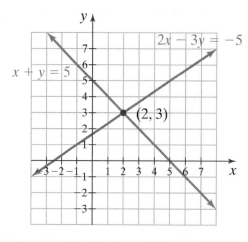

Figure 2 The solution to the system

$$2x - 3y = -5$$
$$x + y = 5$$

is the ordered pair $(2, 3)$.

Work exercises 5–9 in this textbook or in the *MyMathLab*® Study Plan.

OBJECTIVE 3: SOLVING A SYSTEM OF LINEAR EQUATIONS USING THE ELIMINATION METHOD

Another method used to solve a system of two equations is called the elimination method. The elimination method involves adding the two equations together in an attempt to eliminate one of the variables. To accomplish this task, the coefficients of one of the variables must differ only in sign. This can be done by multiplying one or both of the equations by a suitable constant. The elimination method can be summarized in the following five steps.

Solving a System of Equations by the Method of Elimination

Step 1 Choose a variable to eliminate.

Step 2 Multiply one or both equations by an appropriate nonzero constant so that the sum of the coefficients of one of the variables is zero.

Step 3 Add the two equations together to obtain an equation in one variable.

Step 4 Solve the equation in one variable.

Step 5 Substitute the value obtained in step 4 into one of the original equations to solve for the other variable.

The elimination method is used to solve the system in Example 3.

Example 3 Solve a System of Equations Using the Elimination Method

Solve the following system using the method of elimination: $\begin{array}{l} -2x + 5y = 29 \\ 3x + 2y = 4 \end{array}$.

Solution

Step 1 Here we choose to eliminate the variable x.

Step 2 Multiply both sides of the first equation by 3, and multiply both sides of second equation by 2, thus making the coefficient of the x terms in the two equations equal to -6 and 6, respectively.

$$-2x + 5y = 29 \xrightarrow{\text{Multiply by 3}} 3(-2x + 5y) = 3(29) \longrightarrow -6x + 15y = 87$$

$$3x + 2y = 4 \xrightarrow{\text{Multiply by 2}} 2(3x + 2y) = 2(4) \longrightarrow 6x + 4y = 8$$

Step 3   We now add the two new equations together:

$$-6x + 15y = 87$$
$$\underline{6x + 4y = 8}$$
$$19y = 95$$

Step 4   Solve for y:

$$19y = 95$$
$$y = 5 \qquad \textit{Divide both sides by 19.}$$

Step 5   To solve for x, we substitute $y = 5$ into one of the original equations. Here, we choose the first equation:

$$-2x + 5y = 29 \qquad \textit{Write the original first equation.}$$
$$-2x + 5(5) = 29 \qquad \textit{Substitute 5 for y in the first equation.}$$
$$-2x + 25 = 29 \qquad \textit{Simplify.}$$
$$-2x = 4 \qquad \textit{Subtract 25 from both sides.}$$
$$x = -2 \qquad \textit{Divide both sides by } -2.$$

Therefore, the solution is the ordered pair $(-2, 5)$.

Using Technology

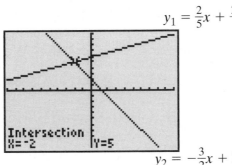

$$y_1 = \frac{2}{5}x + \frac{29}{5}$$

$$y_2 = -\frac{3}{2}x + 2$$

We can solve the system in Example 3 using a graphing utility by first solving both equations for y:

$$-2x + 5y = 29 \longrightarrow y_1 = \frac{2}{5}x + \frac{29}{5}$$

$$3x + 2y = 4 \longrightarrow y_2 = -\frac{3}{2}x + 2$$

Graph y_1 and y_2, and then use the INTERSECT feature to find that the point of intersection is $(-2, 5)$.

Work exercises 10–14 in this textbook or in the *MyMathLab*® Study Plan.

The systems in Examples 2 and 3 had one solution. We now look at an example of a system with no solution, and then solve a system that has infinitely many solutions.

Example 4 Solve a System with No Solution

Solve the system $\begin{array}{l} x - 2y = 11 \\ -2x + 4y = 8 \end{array}$.

Solution We can solve this system using either the substitution method or the elimination method. In this case, we use the elimination method.

Step 1   We choose to eliminate the variable x because it has a **coefficient** of 1.

Step 2   Multiply both sides of the first equation by 2, thus making the coefficient of the x-terms in the two equations equal to 2 and -2, respectively.

$$x - 2y = 11 \xrightarrow{\text{Multiply by 2}} 2x - 4y = 22$$
$$-2x + 4y = 8 \xrightarrow{\text{Leave equation alone}} -2x + 4y = 8$$

7.1 Systems of Linear Equations in Two Variables   **313**

Step 3   We now add the two new equations together:

$$2x - 4y = 22$$
$$\underline{-2x + 4y = 8}$$
$$0 = 30$$

Notice how both variables were eliminated, leaving us with an untrue statement. The number 0 is never equal to 30. This means that there is no solution to this system. This is an example of an **inconsistent** system. You can see in Figure 3 that the graphs of the two linear equations from this system are parallel. Thus, the two lines do not intersect, and the system has no solution.

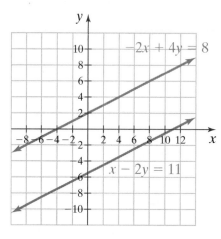

Figure 3 The linear system
$$x - 2y = 11$$
$$-2x + 4y = 8$$ has no solution.

Example 5 Solve a System Having Infinitely Many Solutions

Solve the system $\begin{array}{l} -3x + 6y = 9 \\ x - 2y = -3 \end{array}$.

Solution We can solve this system using either the substitution method or the elimination method. Here, we choose the substitution method.

Step 1   We start by solving the second equation for x:

$$x - 2y = -3 \qquad \textit{Write the second equation.}$$
$$x = 2y - 3 \qquad \textit{Add } 2y \textit{ to both sides.}$$

Step 2   Substitute $2y - 3$ for x in the first equation:

$$-3(2y - 3) + 6y = 9$$

Step 3   Solve for y:

$$-3(2y - 3) + 6y = 9 \qquad \textit{Rewrite the equation from step 2.}$$
$$-6y + 9 + 6y = 9 \qquad \textit{Use the distributive property.}$$
$$9 = 9 \qquad \textit{Combine like terms.}$$

The equation simplifies to $9 = 9$, which is true for *any* value of y. Therefore, this system has infinitely many solutions. Note that we could have multiplied the second equation in the system by -3 to obtain the first equation. Because of this, both equations represent the same line and thus have the same graph (Figure 4). The solution is the set of all ordered pairs that lie on the graph of either equation. In step 1, we wrote x in terms of y as ; thus, the solution to this system is the set of all ordered pairs of the form $(2y - 3, y)$, where y is any real number. For example,

If $y = -2$, then $x = 2(-2) - 3 = -7$, so the ordered pair $(-7, -2)$ is a solution.

If $y = 0$, then $x = 2(0) - 3 = -3$, so the ordered pair $(-3, 0)$ is another solution.

Similarly, we could have solved either equation for y to obtain $y = \frac{1}{2}x + \frac{3}{2}$. So, the solutions can also be expressed as $\left(x, \frac{1}{2}x + \frac{3}{2}\right)$.

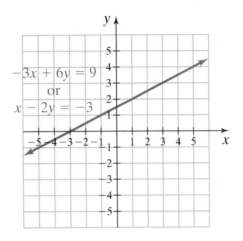

Figure 4 The system $\begin{array}{rcr} -3x + 6y &=& 9 \\ x - 2y &=& -3 \end{array}$ has infinitely many solutions

Work exercises 15–20 in this textbook or in the *MyMathLab*® Study Plan.

OBJECTIVE 4:   SOLVING APPLIED PROBLEMS USING A SYSTEM OF LINEAR EQUATIONS

Applied problems can involve two or more unknown quantities. Sometimes we are able to use a single equation involving one variable to solve such problems as seen in Section 1.2. However, it is often easier to use two variables and create a system of two equations. The following steps are based on the four-step strategy for problem solving that was first introduced in Section 1.2.

Five-Step Strategy for Problem Solving Using Systems of Equations

Step 1   Read the problem several times until you have an understanding of what is being asked. If possible, create diagrams, charts, or tables to assist you.   *Understand the problem.*

Step 2   Choose variables that describe each unknown quantity.   *Devise a plan.*

Step 3   Write a system of equations using the given information and the variables.

Step 4   Carefully solve the system of equations using the method of elimination or substitution.   *Carry out the plan.*

Step 5   Make sure that you have answered the question, and check all answers to ensure they make sense.   *Look back.*

We start with an example that was seen in Section 1.2. This time, we use two variables to solve the problem.

Example 6 Determine the Number of Touchdowns Thrown

Roger Staubach and Terry Bradshaw were both quarterbacks in the National Football League. In 1973, Staubach threw three touchdown passes more than twice the number of touchdown passes thrown by Bradshaw. If the total number of touchdown passes between Staubach and Bradshaw was 33, how many touchdown passes did each player throw?

Solution

Step 1 After carefully reading the problem, we see that we are trying to figure out how many touchdown passes were thrown by each player.

Step 2 Let S = number of touchdowns thrown by Staubach

B = number of touchdowns thrown by Bradshaw

Step 3 The combined number of touchdown passes is 33, so the first equation is

$$\underbrace{S}_{\substack{Staubach's \\ touchdown\ passes}} + \underbrace{B}_{\substack{Bradshaw's \\ touchdown\ passes}} = \underbrace{33}_{\substack{Total\ touchdown \\ passes}}$$

To find the second equation, we use the fact that Staubach threw three more than twice the number of touchdowns thrown by Bradshaw. Therefore, the second equation is $S = 2B + 3$. This equation can be rewritten as $S - 2B = 3$. Therefore, the system of equations is

$$S + B = 33$$
$$S - 2B = 3$$

Step 4 Using the method of elimination and multiplying the first equation by 2, we obtain the system

$$2S + 2B = 66$$
$$S - 2B = 3$$

We can now add the two equations together to eliminate the variable B.

$$2S + 2B = 66$$
$$S - 2B = 3$$
$$3S \qquad = 69$$

Dividing by 3 gives $S = 23$. To find B, replace S with 23 in the original first equation:

$$S + B = 33 \qquad \textit{Write the original first equation.}$$
$$23 + B = 33 \qquad \textit{Replace S with 23.}$$
$$B = 10 \qquad \textit{Subtract 23 from both sides.}$$

Therefore, Staubach threw 23 touchdown passes, and Bradshaw threw 10 touchdown passes.

Step 5 To check, we see that the total number of touchdown passes is $23 + 10 = 33$ and the number of touchdown passes thrown by Staubach is exactly 3 more than twice the number of touchdown passes thrown by Bradshaw.

Example 7 Find the Number of Tickets Sold at a Jazz Festival

During one night at the jazz festival, 2,100 tickets were sold. Adult tickets sold for $12, and child tickets sold for $7. If the receipts totaled $22,100, how many of each type of ticket were sold?

Solution

Step 1 We are asked to determine the number of adult tickets and the number of child tickets sold.

Step 2 Let

$$A = \text{number of adult tickets}$$
$$C = \text{number of child tickets}$$

Step 3 The number of adult tickets plus the number of child tickets totals 2,100, so we obtain our first equation

$$\underbrace{A}_{\substack{Number\ of\ adult \\ tickets}} + \underbrace{C}_{\substack{Number\ of\ child \\ tickets}} = \underbrace{2{,}100}_{\substack{Total\ number\ of \\ tickets}}$$

To find the second equation, we use the fact that the value of an adult ticket is $12 and the value of a child ticket is $7; therefore,

$$\underbrace{12A}_{\substack{Total\ value\ of \\ adult\ tickets}} + \underbrace{7C}_{\substack{Total\ value\ of \\ child\ tickets}} = \underbrace{22{,}100}_{\substack{Total\ value\ of \\ all\ tickets}}$$

Thus, the system of equations is
$$A + C = 2{,}100$$
$$12A + 7C = 22{,}100$$

Step 4 Using the method of elimination and multiplying the first equation by -7, we obtain the system

$$-7A - 7C = -14{,}700$$
$$12A + 7C = 22{,}100$$

Adding the equations together eliminates the variable c:

$$
\begin{array}{rcr}
-7A - 7C &=& -14{,}700 \\
12A + 7C &=& 22{,}100 \\
\hline
5A &=& 7{,}400
\end{array}
$$

Dividing by 5 gives $A = 1{,}480$. To find C, replace A with 1,480 in the original first equation:

$$A + C = 2{,}100 \qquad \textit{Write the original first equation.}$$
$$1{,}480 + C = 2{,}100 \qquad \textit{Replace A with 1,480.}$$
$$C = 620 \qquad \textit{Subtract 1,480 from both sides.}$$

Therefore, there were 1,480 adult tickets sold and 620 child tickets sold.

Step 5 To check, we see that the total number of tickets is $1{,}480 + 620 = 2{,}100$ and the total value of all tickets sold is $(\$12)(1{,}480) + (\$7)(620) = \$22{,}100$.

Following are two more examples. Although the solutions to these examples are not worked out for you in the text, you may view them in detail by clicking on the appropriate video link.

My video summary

Example 8 Mix Beans in a Food Plant

Twin City Foods, Inc., created a 10-lb bean mixture that sells for $5.75 by mixing lima beans and green beans. If lima beans sell for $.70 per pound and green beans sell for $.50 per pound, how many pounds of each bean went into the mixture?

Solution Watch the video to see that the mixture contains $3\frac{3}{4}$ pounds of lima beans and $6\frac{1}{4}$ pounds of green beans.

Example 9 Find the Speed of an Airplane

A small airplane flies from Seattle, Washington, to Portland, Oregon—a distance of 150 miles. Because the pilot encountered a strong headwind, the trip took 1 hour and 15 minutes. On the return flight, the wind is still blowing at the same speed. If the return trip took 45 minutes, what was the average speed of the airplane in still air? What was the speed of the wind?

Solution Watch the video to verify that the speed of the plane is 160 mph and the speed of the wind is 40 mph.

Work exercises 21–30 in this textbook or in the *MyMathLab*® Study Plan.

7.1 Exercises

In Exercises 1–4, two ordered pairs are given. Determine whether each ordered pair is a solution to the given system.

1. $\begin{aligned} -4x - 2y &= -2 \\ 5x + 7y &= 16 \end{aligned}$ Ordered pairs: $(-1, 3)$, $(-3, 7)$

2. $\begin{aligned} \frac{1}{2}x_1 - \frac{1}{3}x_2 &= -\frac{7}{24} \\ -\frac{1}{4}x_1 + \frac{1}{2}x_2 &= \frac{5}{16} \end{aligned}$ Ordered pairs: $\left(-\frac{1}{2}, \frac{1}{4}\right)$, $\left(-\frac{1}{4}, \frac{1}{2}\right)$

3. $\begin{aligned} .3a + .2b &= -.12 \\ -.7a - .1b &= -.05 \end{aligned}$ Ordered pairs: $(.2, -.9)$, $(-.3, .4)$

4. $\begin{aligned} -2x + y &= -1 \\ \frac{2}{3}x - \frac{1}{3}y &= \frac{1}{3} \end{aligned}$ Ordered pairs: $(-2, 5)$, $(3, 5)$

In Exercises 5–9, use the method of *substitution* to solve each system of linear equations.

5. $\begin{aligned} x &= 4 - y \\ 3x - 2y &= -3 \end{aligned}$

6. $\begin{aligned} x + 4y &= 5 \\ 2x - y &= -8 \end{aligned}$

7. $\begin{aligned} 2x - y &= 3 \\ -8x + 3y &= -8 \end{aligned}$

8. $\begin{aligned} 5x + 4y &= -6 \\ 8x - 4y &= -72 \end{aligned}$

9. $\begin{aligned} -\frac{2}{3}x + \frac{1}{2}y &= -\frac{1}{3} \\ -\frac{1}{2}x + \frac{3}{4}y &= -\frac{1}{2} \end{aligned}$

In Exercises 10–14, use the method of *elimination* to solve each system of linear equations

10. $\begin{aligned} x + 14y &= 20 \\ -x + 14y &= 15 \end{aligned}$

11. $\begin{aligned} 3x - 2y &= 20 \\ 2x + 6y &= -38 \end{aligned}$

12. $\begin{aligned} 8x - 6y &= -6 \\ 5x - 7y &= -33 \end{aligned}$

13. $\begin{aligned} .1x - .4y &= -2. \\ .7x - y &= -9.6 \end{aligned}$

14. $\begin{aligned} \frac{7}{16}x - \frac{3}{4}y &= -\frac{5}{2} \\ \frac{3}{4}x + \frac{5}{2}y &= 26 \end{aligned}$

In Exercises 15–20, use the substitution method or the elimination method to solve each system. If the system has infinitely many solutions, express the ordered pairs in terms of x or y as in Example 4.

15. $\begin{aligned} x + y &= 3 \\ -2x - 2y &= 1 \end{aligned}$

16. $\begin{aligned} 3x - 12y &= 6 \\ -2x + 8y &= -4 \end{aligned}$

17. $\begin{aligned} 8x - y &= -13 \\ y &= -8x \end{aligned}$

18. $\begin{aligned} -\frac{1}{2}x + \frac{2}{3}y &= -\frac{1}{3} \\ \frac{3}{2}x - 2y &= 1 \end{aligned}$

19. $\begin{aligned} 8x + 4y &= 5 \\ 2x + y &= -4 \end{aligned}$

20. $\begin{aligned} y &= \frac{5}{7}x - 9 \\ -15x + 21y &= -189 \end{aligned}$

In Exercises 21–30, use a system of equations to solve each problem.

21. Together, teammates Pedro and Ricky got 2,673 base hits last season. Pedro had 281 more hits than Ricky. How many hits did each player have?

22. Benjamin & Associates, a real estate developer, recently built 200 condominiums in McCall, Idaho. The condos were either two-bedroom units or three-bedroom units. If the total number of rooms in the entire complex is 507, how many two-bedroom units are there? How many three-bedroom units are there?

23. On a certain hot summer's day, 492 people used the public swimming pool. The daily prices are $1.50 for children and $2.50 for adults. The receipts for admission totaled $929. How many children and how many adults swam at the public pool that day?

24. Scott invested a total of $3,000 at two separate banks. One bank pays simple interest of 3% per year, whereas the other pays simple interest at a rate of 4% per year. If Scott earned $182.50 in interest during a single year, how much did he have on deposit in each bank?

25. Thursday is ladies' night at the Slurp and Burp Bar and Grill. All adult beverages are $2.50 for men and $1.50 for women. A total of 956 adult beverages were sold last Thursday night. If the Slurp and Burp sold a total of $1,996 in adult beverages last Thursday night, how many adult beverages were sold to women?

26. Farmer Brown planted corn and wheat on his 1,200 acres of land. The cost of planting and harvesting corn (which includes seed, planting, fertilizer, machinery, labor, and other costs) is $285 per acre. The cost of planting and harvesting wheat is $130 per acre. If Farmer Brown's total cost was $245,900, how many acres of corn did he plant?

27. An airplane encountered a headwind during a flight between Joppetown and Jawsburgh that took 3 hours and 36 minutes. The return flight took 3 hours. If the distance from Joppetown to Jawsburgh is 1800 miles, find the airspeed of the plane (the speed of the plane in still air) and the speed of the wind, assuming both remain constant.

28. Gabby's piggy bank contains nickels and quarters worth $9.10. If she has 46 coins in all, how many of each does she have?

29. Ned, the owner of Ned's Nut Shop, sells peanuts for $3.25 per pound and cashews for $6.00 per pound. Ned wants to create a 58.5-lb barrel of mixed nuts by mixing the peanuts and cashews together and sell it for $4.00 per pound. How many pounds of each should Ned use?

30. Angela, the head of the office party planning committee, sent two coworkers, Kevin and Dwight, to the party store to purchase party hats and party whistles. Kevin purchased five packs of party hats and four packs of party whistles and paid a total of $23.01. Dwight purchased four packs of party hats and five packs of party whistles, and paid a total of $20.91. When they returned to the office, Angela informed them that they were supposed to buy a total of five packs of party hats and five packs of party whistles and sent them back to the store to straighten things out. How much should they be refunded?

7.2 Systems of Linear Equations in Three Variables

THINGS TO KNOW

Before working through this section, be sure that you are familiar with the following concepts:

VIDEO   ANIMATION   INTERACTIVE

 1. Solving a System of Linear Equations Using the Substitution Method (Section 7.1)
You Try

 2. Solving a System of Linear Equations Using the Elimination Method (Section 7.1)
You Try

OBJECTIVES

1 Verifying the Solution of a System of Linear Equations

2 Solving a System of Linear Equations Using Gaussian Elimination

3 Using an Augmented Matrix to Solve a System of Linear Equations

4 Applications of Systems of Linear Equations Involving Three Variables

OBJECTIVE 1:   VERIFYING THE SOLUTION OF A SYSTEM OF LINEAR EQUATIONS

In this section, we learn techniques that will enable us to solve systems of three **linear equations** in three variables. We see in Section 7.1 that if a system of two linear equations in two variables has one solution, then the solution can be geometrically represented by the intersection of two lines. Similarly, if a system of three linear equations in three variables has a unique solution, then the solution can be geometrically represented by the intersection of three planes. See Figure 5.

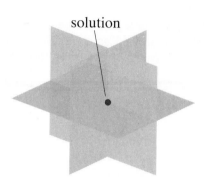

A system of two linear equations in two variables that has a unique solution. The solution can be represented by the point of intersection of two lines.

A system of three linear equations in three variables that has a unique solution. The solution can be represented by the point of intersection of three planes.

Figure 5

The following is an example of a system of three linear equations in three variables:

$$2x + 3y + 4z = 12$$
$$x - 2y + 3z = 0$$
$$-x + y - 2z = -1$$

The solution to this linear system is the set of all *ordered triples* (x, y, z) that satisfy all three equations. In Example 1, we show that the ordered triple $(1, 2, 1)$ is a solution to this linear system.

My video summary

▣ Example 1 Verify the Solution of a System of Linear Equations

Verify that the ordered triple $(1, 2, 1)$ is a solution to the following system of linear equations:

$$2x + 3y + 4z = 12$$
$$x - 2y + 3z = 0$$
$$-x + y - 2z = -1$$

Solution To verify that the ordered triple $(1, 2, 1)$ is a solution, we must substitute $x = 1$, $y = 2$, and $z = 1$ in each of the three equations and determine whether a true statement occurs.

$$2(1) + 3(2) + 4(1) \overset{?}{=} 12 \longrightarrow 2 + 6 + 4 \overset{?}{=} 12 \longrightarrow 12 = 12 \ \sqrt{} \ \text{True}$$

$$(1) - 2(2) + 3(1) \overset{?}{=} 0 \longrightarrow 1 - 4 + 3 \overset{?}{=} 0 \longrightarrow 0 = 0 \ \sqrt{} \ \text{True}$$

$$-(1) + (2) - 2(1) \overset{?}{=} -1 \longrightarrow -1 + 2 - 2 \overset{?}{=} -1 \longrightarrow -1 = -1 \ \sqrt{} \ \text{True}$$

All three statements are true. This implies that the ordered triple $(1, 2, 1)$ is a solution to this system.

Work exercises 1 and 2 in this textbook or in the MyMathLab® Study Plan.

OBJECTIVE 2:   SOLVING A SYSTEM OF LINEAR EQUATIONS USING GAUSSIAN ELIMINATION

In Section 7.1, we used the method of elimination to solve a system of two linear equations by adding a certain multiple of one equation to the other equation, thus eliminating one of the variables. See Example 3 of Section 7.1. If a system of three linear equations in three variables has a unique solution, then we can use a similar technique to eliminate two of the variables of one equation and one of the variables of another equation. Before we learn this technique, consider the following system of equations:

$$\begin{aligned} x - 2y + 3z &= 0 \quad (1) \\ -y + z &= -1 \quad (2) \\ 5z &= 5 \quad (3) \end{aligned}$$

Notice that equation (3) has only one variable, equation (2) contains two variables, and equation (1) has three variables. A system of this form is said to be in **triangular form**. Note that we can easily solve equation (3) by dividing by 5 to get the solution $z = 1$.

Using the value of $z = 1$ in equation (2), we can solve for y:

$$\begin{aligned} -y + z &= -1 & &\text{Write equation (2).} \\ -y + (1) &= -1 & &\text{Substitute 1 for } z. \\ -y &= -2 & &\text{Subtract 1 from both sides.} \\ y &= 2 & &\text{Divide both sides by } -1. \end{aligned}$$

Finally, using $z = 1$ and $y = 2$ in equation (1) gives

$$\begin{aligned} x - 2y + 3z &= 0 & &\text{Write equation (1).} \\ x - 2(2) + 3(1) &= 0 & &\text{Substitute 2 for } y \text{ and 1 for } z. \\ x - 1 &= 0 & &\text{Simplify.} \\ x &= 1 & &\text{Add 1 to both sides.} \end{aligned}$$

Therefore, the solution to this system of equations is the ordered triple $(1, 2, 1)$. This process of substituting the value of z into the second equation to solve for y and then substituting the values of z and y into the first equation to solve for x is called **back substitution**. Our goal when solving a linear system of three equations with three unknowns is to attempt to reduce the system into an **equivalent system** that is in **triangular form** and then use back substitution to solve for each variable. We use a method called **Gaussian elimination**, named after the famous German mathematician Carl Friedrich Gauss, to accomplish this task.

When using Gaussian elimination, there are three algebraic operations that we can use to reduce a system of linear equations into an equivalent system in triangular form. These algebraic operations are called the **elementary row operations** and are listed on the following page.

Elementary Row Operations

The following algebraic operations will result in an equivalent system of linear equations:

1. Interchange any two equations.
2. Multiply any equation by a nonzero constant.
3. Add a multiple of one equation to another equation.

Because Gaussian elimination can be a lengthy process, we use shorthand notation to document the elementary row operation performed at each step. This documentation is important because it will help you "retrace" your steps in case you make a mistake or want to study your work later. We use R_i to describe the ith equation of the system. The shorthand notation is outlined as follows.

Notation Used to Describe Elementary Row Operations

Notation	Meaning
$R_i \Leftrightarrow R_j$	Interchange Rows i and j.
$kR_i \to$ New R_i	k times Row i becomes New Row i.
$kR_i + R_j \to$ New R_j	k times Row i plus Row j becomes New Row j.

Example 2 illustrates how we can use Gaussian elimination to solve a system of three linear equations. You should carefully work through Example 2 and take notes while you watch the video solution.

My video summary

🎞 **Example 2   Use Gaussian Elimination to Solve a System of Three Linear Equations**

For the following system, use elementary row operations to find an equivalent system in triangular form and then use back substitution to solve the system:

$$\begin{aligned} 2x + 3y + 4z &= 12 \\ x - 2y + 3z &= 0 \\ -x + y - 2z &= -1 \end{aligned}$$

Solution Our goal is to use a series of elementary row operations to eliminate the variable x in one of the equations and eliminate the variables x and y in another equation. We typically start by trying to make the first coefficient of the first equation equal to 1. We can accomplish this by interchanging the first two equations. The other four elementary row operations are outlined as follows. You can also watch the video to see each step worked out in detail.

$$\begin{aligned} 2x + 3y + 4z &= 12 \\ x - 2y + 3z &= 0 \\ -x + y - 2z &= -1 \end{aligned} \xrightarrow{R_1 \Leftrightarrow R_2} \begin{aligned} x - 2y + 3z &= 0 \\ 2x + 3y + 4z &= 12 \\ -x + y - 2z &= -1 \end{aligned} \xrightarrow{-2R_1 + R_2 \to \text{New } R_2} \begin{aligned} x - 2y + 3z &= 0 \\ 7y - 2z &= 12 \\ -x + y - 2z &= -1 \end{aligned}$$

$$\xrightarrow{R_1 + R_3 \to \text{New } R_2} \begin{aligned} x - 2y + 3z &= 0 \\ 7y - 2z &= 12 \\ -y + z &= -1 \end{aligned} \xrightarrow{R_2 \Leftrightarrow R_3} \begin{aligned} x - 2y + 3z &= 0 \\ -y + z &= -1 \\ 7y - 2z &= 12 \end{aligned} \xrightarrow{7R_2 + R_3 \to \text{New } R_3} \begin{aligned} x - 2y + 3z &= 0 \\ -y + z &= -1 \\ 5z &= 5 \end{aligned}$$

Now that the linear system is written in triangular form, we can solve the third equation for z and then use back substitution to solve for the other two variables. You should verify that the solution to this system is the ordered triple $(1, 2, 1)$.

Work exercises 3–6 in this textbook or in the MyMathLab® Study Plan.

OBJECTIVE 3: **USING AN AUGMENTED MATRIX TO SOLVE A SYSTEM OF LINEAR EQUATIONS**

In Example 2, the variables x, y, and z were used. We actually could have used any three variables to describe this system. As it turns out, we do not need variables at all to solve a system of linear equations using Gaussian elimination! We need only the coefficients of each variable to perform the elementary row operations. We can simplify the Gaussian elimination process by simplifying the notation. Instead of writing three equations at each step, we only write the coefficients at each step. We accomplish this by writing the coefficients in a rectangular array called an **augmented matrix**. For example, the system from Example 2 can be written using an augmented matrix as follows:

System of Equations from Example 2

$$2x + 3y + 4z = 12$$
$$x - 2y + 3z = 0$$
$$-x + y - 2z = -1$$

Corresponding Augmented Matrix

$$\left[\begin{array}{ccc|c} 2 & 3 & 4 & 12 \\ 1 & -2 & 3 & 0 \\ -1 & 1 & -2 & -1 \end{array}\right]$$

Notice that each row of the corresponding augmented matrix represents the coefficients of each equation. The vertical bar is used to separate the coefficients of each variable from the constant coefficient that appears on the right side of each equation. We can perform the exact same elementary row operations as in Example 2 by writing an augmented matrix at each step. We now illustrate this process. For clarity, the corresponding system of equations is listed below each augmented matrix.

$$\left[\begin{array}{ccc|c} 2 & 3 & 4 & 12 \\ 1 & -2 & 3 & 0 \\ -1 & 1 & -2 & -1 \end{array}\right] \xrightarrow{R_1 \Leftrightarrow R_2} \left[\begin{array}{ccc|c} 1 & -2 & 3 & 0 \\ 2 & 3 & 4 & 12 \\ -1 & 1 & -2 & -1 \end{array}\right] \xrightarrow{-2R_1 + R_2 \to \text{New } R_2} \left[\begin{array}{ccc|c} 1 & -2 & 3 & 0 \\ 0 & 7 & -2 & 12 \\ -1 & 1 & -2 & -1 \end{array}\right] \xrightarrow{R_1 + R_3 \to \text{New } R_2}$$

$$2x + 3y + 4z = 12 \qquad\qquad x - 2y + 3z = 0 \qquad\qquad x - 2y + 3z = 0$$
$$x - 2y + 3z = 0 \qquad\qquad 2x + 3y + 4z = 12 \qquad\qquad 7y - 2z = 12$$
$$-x + y - 2z = -1 \qquad\qquad -x + y - 2z = -1 \qquad\qquad -x + y - 2z = -1$$

$$\left[\begin{array}{ccc|c} 1 & -2 & 3 & 0 \\ 0 & 7 & -2 & 12 \\ 0 & -1 & 1 & -1 \end{array}\right] \xrightarrow{R_2 \Leftrightarrow R_3} \left[\begin{array}{ccc|c} 1 & -2 & 3 & 0 \\ 0 & -1 & 1 & -1 \\ 0 & 7 & -2 & 12 \end{array}\right] \xrightarrow{7R_2 + R_3 \to \text{New } R_3} \left[\begin{array}{ccc|c} 1 & -2 & 3 & 0 \\ 0 & -1 & 1 & -1 \\ 0 & 0 & 5 & 5 \end{array}\right]$$

$$x - 2y + 3z = 0 \qquad\qquad x - 2y + 3z = 0 \qquad\qquad x - 2y + 3z = 0$$
$$7y - 2z = 12 \qquad\qquad -y + z = -1 \qquad\qquad -y + z = -1$$
$$-y + z = -1 \qquad\qquad 7y - 2z = 12 \qquad\qquad 5z = 5$$

This matrix is now in triangular form. (All 0's below the diagonal.)

Once we have a matrix written in triangular form, we can use the last row to solve for z and then use back substitution as before to solve for the remaining variables.

My video summary

▤ Example 3 Solve a Linear System Using an Augmented Matrix (Triangular Form)

Create an augmented matrix and solve the following linear system using Gaussian elimination by writing an equivalent system in triangular form:

$$x + 2y - z = 3$$
$$x - 3y - 2z = 11$$
$$-x - 2y + 2z = -6$$

Solution The augmented matrix that corresponds to the linear system is

$$\begin{bmatrix} 1 & 2 & -1 & 3 \\ 1 & -3 & -2 & 11 \\ -1 & -2 & 2 & -6 \end{bmatrix}$$

We now proceed to use the following elementary row operations:

$$\begin{bmatrix} 1 & 2 & -1 & 3 \\ 1 & -3 & -2 & 11 \\ -1 & -2 & 2 & -6 \end{bmatrix} \xrightarrow{(-1)R_1 + R_2 \to \text{New } R_2} \begin{bmatrix} 1 & 2 & -1 & 3 \\ 0 & -5 & -1 & 8 \\ -1 & -2 & 2 & -6 \end{bmatrix} \xrightarrow{R_1 + R_3 \to \text{New } R_3} \begin{bmatrix} 1 & 2 & -1 & 3 \\ 0 & -5 & -1 & 8 \\ 0 & 0 & 1 & -3 \end{bmatrix}$$

The augmented matrix is now in triangular form. Looking at the last row, we see that $z = -3$. The second row corresponds to the equation $-5y - z = 8$. Therefore,

$-5y - z = 8$	*Use the augmented matrix to write equation (2).*
$-5y - (-3) = 8$	*Substitute -3 for z.*
$-5y + 3 = 8$	*Simplify.*
$-5y = 5$	*Subtract 3 from both sides.*
$y = -1$	*Divide both sides by -5.*

Because $z = -3$ and $y = -1$, we can now use the first row to solve for x. You should verify that $x = 2$. Therefore, the solution to this linear system is the ordered triple $(2, -1, -3)$.

Work exercises 7–12 in this textbook or in the *MyMathLab*® Study Plan.

My video summary

▤ Triangular Form, Row-Echelon Form, and Reduced Row-Echelon Form

When solving a system of three linear equations with three unknowns, only triangular form is needed to find the solution. However, it is often useful to further reduce the augmented matrix into one of two other simpler forms known as row-echelon form or reduced row-echelon form.

We see in Example 3 that the linear system can be reduced into the following triangular form:

Triangular Form

$$\begin{bmatrix} 1 & 2 & -1 & 3 \\ 0 & -5 & -1 & 8 \\ 0 & 0 & 1 & -3 \end{bmatrix}$$

Triangular form only requires zeros below the diagonal.

To reduce this matrix into **row-echelon form** requires that all zeros remain below the diagonal and that all coefficients along the diagonal are 1. If we multiply the second row of the previous matrix by $-\frac{1}{5}$, we get the following equivalent matrix written in **row-echelon form**:

Row-Echelon Form

Row-echelon form requires zeros below the diagonal . . .

$$\begin{bmatrix} 1 & 2 & -1 & 3 \\ 0 & 1 & \frac{1}{5} & -\frac{8}{5} \\ 0 & 0 & 1 & -3 \end{bmatrix}$$

. . . and 1's down the diagonal.

Reduced row-echelon form requires zeros below *and* above the diagonal and 1's down the diagonal. We can reduce the previous matrix into reduced row-echelon form by performing the following three elementary row operations.

$$\begin{bmatrix} 1 & 2 & -1 & 3 \\ 0 & 1 & \frac{1}{5} & -\frac{8}{5} \\ 0 & 0 & 1 & -3 \end{bmatrix} \xrightarrow{(-2)R_2 + R_1 \to \text{New } R_1} \begin{bmatrix} 1 & 0 & -\frac{7}{5} & \frac{31}{5} \\ 0 & 1 & \frac{1}{5} & -\frac{8}{5} \\ 0 & 0 & 1 & -3 \end{bmatrix} \xrightarrow{\left(-\frac{1}{5}\right)R_3 + R_2 \to \text{New } R_2}$$

$$\begin{bmatrix} 1 & 0 & -\frac{7}{5} & \frac{31}{5} \\ 0 & 1 & 0 & -1 \\ 0 & 0 & 1 & -3 \end{bmatrix} \xrightarrow{\left(\frac{7}{5}\right)R_3 + R_1 \to \text{New } R_1} \begin{bmatrix} 1 & 0 & 0 & 2 \\ 0 & 1 & 0 & -1 \\ 0 & 0 & 1 & -3 \end{bmatrix} \begin{matrix} \Rightarrow x = 2 \\ \Rightarrow y = -1 \\ \Rightarrow z = -3 \end{matrix}$$

As you can see from the final matrix, we have now obtained all zeros above and below the diagonal. This final matrix is in **reduced row-echelon form**. The advantage to reducing the augmented matrix into reduced row-echelon form is that you can immediately read the solution by looking down the last column. In the previous reduced row-echelon matrix, you can clearly see that the solution is $x = 2$, $y = -1$, and $z = -3$. The process of reducing a system into **reduced row-echelon form** is called **Gauss-Jordan elimination**. We use Gauss-Jordan elimination from this point forward.

My video summary

🎬 Example 4 Solve a Linear System Using Gauss-Jordan Elimination

Solve the following system using Gauss-Jordan elimination:

$$\begin{aligned} x_1 + x_2 + x_3 &= -1 \\ x_1 + 2x_2 + 4x_3 &= 3 \\ x_1 + 3x_2 + 9x_3 &= 3 \end{aligned}$$

Solution The augmented matrix that corresponds to the linear system is

$$\begin{bmatrix} 1 & 1 & 1 & -1 \\ 1 & 2 & 4 & 3 \\ 1 & 3 & 9 & 3 \end{bmatrix}$$

We need to reduce this matrix into reduced row-echelon form. We must have 1s down the diagonal and zeros everywhere else to the left of the vertical bar. We can accomplish this by performing the following seven elementary row operations. Watch the video for a detailed step-by-step explanation.

$$\begin{bmatrix} 1 & 1 & 1 & | & -1 \\ 1 & 2 & 4 & | & 3 \\ 1 & 3 & 9 & | & 3 \end{bmatrix} \xrightarrow{(-1)R_1 + R_2 \rightarrow \text{New } R_2} \begin{bmatrix} 1 & 1 & 1 & | & -1 \\ 0 & 1 & 3 & | & 4 \\ 1 & 3 & 9 & | & 3 \end{bmatrix} \xrightarrow{(-1)R_1 + R_3 \rightarrow \text{New } R_3}$$

$$\begin{bmatrix} 1 & 1 & 1 & | & -1 \\ 0 & 1 & 3 & | & 4 \\ 0 & 2 & 8 & | & 4 \end{bmatrix} \xrightarrow{(-1)R_2 + R_1 \rightarrow \text{New } R_1} \begin{bmatrix} 1 & 0 & -2 & | & -5 \\ 0 & 1 & 3 & | & 4 \\ 0 & 2 & 8 & | & 4 \end{bmatrix} \xrightarrow{(-2)R_2 + R_3 \rightarrow \text{New } R_3}$$

$$\begin{bmatrix} 1 & 0 & -2 & | & -5 \\ 0 & 1 & 3 & | & 4 \\ 0 & 0 & 2 & | & -4 \end{bmatrix} \xrightarrow{\left(\frac{1}{2}\right)R_3 \rightarrow \text{New } R_3} \begin{bmatrix} 1 & 0 & -2 & | & -5 \\ 0 & 1 & 3 & | & 4 \\ 0 & 0 & 1 & | & -2 \end{bmatrix} \xrightarrow{(-3)R_3 + R_2 \rightarrow \text{New } R_2}$$

$$\begin{bmatrix} 1 & 0 & -2 & | & -5 \\ 0 & 1 & 0 & | & 10 \\ 0 & 0 & 1 & | & -2 \end{bmatrix} \xrightarrow{2R_3 + R_1 \rightarrow \text{New } R_1} \begin{bmatrix} 1 & 0 & 0 & | & -9 \\ 0 & 1 & 0 & | & 10 \\ 0 & 0 & 1 & | & -2 \end{bmatrix} \begin{matrix} \Rightarrow x_1 = -9 \\ \Rightarrow x_2 = 10 \\ \Rightarrow x_3 = -2 \end{matrix}$$

You can see by looking at the last matrix that $x_1 = -9$, $x_2 = 10$, and $x_3 = -2$. Thus, the solution is the ordered triple $(-9, 10, -2)$.

Work exercises 13–18 in this textbook or in the *MyMathLab*® Study Plan.

OBJECTIVE 4: APPLICATIONS OF SYSTEMS OF LINEAR EQUATIONS
INVOLVING THREE VARIABLES

Systems of equations can be used to solve a wide variety of applications. Following are two examples.

My video summary

Example 5 Buy Clothes Online

Wendy ordered 30 T-shirts online for her three children. The small T-shirts cost $4 each, the medium T-shirts cost $5 each, and the large T-shirts were $6 each. She spent $40 more purchasing the large T-shirts than the small T-shirts. Wendy's total bill was $154. How many T-shirts of each size did she buy?

Solution Let S, M, and L represent the number of small, medium, and large T-shirts, respectively. Because a total of 30 T-shirts were purchased, we obtain the first equation

$$S + M + L = 30$$

Because small T-shirts cost $4, we know that the total amount spent on small T-shirts is $4S$. Similarly, the total spent on medium T-shirts is $5M$, and the total spent on large T-shirts is $6L$. Therefore, because the total value of the T-shirts was $154, we obtain the second equation

$$4S + 5M + 6L = 154$$

Finally, Wendy spent $40 more buying large T-shirts than small T-shirts, so $6L = 4S + 40$, which is equivalent to the equation

$$4S - 6L = -40$$

Thus, we obtain the system
$$\begin{aligned} S + M + L &= 30 \\ 4S + 5M + 6L &= 154. \\ 4S \qquad - 6L &= -40 \end{aligned}$$

We now rewrite this system using the corresponding augmented matrix. Note that we must place a 0 in the middle column of the last row of the augmented matrix to represent the coefficient of the variable M of the third equation.

$$\begin{bmatrix} 1 & 1 & 1 & 30 \\ 4 & 5 & 6 & 154 \\ 4 & 0 & 6 & -40 \end{bmatrix}$$

Watch the video to see how to use Gauss-Jordan elimination to rewrite the matrix in the following reduced row-echelon form:

$$\begin{bmatrix} 1 & 0 & 0 & 8 \\ 0 & 1 & 0 & 10 \\ 0 & 0 & 1 & 12 \end{bmatrix}$$

Looking down the last column of this augmented matrix, we see that Wendy purchased 8 small T-shirts, 10 medium T-shirts, and 12 large T-shirts.

Work exercises 19–23 in this textbook or in the *MyMathLab*® Study Plan.

It takes two points to determine the equation of a line. Similarly, it takes three noncollinear points to determine the equation of the quadratic function whose graph passes through the points. Example 6 shows how to find the equation of a quadratic function given three points that lie on its graph. You may want to review quadratic functions from Section 4.1.

Example 6 Use Three Points to Determine a Quadratic Function

Determine the quadratic function whose graph passes through the three points $(1, -9)$, $(-1, -5)$, and $(-3, 7)$.

Solution A quadratic function has the form $f(x) = ax^2 + bx + c$. The points $(1, -9)$, $(-1, -5)$, and $(-3, 7)$ lie on the graph of f so

$$
\begin{aligned}
f(1) &= a(1)^2 + b(1) + c &&= -9 \longrightarrow a + b + c = -9 \\
f(-1) &= a(-1)^2 + b(-1) + c &&= -5 \longrightarrow a - b + c = -5 \\
f(3) &= a(-3)^2 + b(-3) + c &&= 7 \longrightarrow 9a - 3b + c = 7
\end{aligned}
$$

Therefore, we obtain the system and corresponding augmented matrix:

System of Equations	Corresponding Augmented Matrix
$a + b + c = -9$ $a - b + c = -5$ $9a - 3b + c = 7$	$\begin{bmatrix} 1 & 1 & 1 & -9 \\ 1 & -1 & 1 & -5 \\ 9 & -3 & 1 & 7 \end{bmatrix}$

Using Gauss-Jordan elimination, we can rewrite this matrix in the following reduced row-echelon form. (To see each step in detail, click here.)

$$\begin{bmatrix} 1 & 0 & 0 & 1 \\ 0 & 1 & 0 & -2 \\ 0 & 0 & 1 & -8 \end{bmatrix}$$

We can see from the last column that $a = 1$, $b = -2$, and $c = -8$. Therefore, the quadratic function that passes through the points $(1, -9)$, $(-1, -5)$, and $(-3, 7)$ is $f(x) = x^2 - 2x - 8$.

You can see in Figure 6 that the graph of $f(x) = x^2 - 2x - 8$ passes through the three given points.

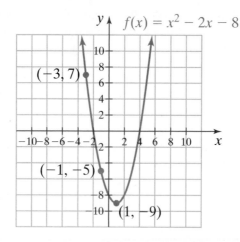

Figure 6 Graph of the quadratic function $f(x) = x^2 - 2x - 8$

Work exercises 24 and 25 in this textbook or in the _MyMathLab_® Study Plan.

7.2 Exercises

In Exercises 1 and 2, two ordered triples are given. Determine whether each ordered triple is a solution of the given system.

1. $(-1, 1, -2), (1, -1, 2)$

$$x + y + z = -2$$
$$-x - 3y - 2z = 2$$
$$2x - 2y + 5z = -14$$

2. $(2, -1, 4), (-2, 1, -4)$

$$\tfrac{1}{2}x + 3y - z = 6$$
$$-x + y + \tfrac{1}{4}z = 2$$
$$x - 4y + z = -10$$

In Exercises 3–6, use Gaussian elimination to solve each linear system by finding an equivalent system in triangular form.

3.
$$x + y + z = 4$$
$$2x - y - 2z = -10$$
$$-x - y + 3z = 8$$

4.
$$x - 2y + z = 6$$
$$2x + y - 3z = -3$$
$$x - 3y + 3z = 10$$

5.
$$-4x_1 + 5x_2 + 9x_3 = -9$$
$$x_1 - 2x_2 + x_3 = 0$$
$$2x_2 - 8x_3 = 8$$

6.
$$2x_1 + 2x_2 + x_3 = 9$$
$$x_1 + x_3 = 4$$
$$4x_2 - 3x_3 = 17$$

In Exercises 7–12, use Gaussian elimination and matrices to solve each system of linear equations. Write your final augmented matrix in triangular form and then solve for each variable using back substitution.

7.
$$x + y + z = 0$$
$$x - 2y + 3z = -7$$
$$-x - y + 4z = 6$$

8.
$$2x - y + z = -6$$
$$x + y - z = 0$$
$$-3x - 3y + 4z = 1$$

9.
$$x_1 + 2x_2 + 3x_3 = 11$$
$$3x_1 + 8x_2 + 5x_3 = 27$$
$$-x_1 + x_2 + 2x_3 = 2$$

10.
$$2x_1 + 3x_3 = 13$$
$$-x_1 + 3x_2 = -8$$
$$2x_1 - x_2 + 4x_3 = 21$$

11.
$$x_1 + x_2 + 10x_3 = 3$$
$$3x_1 - 6x_2 + 6x_3 = -5$$
$$-2x_1 + 3x_2 - 5x_3 = \tfrac{7}{3}$$

12.
$$x_1 - 2x_2 + 2x_3 = 2$$
$$3x_1 + 2x_2 - 2x_3 = -1$$
$$x_1 - x_2 - 2x_3 = 0$$

In Exercises 13–18, use Gauss-Jordan elimination and matrices to solve each system of linear equations.

13.
$$x - y = 7$$
$$y - z = 2$$
$$x + z = 1$$

14.
$$x - 2y + 3z = 0$$
$$4x - 6y + 8z = -4$$
$$-2x - y + z = 7$$

15.
$$-x_1 + 2x_2 - x_3 = -15$$
$$3x_1 + x_2 + 4x_3 = 34$$
$$2x_1 - 4x_2 + 3x_3 = 40$$

16.
$$-2x - y + 3z = 0$$
$$4x + 3y + z = -1$$
$$x + y - 2z = -\tfrac{1}{2}$$

17.
$$x_1 - x_3 = 1$$
$$-x_1 + 2x_2 - 3x_3 = 12$$
$$2x_1 - 4x_2 = -6$$

18.
$$2x_1 + x_2 + 6x_3 = 1$$
$$-x_1 - 3x_3 = -2$$
$$5x_1 + 4x_2 = -7$$

In Exercises 19–25, write a system of linear equations in three variables and then use matrices to solve the system using Gaussian elimination or Gauss-Jordan elimination.

19. Ben was in charge of ordering 35 pizzas for the office party. He ordered three types of pizzas: cheese, supreme, and pepperoni. The cheese pizzas cost $9 each, the pepperoni pizzas cost $12 each, and the supreme pizzas cost $15 each. He spent exactly twice as much on the pepperoni pizzas as he did on the cheese pizzas. If Ben spent a total of $420 on pizza, how many pizzas of each type did he buy?

20. One thousand twenty tickets were sold on opening night of the play "The Music Man" at the Lewiston Civic Theater. The total receipts for the performance were $10,300. Adult tickets cost $12 apiece, children tickets cost $10 apiece, and senior citizen tickets cost $7 apiece. If the combined number of adult and children tickets exceeded twice the number of senior citizen tickets by 170, determine how many tickets of each type were sold.

21. Gary and Larry's Tire Shop sells three types of tires: high performance, ultra performance, and extreme performance. High-performance tires sell for $50/tire. Ultra performance tires sell for $60/tire. Extreme performance tires sell for $80/tire. Last week, the revenue generated by the sale of ultra performance tires exceeded the revenue from the sale of extreme performance tires by $160. If the tire shop sold a total of 238 tires last week for a total revenue of $14,440, how many of each type of tire were sold?

22. Amanda invested a total of $2,200 into three separate accounts that pay 6%, 8%, and 9% annual interest. Amanda has three times as much invested in the account that pays 9% as she does in the account that pays 6%. If the total interest for the year is $178, how much did Amanda invest at each rate?

23. Adam and Murph entered the inaugural Joppetown triathlon, which consisted of running, biking, and swimming. Adam averaged 8 mph during the running segment, 2 mph during the swimming segment, and 15 mph during the biking segment. Murph averaged 6 mph running, 3 mph swimming, and 20 mph biking. Adam's total time was 5 hours and 25 minutes. Murph's total time was 4 hours and 35 minutes. If the total of all three segments is 40 miles, how long was each segment?

24. Determine the quadratic function whose graph passes through the three points $(0, 0)$, $(-4, 36)$, and $(4, 60)$.

25. Determine the quadratic function whose graph passes through the three points $(-1, 2)$, $(3, 10)$, and $(-2, 15)$.

7.3 Inconsistent and Dependent Linear Systems in Three Variables

THINGS TO KNOW

Before working through this section, be sure that you are familiar with the following concept:

VIDEO   ANIMATION   INTERACTIVE

You Try

1. Solving a System of Three Equations in Three Variables Using Gauss-Jordan Elimination (Section 7.2)

OBJECTIVES

1 Solving Linear Systems Having No Solutions

2 Solving Linear Systems Having Infinitely Many Solutions

3 Solving Linear Systems Having Fewer Equations Than Variables

In Section 7.2, we learned how to use Gaussian elimination and Gauss-Jordan elimination to solve systems of linear equations in three variables. Remember that a system of three linear equations in three variables can be geometrically represented by three planes in space. Every system that we encountered in Section 7.2 had a unique solution. (You may want to watch this video describing how to solve a linear system in three variables using Gauss-Jordan elimination.) Systems of equations that have at least one solution are called **consistent systems**. It is possible that a system of three linear equations in three variables has infinitely many solutions. It is also possible for a system of three linear equations in three variables to have no solution at all. Such systems are called **inconsistent systems**. In this section, we concentrate on systems having infinitely many solutions and systems having no solution. Figure 7 illustrates some possibilities of consistent systems and inconsistent systems.

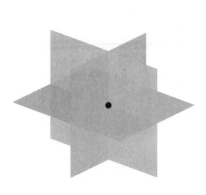

Consistent System
One unique solution

Consistent System
Infinitely many solutions
(Planes intersect at a line.)

Consistent System
Infinitely many solutions
(Three equations describing the same plane.)

Inconsistent System
No solution
(Planes intersect two at a time.)

Inconsistent System
No solution
(Three planes are parallel.)

Inconsistent System
No solution
(Two planes are parallel.)

Figure 7

OBJECTIVE 1: SOLVING LINEAR SYSTEMS HAVING NO SOLUTIONS

You can see in Figure 7 that if the three planes that represent a system of three linear equations in three variables intersect two at a time or if the three planes are parallel, then the system has no solution (no point in common to all three planes) and is said to be **inconsistent**. Example 1 illustrates what happens when we try to solve an inconsistent system.

My video summary

Example 1 Solve System of Equations with No Solution

Use Gauss-Jordan elimination to solve the system:

$$x - y + 2z = 4$$
$$-x + 3y + z = -6$$
$$x + y + 5z = 3$$

Solution Write the system using an augmented matrix and use a series of elementary row operations to attempt to reduce this matrix into **reduced row-echelon form**. Watch the video to see each step worked out in detail.

$$\begin{bmatrix} 1 & -1 & 2 & | & 4 \\ -1 & 3 & 1 & | & -6 \\ 1 & 1 & 5 & | & 3 \end{bmatrix} \xrightarrow{R_1 + R_2 \rightarrow \text{New } R_2} \begin{bmatrix} 1 & -1 & 2 & | & 4 \\ 0 & 2 & 3 & | & -2 \\ 1 & 1 & 5 & | & 3 \end{bmatrix} \xrightarrow{(-1)R_1 + R_3 \rightarrow \text{New } R_3}$$

$$\begin{bmatrix} 1 & -1 & 2 & | & 4 \\ 0 & 2 & 3 & | & -2 \\ 0 & 2 & 3 & | & -1 \end{bmatrix} \xrightarrow{(-1)R_2 + R_3 \rightarrow \text{New } R_3} \begin{bmatrix} 1 & -1 & 2 & | & 4 \\ 0 & 2 & 3 & | & -2 \\ 0 & 0 & 0 & | & 1 \end{bmatrix}$$

Looking at the final augmented matrix, we see that the last row represents the equation $0x + 0y + 0z = 1$ or, simply, $0 = 1$. Obviously, this is an untrue statement (zero can never equal one). Therefore, we say that the system is inconsistent and has no solution.

OBJECTIVE 2: ## SOLVING LINEAR SYSTEMS HAVING INFINITELY MANY SOLUTIONS

A linear system in three variables can have infinitely many solutions; that is, the three planes that geometrically represent the system intersect at infinitely many points. The three planes can intersect at a common line or the three planes can coincide. Example 2 illustrates a system that has infinitely many solutions.

Example 2 Solve System of Equations with Infinitely Many Solutions

Use Gauss-Jordan elimination to solve the system:

$$x + 2y + 3z = 10$$
$$x + y + z = 7$$
$$3x + 2y + z = 18$$

Solution Write the system using an augmented matrix, and use a series of **elementary row operations** to attempt to reduce this matrix into **reduced row-echelon form**. Watch the video to see each step worked out in detail.

$$\begin{bmatrix} 1 & 2 & 3 & | & 10 \\ 1 & 1 & 1 & | & 7 \\ 3 & 2 & 1 & | & 18 \end{bmatrix} \xrightarrow{(-1)R_1 + R_2 \rightarrow \text{New } R_2} \begin{bmatrix} 1 & 2 & 3 & | & 10 \\ 0 & -1 & -2 & | & -3 \\ 3 & 2 & 1 & | & 18 \end{bmatrix} \xrightarrow{(-3)R_1 + R_3 \rightarrow \text{New } R_3}$$

$$\begin{bmatrix} 1 & 2 & 3 & | & 10 \\ 0 & -1 & -2 & | & -3 \\ 0 & -4 & -8 & | & -12 \end{bmatrix} \xrightarrow{(-1)R_2 \rightarrow \text{New } R_2} \begin{bmatrix} 1 & 2 & 3 & | & 10 \\ 0 & 1 & 2 & | & 3 \\ 0 & -4 & -8 & | & -12 \end{bmatrix} \xrightarrow{(-2)R_2 + R_1 \rightarrow \text{New } R_1}$$

$$\begin{bmatrix} 1 & 0 & -1 & | & 4 \\ 0 & 1 & 2 & | & 3 \\ 0 & -4 & -8 & | & -12 \end{bmatrix} \xrightarrow{4R_2 + R_3 \rightarrow \text{New } R_3} \begin{bmatrix} 1 & 0 & -1 & | & 4 \\ 0 & 1 & 2 & | & 3 \\ 0 & 0 & 0 & | & 0 \end{bmatrix}$$

Notice that the last row of the final augmented matrix consists entirely of zeros. This corresponds to the equation $0 = 0$, which is true for every value of x, y, and z. The first two rows correspond to the equations

$$\begin{aligned} x \quad - \quad z &= 4 \\ y + 2z &= 3 \end{aligned}$$

We can solve the first equation for x and the second equation for y to obtain the equations

$$\begin{aligned} x &= 4 + \quad z \\ y &= 3 - 2z \end{aligned}$$

Notice that the variables x and y *depend* on the choice of z. Thus, the system is said to be **dependent**. In this case, the variable z is free to be any real number of our choosing. Consequently, we say that z is a **free variable**. The solutions of this system are all ordered triples of the form $(4 + z, 3 - 2z, z)$. Thus, this system has infinitely many solutions. For example, letting $z = 0$, $z = -1$, and $z = 3$, we get the following ordered triples as solutions to this system:

$$\begin{aligned} z &= \quad 0: (4, 3, 0) \\ z &= -1: (3, 5, -1) \\ z &= \quad 3: (7, -3, 3) \end{aligned}$$

Try substituting these ordered triples into the original system to verify that all three are solutions.

My interactive video summary

📼 Example 3 Determine Whether a System Has No Solution or Infinitely Many Solutions

Each augmented matrix in row reduced form is equivalent to the augmented matrix of a system of linear equations in variables x, y, and z. Determine whether the system is dependent or inconsistent. If the system is dependent, describe the solution as in Example 2.

$$\textbf{a.}\ \left[\begin{array}{ccc|c} 1 & 0 & -2 & 5 \\ 0 & 1 & 3 & -2 \\ 0 & 0 & 0 & 0 \end{array}\right] \quad \textbf{b.}\ \left[\begin{array}{ccc|c} 1 & 0 & 0 & -4 \\ 0 & 1 & 0 & 6 \\ 0 & 0 & 0 & 10 \end{array}\right] \quad \textbf{c.}\ \left[\begin{array}{ccc|c} 1 & 0 & 0 & 3 \\ 0 & 1 & -2 & 4 \\ 0 & 0 & 0 & 0 \end{array}\right]$$

Solution Watch the interactive video to verify that the augmented matrices in parts a and c represent dependent systems, whereas the augmented matrix in part b represents an inconsistent system. The infinite solutions to the system in part a can be described by all ordered triples of the form $(5 + 2z, -2 - 3z, z)$. The infinite solutions to the system in part c can be described by all ordered triples of the form $(3, 4 + 2z, z)$ or $\left(3, y, \frac{1}{2}y - 2\right)$.

Work exercises 1–16 in this textbook or in the *MyMathLab*® Study Plan.

OBJECTIVE 3: ## SOLVING LINEAR SYSTEMS HAVING FEWER EQUATIONS THAN VARIABLES

In every linear system that we have encountered thus far, the number of equations in the system was exactly the same as the number of variables. However, this does not always have to be the case. For example, suppose we encounter a system of linear equations in three variables that has only two equations. We can geometrically illustrate such a system by sketching two planes. There are only three possible scenarios for such a system:

1. The two planes can be distinct parallel planes.

2. The two planes can intersect at a straight line.

3. The two planes can coincide; that is, the two equations describe the same plane. (See Figure 8).

Two planes are parallel.
No solution

Two planes intersect at a line.
Infinitely many solutions

Two planes coincide.
Infinitely many solutions

Figure 8

Example 4 Solve a System Having Fewer Equations Than Variables

Solve the linear system using Gauss-Jordan elimination:

$$\begin{aligned} x + y + z &= 1 \\ 2x - 2y + 6z &= 10 \end{aligned}$$

Solution We write the system using an augmented matrix and obtain the following sequence of equivalent matrices.

$$\begin{bmatrix} 1 & 1 & 1 & | & 1 \\ 2 & -2 & 6 & | & 10 \end{bmatrix} \xrightarrow{-(2)R_1 + R_2 \to \text{New } R_2} \begin{bmatrix} 1 & 1 & 1 & | & 1 \\ 0 & -4 & 4 & | & 8 \end{bmatrix} \xrightarrow{\left(-\frac{1}{4}\right) + R_2 \to \text{New } R_2}$$

$$\begin{bmatrix} 1 & 1 & 1 & | & 1 \\ 0 & 1 & -1 & | & -2 \end{bmatrix} \xrightarrow{(-1)R_2 + R_1 \to \text{New } R_1} \begin{bmatrix} 1 & 0 & 2 & | & 3 \\ 0 & 1 & -1 & | & -2 \end{bmatrix}$$

The last augmented matrix is in row reduced form. We see that the original system is equivalent to the system

$$\begin{aligned} x \quad\;\; + 2z &= 3 \\ y - z &= -2 \end{aligned}$$

We can solve for x and y in terms of z. Thus, z is a free variable, and we find that

$$\begin{aligned} x &= 3 - 2z \\ y &= z - 2 \\ z &= Free \end{aligned}$$

The solution can be written in the form $(3 - 2z, z - 2, z)$.

Work exercises 17–20 in this textbook or in the *MyMathLab*® Study Plan.

7.3 Inconsistent and Dependent Linear Systems in Three Variables   333

7.3   Exercises

In Exercises 1–6, each augmented matrix in row reduced form is equivalent to the augmented matrix of a system of linear equations in variables x, y, and z. Determine whether the system is dependent or inconsistent. If the system is dependent, determine which variable is free and describe the solution as an ordered triple in terms of the free variable. See Example 3.

1. $\begin{bmatrix} 1 & 0 & 0 & | & 6 \\ 0 & 1 & 0 & | & 5 \\ 0 & 0 & 0 & | & -2 \end{bmatrix}$

2. $\begin{bmatrix} 1 & 0 & -2 & | & -5 \\ 0 & 1 & 1 & | & 4 \\ 0 & 0 & 0 & | & 0 \end{bmatrix}$

3. $\begin{bmatrix} 1 & 3 & 0 & | & 5 \\ 0 & 1 & 1 & | & 8 \\ 0 & 0 & 0 & | & 0 \end{bmatrix}$

4. $\begin{bmatrix} 1 & 1 & 2 & | & 8 \\ 0 & 0 & 0 & | & 2 \\ 0 & 0 & 1 & | & 5 \end{bmatrix}$

5. $\begin{bmatrix} 1 & 2 & 0 & | & 1 \\ 0 & 0 & 0 & | & 0 \\ 0 & 0 & 1 & | & -4 \end{bmatrix}$

6. $\begin{bmatrix} 1 & 0 & 2 & | & -6 \\ 0 & 1 & -3 & | & 8 \\ 0 & 0 & 0 & | & 0 \end{bmatrix}$

In Exercises 7–20, use Gauss-Jordan elimination to solve each linear system and determine whether the system has a unique solution, no solution, or infinitely many solutions. If the system has infinitely many solutions, describe the solution as an ordered triple involving a free variable.

7. $\begin{aligned} x - 3y + 5z &= 6 \\ x - 4y + 13z &= 16 \\ x - 2y + 3z &= -5 \end{aligned}$

8. $\begin{aligned} x + y + z &= 3 \\ 2x + y - z &= 4 \\ 3x + y - 3z &= 5 \end{aligned}$

9. $\begin{aligned} 2x - y + z &= -6 \\ x + y - z &= 0 \\ -3x - 3y + 4z &= 1 \end{aligned}$

10. $\begin{aligned} x + y - z &= 3 \\ x + 2y + 2z &= 8 \\ 2x + y - 5z &= 2 \end{aligned}$

11. $\begin{aligned} x + y - z &= 2 \\ x + 2y - 3z &= 6 \\ 2x - 3y + 8z &= -16 \end{aligned}$

12. $\begin{aligned} -x + 2y - z &= -15 \\ 3x + y + 4z &= 34 \\ 2x - 4y + 3z &= 40 \end{aligned}$

13. $\begin{aligned} x + y + z &= 2 \\ 3x + y - 2z &= 8 \\ 2x - y - 8z &= 5 \end{aligned}$

14. $\begin{aligned} x - 2y + 3z &= 0 \\ 4x - 6y + 8z &= -4 \\ -2x - y + z &= 7 \end{aligned}$

15. $\begin{aligned} x + y + z &= 2 \\ 4x + 2y + 3z &= 9 \\ 3x - y + z &= 8 \end{aligned}$

16. $\begin{aligned} x + 4y + 2z &= 5 \\ 3x + 7y + z &= 0 \\ 2x + 5y + z &= 1 \end{aligned}$

17. $\begin{aligned} x + 2y - z &= 11 \\ x + 3y - 2z &= 14 \end{aligned}$

18. $\begin{aligned} x - 2y + z &= -2 \\ -3x + 6y + 3z &= -6 \end{aligned}$

19. $\begin{aligned} x + y + 5z &= 11 \\ 3x + y - z &= 7 \end{aligned}$

20. $\begin{aligned} x - y + z &= 5 \\ 3x + 2y - 2z &= 0 \end{aligned}$

7.4   Matrix Operations

THINGS TO KNOW

Before working through this section, be sure that you are familiar with the following concept:

You Try It

1. Solving a System of Three Equations in Three Variables Using Gauss-Jordan Elimination (Section 7.2)

VIDEO   ANIMATION   INTERACTIVE

OBJECTIVES

1 Understanding the Definition of a Matrix

2 Adding and Subtracting Matrices

3 Scalar Multiplication of Matrices

4 Multiplication of Matrices

5 Applications of Matrix Multiplication

OBJECTIVE 1:   UNDERSTANDING THE DEFINITION OF A MATRIX

My video summary

In Sections 7.2 and 7.3, we introduced matrices to display the coefficients of a system of linear equations in order to simplify the Gaussian elimination (or Gauss-Jordan elimination) process. For example, the linear system

$$2x + 3y + 4z = 12$$
$$x - 2y + 3z = 0$$
$$-x + y - 2z = -1$$

can be represented by the augmented matrix

$$\left[\begin{array}{ccc|c} 2 & 3 & 4 & 12 \\ 1 & -2 & 3 & 0 \\ -1 & 1 & -2 & -1 \end{array}\right]$$

The vertical bar was used to separate the left- and right-hand sides of the equations. We only used this vertical bar for convenience. We could have written this same matrix without the vertical bar as

$$\begin{bmatrix} 2 & 3 & 4 & 12 \\ 1 & -2 & 3 & 0 \\ -1 & 1 & -2 & -1 \end{bmatrix}$$

This matrix has three rows and four columns. Therefore, we say that the *size* of this matrix is 3 by 4 (or 3×4). In general, a matrix is any rectangular array of numbers of any size.

DEFINITION Matrix

A **matrix** is a rectangular array of numbers arranged in a fixed number, m, of rows and a fixed number, n, of columns. The size of a matrix is $m \times n$. The numbers in the array are called entries. Each entry is labeled a_{ij}, where i is the ith row of the entry and j is the jth column of the entry.

$$A = \begin{bmatrix} a_{11} & a_{12} & \cdots & a_{1n} \\ a_{21} & a_{22} & \cdots & a_{2n} \\ \vdots & \vdots & \vdots & \vdots \\ a_{m1} & a_{m2} & \cdots & a_{mn} \end{bmatrix}$$

Two matrices A and B are said to be equal $(A = B)$ provided both matrices are of the same size and the corresponding entries are the same.

We typically use capital letters to name a matrix. Following are some examples of matrices of various sizes:

$$A = \begin{bmatrix} 3 & -1 & 8 \\ 2 & -2 & 9 \\ 5 & 7 & -6 \end{bmatrix}, \quad B = [6 \ -5], \quad C = \begin{bmatrix} 2 & -1 \\ 3 & 1 \end{bmatrix}, \quad D = \begin{bmatrix} 3 & -1 \\ 9 & 6 \\ 2 & 7 \end{bmatrix}, \quad E = \begin{bmatrix} 3 \\ 7 \\ 1 \\ -3 \end{bmatrix}, \quad F = \begin{bmatrix} 1 & 0 & 0 & 0 \\ 0 & 1 & 0 & 0 \\ 0 & 0 & 1 & 0 \\ 0 & 0 & 0 & 1 \end{bmatrix}$$

3×3 Square matrix   Row matrix 2×2 Square matrix 3×2 Matrix   Column matrix   Identity matrix

Matrices A, C, and F are **square matrices** because the number of rows equals the number of columns. (The size of matrix A is 3×3, the size of matrix C is 2×2, and the size of matrix F is 4×4.) Matrix B is called a **row matrix** because it has only one row. Similarly, matrix E is called a **column matrix** because it has only one column. Matrix F is an example of an **identity matrix**. An identity matrix is a square matrix with 1's down the diagonal (from left to right) and zeros everywhere else. (You will see why we call matrix F an identity matrix when we define matrix multiplication.) Each number in a matrix is called an **entry**. We typically use lowercase letters and subscripts to denote each entry. For example, in matrix A, a_{23} refers to the entry of matrix A that is located in the second row and the third column. Therefore, $a_{23} = 9$. Most graphing calculators come equipped with matrix editors. Figure 9 shows matrix A with the entry $a_{23} = 9$ highlighted.

Using Technology

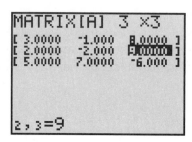

Figure 9 Using a graphing calculator to create a matrix

Work exercises **1–4** in this textbook or in the **MyMathLab** Study Plan.

OBJECTIVE 2: ADDING AND SUBTRACTING MATRICES

Just like with real numbers, we can add and subtract two matrices. To add or subtract matrices, they must be the *same size*. To add matrices, we simply add the corresponding entries. Similarly, to subtract matrices, we subtract the corresponding entries. Example 2 illustrates the addition and subtraction of two 3×4 matrices.

My video summary

▣ Example 1 Add and Subtract Matrices

Let $A = \begin{bmatrix} 2 & -1 & 0 & 3 \\ -1 & 4 & -2 & 5 \\ 5 & -3 & 7 & 9 \end{bmatrix}$ and $B = \begin{bmatrix} -3 & -2 & 6 & 1 \\ 0 & 4 & -5 & 1 \\ -1 & 4 & 2 & 7 \end{bmatrix}$.

a. Find matrix $A + B$.
b. Find matrix $B + A$.
c. Find matrix $A - B$.

Solution

a. Adding each corresponding entry of matrix A and matrix B, we obtain the matrix

$$A + B = \begin{bmatrix} (2 + (-3)) & (-1 + (-2)) & (0 + 6) & (3 + 1) \\ (-1 + 0) & (4 + 4) & (-2 + (-5)) & (5 + 1) \\ (5 + -1) & (-3 + 4) & (7 + 2) & (9 + 7) \end{bmatrix} = \begin{bmatrix} -1 & -3 & 6 & 4 \\ -1 & 8 & -7 & 6 \\ 4 & 1 & 9 & 16 \end{bmatrix}$$

b.

$$B + A = \begin{bmatrix} (-3 + 2) & (-2 + (-1)) & (6 + 0) & (1 + 3) \\ (0 + (-1)) & (4 + 4) & (-5 + (-2)) & (1 + 5) \\ (-1 + 5) & (4 + (-3)) & (2 + 7) & (7 + 9) \end{bmatrix} = \begin{bmatrix} -1 & -3 & 6 & 4 \\ -1 & 8 & -7 & 6 \\ 4 & 1 & 9 & 16 \end{bmatrix}$$

c. Subtracting each corresponding entry of matrix B from matrix A, we obtain the matrix

$$A - B = \begin{bmatrix} (2 - (-3)) & (-1 - (-2)) & (0 - 6) & (3 - 1) \\ (-1 - 0) & (4 - 4) & (-2 - (-5)) & (5 - 1) \\ (5 - (-1)) & (-3 - 4) & (7 - 2) & (9 - 7) \end{bmatrix} = \begin{bmatrix} 5 & 1 & -6 & 2 \\ -1 & 0 & 3 & 4 \\ 6 & -7 & 5 & 2 \end{bmatrix}$$

Notice in Example 1 that $A + B = B + A$. This is no coincidence. Matrices, like real numbers, are said to be **commutative** under the operation of addition. That is, the order in which we add two matrices does not matter. However, matrix subtraction is *not* commutative. Matrices are also **associative** under addition, which means $(A + B) + C = A + (B + C)$ provided that A, B, and C are the same size.

Work exercises 5–7 in this textbook or in the *MyMathLab*® Study Plan.

OBJECTIVE 3:   SCALAR MULTIPLICATION OF MATRICES

We can multiply any matrix by a real number. We call this real number a **scalar** and this operation is called scalar multiplication. Scalar multiplication is performed by multiplying every entry in the matrix by that scalar. For example, suppose matrix A is given by

$$A = \begin{bmatrix} a_{11} & a_{12} & \cdots & a_{1n} \\ a_{21} & a_{22} & \cdots & a_{2n} \\ \vdots & \vdots & \vdots & \vdots \\ a_{m1} & a_{m2} & \cdots & a_{mn} \end{bmatrix}$$

then if c is any real number, we define matrix cA as

$$cA = \begin{bmatrix} ca_{11} & ca_{12} & \cdots & ca_{1n} \\ ca_{21} & ca_{22} & \cdots & ca_{2n} \\ \vdots & \vdots & \vdots & \vdots \\ ca_{m1} & ca_{m2} & \cdots & ca_{mn} \end{bmatrix}$$

My video summary

▦ Example 2 Scalar Multiplication and Addition of Matrices

If $A = \begin{bmatrix} -4 & 0 \\ 3 & 1 \end{bmatrix}$ and $B = \begin{bmatrix} 6 & 3 \\ -3 & -9 \end{bmatrix}$, find the following matrices: $-2A$, $\frac{1}{3}B$, and $-2A + \frac{1}{3}B$.

Solution Work through the video to verify that

$$-2A = \begin{bmatrix} 8 & 0 \\ -6 & -2 \end{bmatrix}, \frac{1}{3}B = \begin{bmatrix} 2 & 1 \\ -1 & -3 \end{bmatrix}, \text{ and } -2A + \frac{1}{3}B = \begin{bmatrix} 10 & 1 \\ -7 & -5 \end{bmatrix}$$

Figure 10 shows how a graphing utility can be used to perform scalar multiplication and matrix addition.

Using Technology

Figure 10 Using a graphing utility to compute matrix $-2A + \frac{1}{3}B$

Work exercises 8–10 in this textbook or in the **MyMathLab**® Study Plan.

OBJECTIVE 4:   MULTIPLICATION OF MATRICES

My video summary

 Unlike with real numbers, we cannot always multiply two matrices. The product of two matrices is defined only if the number of columns of the first matrix is equal to the number of rows of the second matrix. If the number of columns of the first matrix does not equal the number of rows of the second matrix, then we say that the product matrix is not defined.

Matrix Multiplication

If A is an $m \times n$ matrix and B is an $n \times q$ matrix, then there exists an $m \times q$ product matrix AB. The element in the ith row and jth column of matrix AB is equal to the sum of the products of the corresponding entries from row i of matrix A and column j of matrix B.

We now illustrate matrix multiplication by multiplying the 2×3 and 3×2 matrices. Watch the video to see every step of the multiplication process.

$$
\overset{A}{\begin{bmatrix} 1 & 1 & 2 \\ 2 & 4 & -3 \end{bmatrix}} \overset{B}{\begin{bmatrix} 0 & -5 \\ 4 & -1 \\ 1 & 3 \end{bmatrix}} = \begin{bmatrix} (1)(0)+(1)(4)+(2)(1) & (1)(-5)+(1)(-1)+(2)(3) \\ (2)(0)+(4)(4)+(-3)(1) & (2)(-5)+(4)(-1)+(-3)(3) \end{bmatrix} = \overset{AB}{\begin{bmatrix} 6 & 0 \\ 13 & -23 \end{bmatrix}}
$$

$$2 \times 3 \qquad 3 \times 2 \qquad\qquad\qquad\qquad\qquad\qquad\qquad\qquad\qquad 2 \times 2$$

Same

Size of AB is 2×2

Notice that matrix AB has the same number of rows as matrix A and the same number of columns as matrix B. This will always be the case. In other words, if we multiply an $m \times n$ matrix by an $n \times q$ matrix, then the size of the resultant matrix will be $m \times q$.

My interactive video summary

Example 3 Matrix Multiplication

Given matrices $A = \begin{bmatrix} 1 & 1 & 2 \\ 2 & 4 & -3 \\ 3 & 6 & -5 \end{bmatrix}$, $B = \begin{bmatrix} 0 & -1 & 1 \\ -2 & 1 & 3 \\ 4 & 5 & -3 \end{bmatrix}$, and $C = \begin{bmatrix} 1 & 0 & 0 \\ 0 & 1 & 0 \\ 0 & 0 & 1 \end{bmatrix}$,

find the products AB, BA, and AC.

Solution Watch the interactive video to verify that

$$AB = \begin{bmatrix} 6 & 10 & -2 \\ -20 & -13 & 23 \\ -32 & -22 & 36 \end{bmatrix}, BA = \begin{bmatrix} 1 & 2 & -2 \\ 9 & 20 & -22 \\ 5 & 6 & 8 \end{bmatrix}, \text{ and } AC = \begin{bmatrix} 1 & 1 & 2 \\ 2 & 4 & -3 \\ 3 & 6 & -5 \end{bmatrix}$$

Figure 11 illustrates how a graphing utility can be used to perform matrix multiplication.

Using Technology

Figure 11   Using a graphing utility to compute matrices AB and BA

Work exercises 11–20 in this textbook or in the *MyMathLab*® Study Plan.

In Example 3, notice that $AB \neq BA$. This implies that matrix multiplication is not commutative. Although matrix multiplication is not commutative in general, it is possible to find two matrices that commute under the operation of multiplication. For example, consider matrix A and matrix C from Example 3. We can easily verify that

$$AC = \begin{bmatrix} 1 & 1 & 2 \\ 2 & 4 & -3 \\ 3 & 6 & -5 \end{bmatrix} \begin{bmatrix} 1 & 0 & 0 \\ 0 & 1 & 0 \\ 0 & 0 & 1 \end{bmatrix} = \begin{bmatrix} 1 & 1 & 2 \\ 2 & 4 & -3 \\ 3 & 6 & -5 \end{bmatrix} = A \text{ and}$$

$$CA = \begin{bmatrix} 1 & 0 & 0 \\ 0 & 1 & 0 \\ 0 & 0 & 1 \end{bmatrix} \begin{bmatrix} 1 & 1 & 2 \\ 2 & 4 & -3 \\ 3 & 6 & -5 \end{bmatrix} = \begin{bmatrix} 1 & 1 & 2 \\ 2 & 4 & -3 \\ 3 & 6 & -5 \end{bmatrix} = A.$$

Notice that C is an **identity matrix**. We will typically use the notation I_n to denote an $n \times n$ identity matrix. For example,

$$I_2 = \begin{bmatrix} 1 & 0 \\ 0 & 1 \end{bmatrix}, I_3 = \begin{bmatrix} 1 & 0 & 0 \\ 0 & 1 & 0 \\ 0 & 0 & 1 \end{bmatrix}, I_4 = \begin{bmatrix} 1 & 0 & 0 & 0 \\ 0 & 1 & 0 & 0 \\ 0 & 0 & 1 & 0 \\ 0 & 0 & 0 & 1 \end{bmatrix}, \text{ and so on.}$$

Whenever an $m \times n$ matrix A is multiplied by the identity matrix I_n, the resultant matrix is always A. Thus, we have the following identity property for matrix multiplication. This property is extremely important to understand and is revisited in Section 7.5.

Identity Property for Matrix Multiplication

For any $m \times n$ matrix A, $I_m A = A$ and $AI_n = A$. If A is an $n \times n$ square matrix, then $AI_n = I_n A = A$.

Example 4 Multiplication of an Identity Matrix

If $A = \begin{bmatrix} 2 & -1 \\ 5 & 6 \\ 0 & -3 \end{bmatrix}$, find appropriate identity matrices I_m and I_n and verify that $I_m A = A$ and $AI_n = A$.

Solution A is a 3×2 matrix, so $m = 3$ and $n = 2$. Therefore,

$$I_m = I_3 = \begin{bmatrix} 1 & 0 & 0 \\ 0 & 1 & 0 \\ 0 & 0 & 1 \end{bmatrix} \text{ and } I_n = I_2 = \begin{bmatrix} 1 & 0 \\ 0 & 1 \end{bmatrix}. \text{ Thus,}$$

$$I_m A = I_3 A = \begin{bmatrix} 1 & 0 & 0 \\ 0 & 1 & 0 \\ 0 & 0 & 1 \end{bmatrix} \begin{bmatrix} 2 & -1 \\ 5 & 6 \\ 0 & -3 \end{bmatrix} = \begin{bmatrix} 2 & -1 \\ 5 & 6 \\ 0 & -3 \end{bmatrix} = A$$

$$\text{and } AI_n = AI_2 = \begin{bmatrix} 2 & -1 \\ 5 & 6 \\ 0 & -3 \end{bmatrix} \begin{bmatrix} 1 & 0 \\ 0 & 1 \end{bmatrix} = \begin{bmatrix} 2 & -1 \\ 5 & 6 \\ 0 & -3 \end{bmatrix} = A.$$

Work exercises 21–23 in this textbook or in the *MyMathLab*® Study Plan.

Summary of Matrix Operations

Assume matrix addition and matrix multiplication are defined for matrices A, B, and C and let k be a scalar, then

1. $A + B = B + A$
2. $(A + B) + C = A + (B + C)$
3. $k(A + B) = kA + kB$
4. $(AB)C = A(BC)$
5. $A(B + C) = AB + AC$
6. $(B + C)A = BA + CA$
7. $k(AB) = (kA)B = A(kB)$
8. If A is an $m \times n$ matrix, then $I_m A = A$ and $AI_n = A$.

OBJECTIVE 5:   APPLICATIONS OF MATRIX MULTIPLICATION

Matrices can be used as a convenient way to express data. For example, suppose that a certain sporting goods manufacturer has manufacturing plants in Atlanta, Boise, Columbus, and Detroit. The number of each type of ball produced (in thousands) during the month of June last year is represented by matrix A.

	Atlanta	Boise	Columbus	Detroit	
Basketballs	8	5	4.5	10	
Footballs	7	3	2	8	$= A$
Golf balls	10	0	8	15	

We can use this matrix to determine the number of each ball that was produced in each plant. For example, the Detroit plant produced 10,000 basketballs last June.

We can deduce that the Boise plant does not produce golf balls (or perhaps the golf ball portion of the plant was shut down during June). Now, suppose that during the month of June, the selling price of a basketball was $25, footballs sold for $20, and golf balls sold for $1.50. We can calculate the total revenue at each plant by multiplying matrix A on the left by the 1×3 **row matrix** $[25 \quad 20 \quad 1.5] = B$, which represents the selling price of each ball. Thus, we can calculate the revenue matrix BA:

$$BA = [25 \quad 20 \quad 1.5] \begin{bmatrix} 8 & 5 & 4.5 & 10 \\ 7 & 3 & 2 & 8 \\ 10 & 0 & 8 & 15 \end{bmatrix} \begin{matrix} \text{Atlanta} & \text{Boise} & \text{Columbus} & \text{Detroit} \\ = [355 & 185 & 164.5 & 432.5] \end{matrix}$$

Therefore, we see that total revenue generated by each manufacturing plant in the month of June was

Atlanta:	$355,000
Boise:	$185,000
Columbus:	$164,500
Detroit:	$432,500

Work exercises 24–26 in this textbook or in the *MyMathLab*® Study Plan.

7.4 Exercises

1. $A = \begin{bmatrix} 2 & -5 & 7 & 9 \\ 1 & 3 & -4 & 11 \\ 8 & 6 & 1 & 0 \end{bmatrix}$

a. What is the size of matrix A?

b. Determine whether A is a square matrix, row matrix, column matrix, or none of these.

2. $A = \begin{bmatrix} 1 & 0 & 5 \\ 2 & 1 & 0 \\ 0 & -1 & 1 \end{bmatrix}$

a. What is the size of matrix A?

b. Determine whether A is a square matrix, row matrix, column matrix, or none of these.

3. $A = \begin{bmatrix} 4 & -2 & \pi & \sqrt{3} \end{bmatrix}$

a. What is the size of matrix A?

b. Determine whether A is a square matrix, row matrix, column matrix, or none of these.

4. $A = \begin{bmatrix} 3.9 \\ -1.2 \\ 5.3 \\ -2.7 \end{bmatrix}$

a. What is the size of matrix A?

b. Determine whether A is a square matrix, row matrix, column matrix, or none of these.

In Exercises 5–10, use matrices A and B to find the indicated matrix:

$$A = \begin{bmatrix} 1 & 0 & -2 \\ 3 & 1 & 4 \\ 5 & -1 & -5 \end{bmatrix}, B = \begin{bmatrix} 3 & 4 & -1 \\ 5 & 0 & 2 \\ 3 & 4 & -5 \end{bmatrix}$$

5. $A + B$　　**6.** $A - B$　　**7.** $B - A$　　**8.** $2A$　　**9.** $\frac{1}{2}B$　　**10.** $2A - \frac{1}{2}B$

In Exercises 11–14, let $(m \times n)$ denote a matrix of size $m \times n$. Find the size of the following products, or state that the product is not defined.

11. $(2 \times 3)(3 \times 2)$　　**12.** $(3 \times 4)(4 \times 5)$　　**13.** $(2 \times 3)(2 \times 3)$　　**14.** $(1 \times 4)(4 \times 1)$

In Exercises 15–20, use matrices A, B, C, and D to find the indicated matrix:

$$A = \begin{bmatrix} -1 & 2 \\ 3 & 5 \end{bmatrix}, B = \begin{bmatrix} 1 & 3 & -1 \\ 4 & -2 & 0 \end{bmatrix}, C = \begin{bmatrix} 1 & 0 & -1 \\ 4 & 0 & 3 \\ 2 & -1 & -2 \end{bmatrix}, D = \begin{bmatrix} 1 & -2 \\ -3 & -4 \end{bmatrix}$$

15. AB **16.** BC **17.** $AB + BC$ **18.** A^2 **19.** ABC **20.** $(A + D)B$

In Exercises 21–23, matrix A is given. Find appropriate identity matrices I_m and I_n such that $I_m A = A$ and $AI_n = A$.

21. $A = \begin{bmatrix} 5 & 1 & -1 \\ 3 & -2 & 4 \end{bmatrix}$

22. $A = \begin{bmatrix} 2 & -1 & 0 & 3 \\ -1 & 4 & -2 & 5 \\ 5 & -3 & 7 & 9 \end{bmatrix}$

23. $A = \begin{bmatrix} 3 & -1 \\ 9 & 6 \\ 2 & 7 \\ 0 & 4 \\ 1 & -3 \end{bmatrix}$

24. The youth from a local church are having a breakfast fund-raising event. They are planning on serving biscuits, pancakes, and waffles. The ingredients for one batch of each are as follows:

Biscuits: 3 cups baking mix, 1 egg, 1 cup milk, $\frac{1}{2}$ tablespoon cooking oil

Pancakes: 2 cups baking mix, 2 eggs, $\frac{2}{3}$ cup milk

Waffles: $2\frac{1}{2}$ cups baking mix, 1 egg, 1 cup milk, 2 teaspoons cooking oil

The ingredients can be summarized by matrix A.

$$\begin{array}{c} \\ \text{Biscuits} \\ \text{Pancakes} \\ \text{Waffles} \end{array} \begin{array}{cccc} \text{Mix} & \text{Eggs} & \text{Milk} & \text{Oil} \\ \begin{bmatrix} 3 & 1 & 1 & \frac{1}{2} \\ 2 & 2 & \frac{2}{3} & 0 \\ 2\frac{1}{2} & 1 & 1 & 2 \end{bmatrix} = A \end{array}$$

It was determined that they were going to need 16 batches of biscuits, 30 batches of pancakes, and 40 batches of waffles, which is summarized by matrix B.

$$\begin{array}{c} \\ \text{Batches} \end{array} \begin{array}{ccc} \text{Biscuits} & \text{Pancakes} & \text{Waffles} \\ [16 & 30 & 40] = B \end{array}$$

a. Calculate matrix BA.

b. Interpret the meaning of each entry of matrix BA.

25. A lawn-mower manufacturer produces two types of lawn-mowers, a standard model and the self-propelled

model. The production of each model requires assembly and quality assurance testing. The number of hours required for each model are given by the following matrix.

$$\begin{array}{c} \\ \text{Standard} \\ \text{Self-propelled} \end{array} \begin{array}{cc} \text{Assembly} & \text{Testing} \\ \begin{bmatrix} 2 & .5 \\ 2.5 & 1 \end{bmatrix} = A \end{array}$$

The company has three manufacturing plants located in New York, Los Angeles, and Beijing, China. The hourly labor rates (in dollars) for each plant is given by the following matrix.

$$\begin{array}{c} \\ \text{Assembly} \\ \text{Testing} \end{array} \begin{array}{ccc} \text{New York} & \text{Los Angeles} & \text{Beijing} \\ \begin{bmatrix} \$12 & \$10 & \$4 \\ \$9 & \$8 & \$2 \end{bmatrix} = B \end{array}$$

a. Calculate matrix AB.

b. If ab_{ij} represents the entry of matrix AB in row i and column j, then interpret the meaning of the entries ab_{21}, ab_{12}, and ab_{23}.

26. A small clothing store has locations in Buffalo, Rochester, and Syracuse. The store sells only three items: sport coats, slacks, and shirts. During the first quarter, the Buffalo branch sold 100 sport coats, 120 pairs of slacks, and 250 shirts. The Rochester store sold 110 sport coats, 100 slacks, and 300 shirts. The Syracuse store sold 180 sport coats, 200 slacks, and 320 shirts. At each store, sport coats sell for $100, slacks sell for $50, and shirts sell for $35. Use matrix multiplication to get a matrix that shows the total revenue for each store. What was the total revenue of all three stores combined?

7.5 Inverses of Matrices and Matrix Equations

THINGS TO KNOW

Before working through this section, be sure that you are familiar with the following concepts:

VIDEO   ANIMATION   INTERACTIVE

You Try   **1.**   Solving a System of Three Equations in Three Variables Using Gauss-Jordan Elimination (Section 7.2)

VIDEO   ANIMATION   INTERACTIVE

2. Scalar Multiplication of Matrices
(Section 7.4)

You Try

3. Multiplication of Matrices
(Section 7.4)

You Try

4. Applications of Matrix Multiplication
(Section 7.4)

You Try

OBJECTIVES

1 Understanding the Definition of an Inverse Matrix

2 Finding the Inverse of a 2 × 2 Matrix Using a Formula

3 Finding the Inverse of an Invertible Square Matrix

4 Solving Systems of Equations Using an Inverse Matrix

OBJECTIVE 1:   UNDERSTANDING THE DEFINITION OF AN INVERSE MATRIX

In Section 7.4 we introduced the **identity matrix** and the **identity property for matrix multiplication.** Recall, if $A = \begin{bmatrix} 4 & 6 \\ -1 & -2 \end{bmatrix}$, then

$$AI_2 = \begin{bmatrix} 4 & 6 \\ -1 & -2 \end{bmatrix}\begin{bmatrix} 1 & 0 \\ 0 & 1 \end{bmatrix} = \begin{bmatrix} 4 & 6 \\ -1 & -2 \end{bmatrix} = A \quad \text{and}$$

$$I_2A = \begin{bmatrix} 1 & 0 \\ 0 & 1 \end{bmatrix}\begin{bmatrix} 4 & 6 \\ -1 & -2 \end{bmatrix} = \begin{bmatrix} 4 & 6 \\ -1 & -2 \end{bmatrix} = A.$$

Now, suppose there exists a 2 × 2 matrix B such that $AB = BA = I_2$. For example, if $A = \begin{bmatrix} 4 & 6 \\ -1 & -2 \end{bmatrix}$ and $B = \begin{bmatrix} 1 & 3 \\ -\frac{1}{2} & -2 \end{bmatrix}$, then

$$AB = \begin{bmatrix} 4 & 6 \\ -1 & -2 \end{bmatrix}\begin{bmatrix} 1 & 3 \\ -\frac{1}{2} & -2 \end{bmatrix} = \begin{bmatrix} 4-3 & 12-12 \\ -1+1 & -3+4 \end{bmatrix} = \begin{bmatrix} 1 & 0 \\ 0 & 1 \end{bmatrix} = I_2 \quad \text{and}$$

$$BA = \begin{bmatrix} 1 & 3 \\ -\frac{1}{2} & -2 \end{bmatrix}\begin{bmatrix} 4 & 6 \\ -1 & -2 \end{bmatrix} = \begin{bmatrix} 4-3 & 6-6 \\ -2+2 & -3+4 \end{bmatrix} = \begin{bmatrix} 1 & 0 \\ 0 & 1 \end{bmatrix} = I_2$$

We say that matrix B is the **multiplicative inverse** of matrix A. We typically use the notation A^{-1} to denote the multiplicative inverse of matrix A.

DEFINITION Multiplicative Inverse of a Matrix

Let A be an $n \times n$ square matrix. If there exists an $n \times n$ matrix A^{-1} such that $AA^{-1} = A^{-1}A = I_n$, then A^{-1} is called the **multiplicative inverse** of matrix A.

Note: For the remainder of this section, the term *inverse* implies *multiplicative inverse*.

My video summary

▦ Example 1 Verify the Inverse of a Square Matrix

Verify that $A = \begin{bmatrix} 1 & 1 & 2 \\ 2 & 4 & -3 \\ 3 & 6 & -5 \end{bmatrix}$ and $B = \begin{bmatrix} 2 & -17 & 11 \\ -1 & 11 & -7 \\ 0 & 3 & -2 \end{bmatrix}$ are inverse matrices.

Solution To verify that the two matrices are inverses, we must show that $AB = BA = I_3$.

$$AB = \begin{bmatrix} 1 & 1 & 2 \\ 2 & 4 & -3 \\ 3 & 6 & -5 \end{bmatrix}\begin{bmatrix} 2 & -17 & 11 \\ -1 & 11 & -7 \\ 0 & 3 & -2 \end{bmatrix} = \begin{bmatrix} 1 & 0 & 0 \\ 0 & 1 & 0 \\ 0 & 0 & 1 \end{bmatrix} = I_3$$

$$BA = \begin{bmatrix} 2 & -17 & 11 \\ -1 & 11 & -7 \\ 0 & 3 & -2 \end{bmatrix}\begin{bmatrix} 1 & 1 & 2 \\ 2 & 4 & -3 \\ 3 & 6 & -5 \end{bmatrix} = \begin{bmatrix} 1 & 0 & 0 \\ 0 & 1 & 0 \\ 0 & 0 & 1 \end{bmatrix} = I_3$$

Because $AB = BA = I_3$, we have verified that matrix A is the inverse of matrix B and vice versa. Watch the video to see each step of the multiplication process.

Work exercises **1–4** in this textbook or in the ***MyMathLab*** ® Study Plan.

OBJECTIVE 2:   FINDING THE INVERSE OF A 2 × 2 MATRIX USING A FORMULA

The definition of the inverse of a matrix indicates that a matrix must be square in order for it to have an inverse. However, not every square matrix has an inverse. If a square matrix has an inverse, we say that the matrix is **invertible**. If a square matrix does not have an inverse, we say that the matrix is **singular**. There is an easy way to determine whether a 2 × 2 matrix is invertible or singular. Every square matrix has a number associated with it called the **determinant**. (We learn more about determinants in Section 7.6.) The formula for the determinant of a 2 × 2 matrix is as follows.

Determinant of a 2 × 2 Matrix

Let $A = \begin{bmatrix} a & b \\ c & d \end{bmatrix}$. The determinant of matrix A is denoted as $|A|$ or $\begin{vmatrix} a & b \\ c & d \end{vmatrix}$ and is defined by $|A| = \begin{vmatrix} a & b \\ c & d \end{vmatrix} = ad - cb$. If $|A| \neq 0$, then matrix A is invertible.

You can see that if $|A| \neq 0$, then matrix A has an inverse. For example, the matrix $A = \begin{bmatrix} 4 & 6 \\ -1 & -2 \end{bmatrix}$ is invertible because $a = 4, b = 6, c = -1, d = -2$, and $|A| = \begin{vmatrix} 4 & 6 \\ -1 & -2 \end{vmatrix} = (4)(-2) - (-1)(6) = -8 + 6 = -2 \neq 0$. Following is a formula for determining the inverse of a 2 × 2 matrix.

My video summary

Formula for Determining the Inverse of a 2 × 2 Matrix

Let $A = \begin{bmatrix} a & b \\ c & d \end{bmatrix}$. If $|A| \neq 0$, then A is invertible and $A^{-1} = \dfrac{1}{|A|}\begin{bmatrix} d & -b \\ -c & a \end{bmatrix}$.

See the video proof.

Example 2 Find the Inverse of a 2 × 2 Matrix

Determine whether the following matrices are invertible or singular. If the matrix is invertible, find its inverse and then verify by using matrix multiplication.

a. $A = \begin{bmatrix} 2 & 3 \\ -4 & -6 \end{bmatrix}$ **b.** $B = \begin{bmatrix} 2 & -1 \\ 4 & 3 \end{bmatrix}$

Solution

a. $|A| = \begin{vmatrix} 2 & 3 \\ -4 & -6 \end{vmatrix} = (2)(-6) - (-4)(3) = -12 + 12 = 0.$

The determinant is zero, so matrix A is singular.

b. $|B| = \begin{vmatrix} 2 & -1 \\ 4 & 3 \end{vmatrix} = (2)(3) - (4)(-1) = 6 + 4 = 10.$

The determinant is nonzero, so matrix B is invertible.

$$B^{-1} = \frac{1}{10}\begin{bmatrix} 3 & 1 \\ -4 & 2 \end{bmatrix} = \begin{bmatrix} \frac{3}{10} & \frac{1}{10} \\ -\frac{2}{5} & \frac{1}{5} \end{bmatrix}$$

We can verify that this is the inverse matrix by showing that $BB^{-1} = B^{-1}B = I_2$. Watch the video to see the solution worked out in detail.

Using Technology

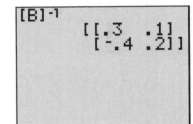

Using the matrix editor feature of a graphing utility, we can attempt to find the inverse of the matrices from Example 2. You can see in the first screenshot that an error occurs because matrix A is singular.

Work exercises 5–8 in this textbook or in the *MyMathLab* Study Plan.

OBJECTIVE 3:   FINDING THE INVERSE OF AN INVERTIBLE SQUARE MATRIX

Before we learn how to find the inverse of any square invertible matrix, we start by showing an alternative way of finding the inverse of a 2 × 2 invertible matrix and then generalize this technique for any invertible $n \times n$ matrix. In Example 2, we saw that $B = \begin{bmatrix} 2 & -1 \\ 4 & 3 \end{bmatrix}$ was invertible.

Thus, there exists an inverse matrix B^{-1} such that $BB^{-1} = B^{-1}B = I_2$. If we let $B^{-1} = \begin{bmatrix} x & y \\ z & w \end{bmatrix}$, then

$$\underbrace{\begin{bmatrix} 2 & -1 \\ 4 & 3 \end{bmatrix}}_{B} \underbrace{\begin{bmatrix} x & y \\ z & w \end{bmatrix}}_{B^{-1}} = \underbrace{\begin{bmatrix} 1 & 0 \\ 0 & 1 \end{bmatrix}}_{I_2}.$$

Multiplying BB^{-1}, we get

$$\underbrace{\begin{bmatrix} 2x - z & 2y - w \\ 4x + 3z & 4y + 3w \end{bmatrix}}_{BB^{-1}} = \underbrace{\begin{bmatrix} 1 & 0 \\ 0 & 1 \end{bmatrix}}_{I_2}.$$

We can compare the corresponding entries of the previous two matrices to obtain the following two systems of equations:

$$\begin{array}{cc} 2x - z = 1 & 2y - w = 0 \\ 4x + 3z = 0 & \text{and} \quad 4y + 3w = 1 \end{array}$$

These two systems correspond to the following two augmented matrices:

$$\left[\begin{array}{cc|c} 2 & -1 & 1 \\ 4 & 3 & 0 \end{array}\right] \quad \text{and} \quad \left[\begin{array}{cc|c} 2 & -1 & 0 \\ 4 & 3 & 1 \end{array}\right]$$

Because the coefficients of the left-hand side of the previous two matrices are exactly the same, we can combine the two augmented matrices into the following augmented matrix:

$$\left[\begin{array}{cc|c} 2 & -1 & 1 \\ 4 & 3 & 0 \end{array}\right] \quad \left[\begin{array}{cc|c} 2 & -1 & 0 \\ 4 & 3 & 1 \end{array}\right]$$

$$\left[\begin{array}{cc|cc} 2 & -1 & 1 & 0 \\ 4 & 3 & 0 & 1 \end{array}\right]$$

To solve for each variable, we use Gauss-Jordan elimination to write the left-hand side of the augmented matrix in **reduced row-echelon form**. Click here to see how the previous augmented matrix can be reduced to the matrix

$$\left[\begin{array}{cc|cc} 1 & 0 & \frac{3}{10} & \frac{1}{10} \\ 0 & 1 & -\frac{2}{5} & \frac{1}{5} \end{array}\right]$$

Therefore, we see that $x = \frac{3}{10}$, $y = \frac{1}{10}$, $z = -\frac{2}{5}$, and $w = \frac{1}{5}$. Because $B^{-1} = \begin{bmatrix} x & y \\ z & w \end{bmatrix}$, then $B^{-1} = \begin{bmatrix} \frac{3}{10} & \frac{1}{10} \\ -\frac{2}{5} & \frac{1}{5} \end{bmatrix}$.

You can see that this is precisely what we obtained for the inverse in Example 2. Note that we started with the augmented matrix

$$\underbrace{\left[\begin{array}{cc|cc} 2 & -1 \\ 4 & 3 \end{array}\right.}_{B} \underbrace{\left.\begin{array}{cc} 1 & 0 \\ 0 & 1 \end{array}\right]}_{I_2},$$

then used Gauss-Jordan elimination to end up with the augmented matrix

$$\underbrace{\begin{bmatrix} 1 & 0 \\ 0 & 1 \end{bmatrix}}_{I_2} \left| \underbrace{\begin{array}{cc} \frac{3}{10} & \frac{1}{10} \\ -\frac{2}{5} & \frac{1}{5} \end{array}}_{B^{-1}} \right].$$

This technique can be used to find the inverse of any invertible square matrix. The process is summarized as follows.

Steps for Finding the Multiplicative Inverse of an Invertible Square Matrix

If A is an $n \times n$ invertible matrix, then the multiplicative inverse of A can be obtained by following these steps:

Step 1   Form the augmented matrix $\left[A | I_n \right]$.

Step 2   Use Gauss-Jordan elimination to reduce A into the identity matrix I_n.

Step 3   The new augmented matrix is of the form $\left[I_n | A^{-1} \right]$, where A^{-1} is the multiplicative inverse of A.

My video summary

Example 3 Find the Inverse of a 3 × 3 Matrix

If possible, determine the inverse of $A = \begin{bmatrix} 1 & 1 & 2 \\ 2 & 4 & -3 \\ 3 & 6 & -5 \end{bmatrix}$.

Solution To attempt to find the inverse of matrix A, we create the augmented matrix by attaching the identity matrix, I_3, to the right-hand side.

$$\left[\underbrace{\begin{array}{ccc} 1 & 1 & 2 \\ 2 & 4 & -3 \\ 3 & 6 & -5 \end{array}}_{A} \left| \underbrace{\begin{array}{ccc} 1 & 0 & 0 \\ 0 & 1 & 0 \\ 0 & 0 & 1 \end{array}}_{I_3} \right. \right]$$

We now use Gauss-Jordan elimination to attempt to write the matrix on the left-hand side as I_3. Watch this video to see each step of the Gauss-Jordan elimination process. The reduced augmented matrix is

$$\left[\underbrace{\begin{array}{ccc} 1 & 0 & 0 \\ 0 & 1 & 0 \\ 0 & 0 & 1 \end{array}}_{I_3} \left| \underbrace{\begin{array}{ccc} 2 & -17 & 11 \\ -1 & 11 & -7 \\ 0 & 3 & -2 \end{array}}_{A^{-1}} \right. \right].$$

Therefore, $A^{-1} = \begin{bmatrix} 2 & -17 & 11 \\ -1 & 11 & -7 \\ 0 & 3 & -2 \end{bmatrix}$. You should verify that $AA^{-1} = A^{-1}A = I_3$.

Example 4 Example of a 3 × 3 Singular Matrix

If possible, determine the inverse of $A = \begin{bmatrix} 5 & 0 & -1 \\ 1 & -3 & -2 \\ 0 & 5 & 3 \end{bmatrix}$.

Solution To attempt to find the inverse of matrix A, we create the augmented matrix by attaching the identity matrix, I_3, to the right-hand side.

$$\begin{bmatrix} 5 & 0 & -1 & | & 1 & 0 & 0 \\ 1 & -3 & -2 & | & 0 & 1 & 0 \\ 0 & 5 & 3 & | & 0 & 0 & 1 \end{bmatrix}$$
$$\underbrace{\qquad}_{A} \qquad \underbrace{\qquad}_{I_3}$$

We now use a series of row operations to try to obtain I_3 on the left-hand side:

$$\begin{bmatrix} 5 & 0 & -1 & | & 1 & 0 & 0 \\ 1 & -3 & -2 & | & 0 & 1 & 0 \\ 0 & 5 & 3 & | & 0 & 0 & 1 \end{bmatrix} \xrightarrow{R_1 \Leftrightarrow R_2} \begin{bmatrix} 1 & -3 & -2 & | & 0 & 1 & 0 \\ 5 & 0 & -1 & | & 1 & 0 & 0 \\ 0 & 5 & 3 & | & 0 & 0 & 1 \end{bmatrix} \xrightarrow{-5R_1 + R_2 \to \text{New } R_2}$$

$$\begin{bmatrix} 1 & -3 & -2 & | & 0 & 1 & 0 \\ 0 & 15 & 9 & | & 1 & -5 & 0 \\ 0 & 5 & 3 & | & 0 & 0 & 1 \end{bmatrix} \xrightarrow{\left(\frac{1}{15}\right) R_2 \to \text{New } R_2} \begin{bmatrix} 1 & -3 & -2 & | & 0 & 1 & 0 \\ 0 & 1 & \frac{3}{5} & | & \frac{1}{15} & -\frac{1}{3} & 0 \\ 0 & 5 & 3 & | & 0 & 0 & 1 \end{bmatrix} \xrightarrow{(-5)R_2 + R_3 \to \text{New } R_3}$$

$$\begin{bmatrix} 1 & -3 & -2 & | & 0 & 1 & 0 \\ 0 & 1 & \frac{3}{5} & | & \frac{1}{15} & -\frac{1}{3} & 0 \\ 0 & 0 & 0 & | & -\frac{1}{3} & \frac{5}{3} & 1 \end{bmatrix}$$

CAUTION Because the last row of the matrix on the left-hand side is all zeros, there is no way to obtain the identity matrix. Therefore, matrix A does not have an inverse and thus is singular. In Section 7.6, we learn another way to determine whether a square matrix is invertible.

Work exercises 9–14 in this textbook or in the *MyMathLab*® Study Plan.

OBJECTIVE 4: SOLVING SYSTEMS OF EQUATIONS USING AN INVERSE MATRIX

My video summary

Suppose we are given the linear equation $ax = b$, where $a \neq 0$. If $a^{-1} = \frac{1}{a}$ (the multiplicative inverse of a), then we can solve for x as follows:

$$ax = b$$
$$a^{-1}ax = a^{-1}b \qquad \textit{Multiply both sides by the multiplicative inverse of a.}$$
$$\frac{1}{a}ax = a^{-1}b \qquad a^{-1} = \frac{1}{a}$$
$$x = a^{-1}b \qquad \frac{1}{a} \cdot a = 1$$

We can use this same technique to solve systems of n linear equations with n variables as follows.

Suppose we are given the linear system

$$a_{11}x_1 + a_{12}x_2 + a_{13}x_3 + \cdots a_{1n}x_n = b_1.$$
$$a_{21}x_1 + a_{22}x_2 + a_{23}x_3 + \cdots a_{2n}x_n = b_2$$
$$\vdots \qquad \vdots \qquad \vdots \qquad \vdots \qquad \vdots$$
$$a_{n1}x_1 + a_{n2}x_2 + a_{n3}x_3 + \cdots a_{nn}x_n = b_n$$

This system is equivalent to the matrix equation

$$\begin{bmatrix} a_{11} & a_{12} & a_{13} & \cdots & a_{1n} \\ a_{21} & a_{22} & a_{23} & \cdots & a_{2n} \\ \vdots & \vdots & \vdots & & \vdots \\ a_{n1} & a_{n2} & a_{n3} & & a_{nn} \end{bmatrix} \begin{bmatrix} x_1 \\ x_2 \\ \vdots \\ x_n \end{bmatrix} = \begin{bmatrix} b_1 \\ b_2 \\ \vdots \\ b_n \end{bmatrix}.$$

If we let $A = \underbrace{\begin{bmatrix} a_{11} & a_{12} & a_{13} & \cdots & a_{1n} \\ a_{21} & a_{22} & a_{23} & \cdots & a_{2n} \\ \vdots & \vdots & \vdots & & \vdots \\ a_{n1} & a_{n2} & a_{n3} & & a_{nn} \end{bmatrix}}_{\text{Coefficient Matrix}}$, $X = \underbrace{\begin{bmatrix} x_1 \\ x_2 \\ \vdots \\ x_n \end{bmatrix}}_{\substack{\text{Variable} \\ \text{Matrix}}}$, and $B = \underbrace{\begin{bmatrix} b_1 \\ b_2 \\ \vdots \\ b_n \end{bmatrix}}_{\substack{\text{Constant} \\ \text{Matrix}}}$, then

the previous matrix equation can be written as $AX = B$. Matrix A is called the **coefficient matrix**, matrix X is called the **variable matrix**, and matrix B is called the **constant matrix**. If A is invertible, then we can determine the entries of matrix X by multiplying both sides of the matrix equation *on the left* by A^{-1}:

$$AX = B \qquad \text{*Write the matrix equation.*}$$
$$A^{-1}AX = A^{-1}B \qquad \text{*Multiply both sides of the equation on the left by A^{-1}.*}$$
$$I_nX = A^{-1}B \qquad \text{*Definition of an inverse matrix $(A^{-1}A = I_n)$.*}$$
$$X = A^{-1}B \qquad \text{*Identity property for matrix multiplication $(I_nX = X)$.*}$$

CAUTION Because matrix multiplication is **not commutative**, we must be careful when multiplying both sides of a matrix equation by another matrix. Notice that when we multiplied both sides of the previous equation by A^{-1}, we had to position A^{-1} on the left of each side.

Note: If $AX = B$, then $A^{-1}AX \neq BA^{-1}$.

My video summary

Example 5   Solve a System of Equations Using an Inverse Matrix

Solve the following system of linear equations using an inverse matrix:

$$-3x + 2y = 4$$
$$5x - 4y = 9$$

Solution Let $A = \begin{bmatrix} -3 & 2 \\ 5 & -4 \end{bmatrix}$, $X = \begin{bmatrix} x \\ y \end{bmatrix}$, and $B = \begin{bmatrix} 4 \\ 9 \end{bmatrix}$. The linear system is equivalent to the matrix equation $AX = B$:

$$\overbrace{\begin{bmatrix} -3 & 2 \\ 5 & -4 \end{bmatrix}}^{A} \overbrace{\begin{bmatrix} x \\ y \end{bmatrix}}^{X} = \overbrace{\begin{bmatrix} 4 \\ 9 \end{bmatrix}}^{B}$$

We now calculate A^{-1} using the formula for finding the inverse of a 2×2 invertible matrix:

$$|A| = \begin{vmatrix} -3 & 2 \\ 5 & -4 \end{vmatrix} = (-3)(-4) - (5)(2) = 2, \text{ so } A^{-1} = \frac{1}{2}\begin{bmatrix} -4 & -2 \\ -5 & -3 \end{bmatrix} = \begin{bmatrix} -2 & -1 \\ -\frac{5}{2} & -\frac{3}{2} \end{bmatrix}$$

To solve for X, we must multiply both sides of the matrix equation on the left by A^{-1}.

7.5 Inverses of Matrices and Matrix Equations   **349**

$$\overbrace{\begin{bmatrix} -3 & 2 \\ 5 & -4 \end{bmatrix}}^{A} \overbrace{\begin{bmatrix} x \\ y \end{bmatrix}}^{X} = \overbrace{\begin{bmatrix} 4 \\ 9 \end{bmatrix}}^{B}$$

Write the matrix equation $AX = B$.

$$\overbrace{\begin{bmatrix} -2 & -1 \\ -\frac{5}{2} & -\frac{3}{2} \end{bmatrix}}^{A^{-1}} \overbrace{\begin{bmatrix} -3 & 2 \\ 5 & -4 \end{bmatrix}}^{A} \overbrace{\begin{bmatrix} x \\ y \end{bmatrix}}^{X} = \overbrace{\begin{bmatrix} -2 & -1 \\ -\frac{5}{2} & -\frac{3}{2} \end{bmatrix}}^{A^{-1}} \overbrace{\begin{bmatrix} 4 \\ 9 \end{bmatrix}}^{B}$$

Multiply both sides of the equation on the left by A^{-1}.

$$\underbrace{\begin{bmatrix} 1 & 0 \\ 0 & 1 \end{bmatrix}}_{I_2} \begin{bmatrix} x \\ y \end{bmatrix} = \begin{bmatrix} -17 \\ -\frac{47}{2} \end{bmatrix}$$

Definition of an inverse matrix $(A^{-1}A = I_n)$.

$$\begin{bmatrix} x \\ y \end{bmatrix} = \begin{bmatrix} -17 \\ -\frac{47}{2} \end{bmatrix}$$

Identity property for matrix multiplication $(I_n X = X)$.

So, $x = -17$ and $y = -\frac{47}{2}$.

Example 6 Solve a System of Equations Using an Inverse Matrix

Solve the following system of linear equations using an inverse matrix:

$$\begin{aligned} 2x + 4y + z &= 3 \\ -x + y - z &= 6 \\ x + 4y \phantom{{}+z} &= 7 \end{aligned}$$

Solution The linear system can be written as the following matrix equation:

$$\underbrace{\begin{bmatrix} 2 & 4 & 1 \\ -1 & 1 & -1 \\ 1 & 4 & 0 \end{bmatrix}}_{A} \underbrace{\begin{bmatrix} x \\ y \\ z \end{bmatrix}}_{X} = \underbrace{\begin{bmatrix} 3 \\ 6 \\ 7 \end{bmatrix}}_{B}$$

The solution can be obtained by multiplying both sides of this matrix equation by A^{-1} to obtain $X = A^{-1}B$. To find the inverse matrix, we create the augmented matrix by attaching the identity matrix, I_3, to the right-hand side.

$$\left[\underbrace{\begin{array}{ccc} 2 & 4 & 1 \\ -1 & 1 & -1 \\ 1 & 4 & 0 \end{array}}_{A} \middle| \underbrace{\begin{array}{ccc} 1 & 0 & 0 \\ 0 & 1 & 0 \\ 0 & 0 & 1 \end{array}}_{I_3} \right]$$

After a series of row operations, we obtain the following augmented matrix:

$$\left[\underbrace{\begin{array}{ccc} 1 & 0 & 0 \\ 0 & 1 & 0 \\ 0 & 0 & 1 \end{array}}_{A} \middle| \underbrace{\begin{array}{ccc} -4 & -4 & 5 \\ 1 & 1 & -1 \\ 5 & 4 & -6 \end{array}}_{A^{-1}} \right]$$

(Click here to see every step of this row reduction process.)

Therefore, $A^{-1} = \begin{bmatrix} -4 & -4 & 5 \\ 1 & 1 & -1 \\ 5 & 4 & -6 \end{bmatrix}$.

The solution to the system of equations is

$$X = A^{-1}B$$

$$= \begin{bmatrix} -4 & -4 & 5 \\ 1 & 1 & -1 \\ 5 & 4 & -6 \end{bmatrix} \begin{bmatrix} 3 \\ 6 \\ 7 \end{bmatrix}$$

$$= \begin{bmatrix} -1 \\ 2 \\ -3 \end{bmatrix}$$

Thus, $x = -1$, $y = 2$, and $z = -3$.

Using Technology

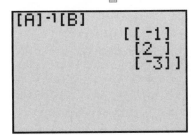

A graphing utility can be used to solve the linear system from Example 6. First, use the matrix editor feature to input the entries of the coefficient matrix A and the constant matrix B. Then, find the product $A^{-1}B$.

Work exercises 15–21 in this textbook or in the *MyMathLab*® Study Plan.

7.5 Exercises

In Exercises 1–4, determine whether matrix B is the inverse of matrix A by finding the product AB.

1. $A = \begin{bmatrix} -3 & 1 \\ -5 & 2 \end{bmatrix}$, $B = \begin{bmatrix} -2 & 1 \\ -5 & 3 \end{bmatrix}$

2. $A = \begin{bmatrix} 2 & -4 \\ 1 & -3 \end{bmatrix}$, $B = \begin{bmatrix} \frac{3}{2} & -2 \\ \frac{1}{2} & -1 \end{bmatrix}$

3. $A = \begin{bmatrix} -28 & -13 & 3 \\ 2 & 1 & 0 \\ -7 & -3 & 1 \end{bmatrix}$, $B = \begin{bmatrix} 1 & 4 & -3 \\ -2 & -7 & 6 \\ 1 & 7 & -2 \end{bmatrix}$

4. $A = \begin{bmatrix} -5 & 7 & 8 \\ -5 & 8 & 9 \\ -5 & 9 & 8 \end{bmatrix}$, $B = \begin{bmatrix} -\frac{17}{10} & \frac{8}{5} & -\frac{1}{10} \\ -\frac{1}{2} & 0 & \frac{1}{2} \\ -\frac{1}{2} & 1 & -\frac{1}{2} \end{bmatrix}$

In Exercises 5–8, determine whether the given 2×2 matrix is invertible or singular. If it is invertible, find the inverse matrix.

5. $\begin{bmatrix} 2 & 1 \\ 5 & 3 \end{bmatrix}$

6. $\begin{bmatrix} -5 & 4 \\ -2 & 2 \end{bmatrix}$

7. $\begin{bmatrix} -3 & 6 \\ 4 & -8 \end{bmatrix}$

8. $\begin{bmatrix} \frac{1}{3} & \frac{5}{4} \\ \frac{6}{5} & \frac{9}{2} \end{bmatrix}$

In Exercises 9–14, find the inverse of each matrix if possible.

9. $\begin{bmatrix} 1 & -1 \\ 2 & 0 \end{bmatrix}$

10. $\begin{bmatrix} 1 & -6 & -2 \\ 0 & -1 & 0 \\ 2 & -11 & -3 \end{bmatrix}$

11. $\begin{bmatrix} 1 & 3 & 2 \\ 2 & -1 & 4 \\ 3 & 2 & 6 \end{bmatrix}$

12. $\begin{bmatrix} 1 & 4 & -3 \\ -2 & -7 & 6 \\ 1 & 7 & -2 \end{bmatrix}$

13. $\begin{bmatrix} 1 & 1 & 3 \\ -1 & -1 & -2 \\ -1 & 3 & 5 \end{bmatrix}$

14. $\begin{bmatrix} 1 & 0 & -2 & 0 \\ 0 & 1 & 0 & -5 \\ -4 & 0 & 9 & 0 \\ 0 & 2 & 1 & -9 \end{bmatrix}$

In Exercises 15–21, solve each linear system using an inverse matrix.

15. $\begin{aligned} -x + y &= -5 \\ -2x + 3y &= -13 \end{aligned}$

16. $\begin{aligned} -3x + 4y &= 3 \\ x - 2y &= -\frac{4}{3} \end{aligned}$

17. $\begin{aligned} -\frac{3}{10}x - \frac{1}{10}y &= -3 \\ \frac{2}{5}x - \frac{1}{5}y &= 7 \end{aligned}$

18. $\begin{aligned} x + y + 3z &= 11 \\ -x - y - 2z &= -7 \\ x - 3y - 5z &= -13 \end{aligned}$

19. $\begin{aligned} x + 4y - 3z &= -5 \\ -2x - 7y + 6z &= 11 \\ x + 7y - 2z &= 1 \end{aligned}$

20. $\begin{aligned} x + y - 2z &= -4 \\ x + 3y + 4z &= 8 \\ x + 2y - z &= 2 \end{aligned}$

21. $\begin{aligned} x - y + 2z + 3w &= 2 \\ 2x - y + 6z + 5w &= 6 \\ 3x - y + 9z + 6w &= 9 \\ 2x - 2y + 4z + 7w &= 5 \end{aligned}$

7.6 Determinants and Cramer's Rule

THINGS TO KNOW

Before working through this section, be sure that you are familiar with the following concepts:

VIDEO ANIMATION INTERACTIVE

You Try It

1. Solving a System of Three Equations in Three Variables Using Gauss-Jordan Elimination (Section 7.2)

You Try It

2. Finding the Inverse of a 2×2 Matrix Using a Formula (Section 7.5)

You Try It

3. Solving Systems of Equations Using an Inverse Matrix (Section 7.5)

OBJECTIVES

1 Using Cramer's Rule to Solve Linear Systems in Two Variables

2 Calculating the Determinant of an $n \times n$ Matrix by Expansion of Minors

3 Using Cramer's Rule to Solve Linear Systems in n Variables

OBJECTIVE 1: USING CRAMER'S RULE TO SOLVE LINEAR SYSTEMS IN TWO VARIABLES

In this section, we learn how determinants can be used to solve systems of linear equations. Let's start by considering the following system of linear equations:

$$ax + by = m$$
$$cx + dy = n$$

In Section 7.5, we saw that this linear system can be written in matrix form as

$$\underbrace{\begin{bmatrix} a & b \\ c & d \end{bmatrix}}_{A} \underbrace{\begin{bmatrix} x \\ y \end{bmatrix}}_{X} = \underbrace{\begin{bmatrix} m \\ n \end{bmatrix}}_{B}$$

We now define two new matrices, D_x and D_y. Matrix D_x is created by replacing the first column of the **coefficient matrix** by the entries of the **constant matrix**. So, $D_x = \begin{bmatrix} m & b \\ n & d \end{bmatrix}$. Similarly, matrix D_y is created by replacing the second column of the **coefficient matrix** by the entries of the **constant matrix**. Therefore, $D_y = \begin{bmatrix} a & m \\ c & n \end{bmatrix}$. Recall that the determinant of the coefficient matrix is defined as

$$|A| = \begin{vmatrix} a & b \\ c & d \end{vmatrix} = ad - cb. \text{ So, } |D_x| = \begin{vmatrix} m & b \\ n & d \end{vmatrix} = md - nb \text{ and } |D_y| = \begin{vmatrix} a & m \\ c & n \end{vmatrix} = an - cm.$$

If $|A| \neq 0$, then the linear system has a unique solution where x and y can be found by the following formula:

$$x = \frac{|D_x|}{|A|} \text{ and } y = \frac{|D_y|}{|A|}$$

This technique of using determinants to solve a linear system is known as **Cramer's rule**.

Cramer's Rule for a System of Linear Equations in Two Variables

For the linear system of equations $\begin{array}{l} ax + by = m \\ cx + dy = n \end{array}$, let $|A| = \begin{vmatrix} a & b \\ c & d \end{vmatrix}$,

$|D_x| = \begin{vmatrix} m & b \\ n & d \end{vmatrix}$, and $|D_y| = \begin{vmatrix} a & m \\ c & n \end{vmatrix}$. If $|A| \neq 0$, then there is a unique solution to this system given by

$$x = \frac{|D_x|}{|A|} \text{ and } y = \frac{|D_y|}{|A|}.$$

My video summary

The proof of Cramer's rule is actually not too difficult. We can prove Cramer's rule for a system of linear equations in two variables by solving the linear system using the **method of elimination** that was discussed in Section 7.1. Click here to see the video proof of Cramer's rule.

Example 1 Use Cramer's Rule to Solve a Linear System in Two Variables

Use Cramer's rule to solve the following system:

$$\begin{array}{rcr} 2x - 2y & = & 4 \\ 3x + y & = & -3 \end{array}$$

Solution For this particular system,

$$|A| = \begin{vmatrix} 2 & -2 \\ 3 & 1 \end{vmatrix} = (2)(1) - (3)(-2) = 8,$$

$$|D_x| = \begin{vmatrix} 4 & -2 \\ -3 & 1 \end{vmatrix} = (4)(1) - (-3)(-2) = -2, \text{ and}$$

$$|D_y| = \begin{vmatrix} 2 & 4 \\ 3 & -3 \end{vmatrix} = (2)(-3) - (3)(4) = -18.$$

Therefore, by Cramer's rule, the solution is

$$x = \frac{|D_x|}{|A|} = \frac{-2}{8} = -\frac{1}{4} \text{ and } y = \frac{|D_y|}{|A|} = \frac{-18}{8} = -\frac{9}{4}$$

Work exercises 1–4 in this textbook or in the *MyMathLab*® Study Plan.

Cramer's rule can actually be used to solve larger systems of n equations with n unknowns. Before we learn how to extend the use of Cramer's rule for linear systems of more than two equations, we must first learn how to find the determinant of larger square matrices.

OBJECTIVE 2:   CALCULATING THE DETERMINANT OF AN $n \times n$ MATRIX BY EXPANSION OF MINORS

Before we learn how to calculate the determinant of any square matrix, we first learn how to calculate the determinant of a 3×3 matrix. For every entry of a 3×3 matrix, there is a 2×2 matrix associated with it. This 2×2 matrix is obtained by deleting the row and the column in which this entry appears. The determinants of these 2×2 matrices are called **minors**. In general, if a_{ij} is the entry in the ith row and jth column of a 3×3 matrix, then matrix M_{ij} is obtained by deleting the ith row and the jth column of this matrix. The determinant $|M_{ij}|$ is the minor corresponding to entry a_{ij}. For example, consider the following 3×3 matrix:

$$\begin{bmatrix} 2 & -17 & 11 \\ -1 & 11 & -7 \\ 0 & 3 & -2 \end{bmatrix}$$

To find the minor for the entry $a_{11} = 2$, we delete the first row and the first column to obtain a 2×2 matrix.

$$\begin{bmatrix} \cancel{2} & \cancel{-17} & \cancel{11} \\ -1 & 11 & -7 \\ 0 & 3 & -2 \end{bmatrix}$$ The 2×2 matrix associated with the entry $a_{11} = 2$ is

$$\begin{bmatrix} M_{11} \end{bmatrix} = \begin{bmatrix} 11 & -7 \\ 3 & -2 \end{bmatrix}.$$

Therefore, the minor for $a_{11} = 2$ is $|M_{11}| = \begin{vmatrix} 11 & -7 \\ 3 & -2 \end{vmatrix} = -1$. Similarly, the minor for the entry $a_{12} = -17$ is found by deleting the first row and second column.

$$\begin{bmatrix} \cancel{2} & \cancel{-17} & \cancel{11} \\ -1 & 11 & -7 \\ 0 & 3 & -2 \end{bmatrix}$$ The 2×2 matrix associated with the entry $a_{12} = -17$ is

$$\begin{bmatrix} M_{12} \end{bmatrix} = \begin{bmatrix} -1 & -7 \\ 0 & -2 \end{bmatrix}.$$

Thus, the minor for the entry $a_{12} = -17$ is $|M_{12}| = \begin{vmatrix} -1 & -7 \\ 0 & -2 \end{vmatrix} = (-1)(-2) - (0)(-7) = 2$.

To find the determinant of a 3×3 matrix, we use a technique called **expansion by minors**.

Determinant of a 3×3 Matrix Using Expansion by Minors

Let $A = \begin{bmatrix} a_{11} & a_{12} & a_{13} \\ a_{21} & a_{22} & a_{23} \\ a_{31} & a_{32} & a_{33} \end{bmatrix}$, then $|A| = a_{11}|M_{11}| - a_{12}|M_{12}| + a_{13}|M_{13}|$.

My video summary

Example 2 Calculate the Determinant of a 3 × 3 Matrix

Calculate $|A|$ if $A = \begin{bmatrix} 2 & -17 & 11 \\ -1 & 11 & -7 \\ 0 & 3 & -2 \end{bmatrix}$.

Solution $|A| = a_{11}|M_{11}| - a_{12}|M_{12}| + a_{13}|M_{13}|$

$$= 2\begin{vmatrix} 11 & -7 \\ 3 & -2 \end{vmatrix} - (-17)\begin{vmatrix} -1 & -7 \\ 0 & -2 \end{vmatrix} + 11\begin{vmatrix} -1 & 11 \\ 0 & 3 \end{vmatrix}$$

$$= 2(-1) + 17(2) + 11(-3)$$

$$= -1$$

Using Technology

We can use a graphing utility to find the determinant of the matrix from Example 2.

In Example 2, we calculated the determinant by expanding the minors across the first row. Actually, we can find the determinant using the expansions of minors across any row or down any column. We have to be careful to properly alternate the signs of the coefficients of the minors. The coefficient, a_{ij}, of each minor must be preceded by a "+" or "−", which can be determined by the formula $(-1)^{i+j}$. If the sum $i + j$ is odd, then the sign preceding the minor is $(-1)^{i+j} =$ "−". Likewise, if the sum $i + j$ is even, the sign preceding the coefficient is $(-1)^{i+j} =$ "+". The signs can be summarized in the following matrix:

$$\begin{bmatrix} + & - & + \\ - & + & - \\ + & - & + \end{bmatrix}$$

My video summary

Example 3 Calculate the Determinant of a 3 × 3 Matrix by Expansion of Minors Down the First Column

Find $|A|$ using the matrix from Example 2 by expanding the minors down the first column.

Solution Expanding the minors down the first column, we get

$$|A| = \begin{bmatrix} 2 & -17 & 11 \\ -1 & 11 & -7 \\ 0 & 3 & -2 \end{bmatrix}$$

$$= 2\begin{vmatrix} 11 & -7 \\ 3 & -2 \end{vmatrix} - (-1)\begin{vmatrix} -17 & 11 \\ 3 & -2 \end{vmatrix} + 0\begin{vmatrix} -17 & 11 \\ 11 & -7 \end{vmatrix}$$

$$= 2(-1) + 1(1) + 0$$

$$= -1$$

Notice that we obtained the same value as we did in Example 2. When calculating determinants, it is typically easier to expand by minors across rows or down columns that contain zeros.

In Example 4 of Section 7.5, we found that the matrix $A = \begin{bmatrix} 5 & 0 & -1 \\ 1 & -3 & -2 \\ 0 & 5 & 3 \end{bmatrix}$

did not have an inverse (i.e., the matrix is **singular**). The determinant of this matrix can be found by expanding the minors down the first column:

$$|A| = \begin{bmatrix} 5 & 0 & -1 \\ 1 & -3 & -2 \\ 0 & 5 & 3 \end{bmatrix} = 5 \begin{vmatrix} -3 & -2 \\ 5 & 3 \end{vmatrix} - 1 \begin{vmatrix} 0 & -1 \\ 5 & 3 \end{vmatrix} = 5(1) - 1(5) = 0$$

The fact that the determinant of this matrix is zero is no coincidence, as stated in the following theorem.

Theorem

A square matrix A is invertible if and only if $|A| \neq 0$. (If $|A| = 0$, then A is a singular matrix.)

We can calculate the determinant of any $n \times n$ matrix in exactly the same way as with a 3×3 matrix by expansion of minors across any row or down any column. We have to make sure to alternate the signs of each coefficient properly. We define the determinant of any square matrix as follows.

Determinant of a Square Matrix

Let $A = \begin{bmatrix} a_{11} & a_{12} & \cdots & a_{1n} \\ a_{21} & a_{22} & \cdots & a_{2n} \\ \vdots & \vdots & \ddots & \vdots \\ a_{n1} & a_{n2} & \cdots & a_{nn} \end{bmatrix}$, then for each entry a_{ij} there is a corresponding

minor $|M_{ij}|$, which is the determinant of a $(n-1) \times (n-1)$ matrix. The determinant $|A|$ is equal to the sum of the products of the entries of any row or column and the corresponding minors. (Each coefficient must be preceded by $(-1)^{i+j}$.)

My video summary

⊞ Example 4 Calculate the Determinant of a 4 × 4 Matrix

Calculate $|A|$ if $A = \begin{bmatrix} 8 & 0 & 1 & 10 \\ -1 & 0 & 0 & -5 \\ 5 & 2 & -2 & 4 \\ -9 & 0 & 0 & -12 \end{bmatrix}$.

Solution Because the second column only has one nonzero entry, we calculate $|A|$ by expanding down the second column. The first entry of column 2 is $a_{12} = 0$. Note that $i = 1$ and $j = 2$, so $i + j = 1 + 2 = 3$, which is an odd number. Therefore, $(-1)^{i+j} = (-1)^{1+2} = (-1)^3 = -1$. The coefficient of the first term must be preceded by a negative sign. The signs then alternate.

$$|A| = -0 \cdot |M_{12}| + 0 \cdot |M_{22}| - 2 \cdot |M_{32}| + 0 \cdot |M_{42}|$$
$$= -2 \cdot |M_{32}|$$

$$= -2 \cdot \begin{vmatrix} 8 & 1 & 10 \\ -1 & 0 & -5 \\ -9 & 0 & -12 \end{vmatrix}$$ *Use expansion by minors down the second column.*

$$= -2 \cdot (-1) \begin{vmatrix} -1 & -5 \\ -9 & -12 \end{vmatrix}$$

$$= -2 \cdot (-1)(12 - 45)$$

$$= -66$$

Watch the video to see this solution worked out in detail.

Work exercises 5–12 in this textbook or in the *MyMathLab*® Study Plan.

OBJECTIVE 3: USING CRAMER'S RULE TO SOLVE LINEAR SYSTEMS IN n VARIABLES

Now that we know how to compute the determinant of an $n \times n$ matrix, we can extend the use of Cramer's rule to solve a system of linear equations in n variables.

Cramer's Rule for a System of Linear Equations in n Variables

Given the linear system

$$a_{11}x_1 + a_{12}x_2 + a_{13}x_3 + \cdots a_{1n}x_n = b_1$$
$$a_{21}x_1 + a_{22}x_2 + a_{23}x_3 + \cdots a_{2n}x_n = b_2$$
$$\vdots \qquad \vdots \qquad \vdots \qquad \vdots \qquad \vdots$$
$$a_{n1}x_1 + a_{n2}x_2 + a_{n3}x_3 + \cdots a_{nn}x_n = b_n$$

, or equivalently

$$\underbrace{\begin{bmatrix} a_{11} & a_{12} & a_{13} & \cdots & a_{1n} \\ a_{21} & a_{22} & a_{23} & \cdots & a_{2n} \\ \vdots & \vdots & \vdots & & \vdots \\ a_{n1} & a_{n2} & a_{n3} & & a_{nn} \end{bmatrix}}_{\substack{\text{Coefficient} \\ \text{Matrix } A}} \underbrace{\begin{bmatrix} x_1 \\ x_2 \\ \vdots \\ x_n \end{bmatrix}}_{\substack{\text{Variable} \\ \text{Matrix } X}} = \underbrace{\begin{bmatrix} b_1 \\ b_2 \\ \vdots \\ b_n \end{bmatrix}}_{\substack{\text{Constant} \\ \text{Matrix } B}}$$

if $|D_{x_i}| = \underbrace{\begin{bmatrix} a_{11} & a_{12} & \cdots & b_1 & \cdots & a_{1n} \\ a_{21} & a_{22} & \cdots & b_2 & \cdots & a_{2n} \\ \vdots & \vdots & \cdots & \vdots & & \vdots \\ a_{n1} & a_{n2} & \cdots & b_n & & a_{nn} \end{bmatrix}}_{\substack{\text{Replace the }i\text{th column of }A\text{ with the} \\ \text{coefficients of the constant matrix.}}}$ and if $|A| \neq 0$,

then there is a unique solution to the linear system given by $x_i = \dfrac{|D_{x_i}|}{|A|}$.

My interactive video summary

▶ **Example 5 Use Cramer's Rule to Solve a Linear System in Three Variables**

Use Cramer's rule to solve the system
$$\begin{aligned} x + y + z &= 2 \\ x - 2y + z &= 5. \\ 2x + y - z &= -1 \end{aligned}$$

Solution We must calculate the following four determinants:

$$|A| = \begin{vmatrix} 1 & 1 & 1 \\ 1 & -2 & 1 \\ 2 & 1 & -1 \end{vmatrix}, \ |D_x| = \begin{vmatrix} 2 & 1 & 1 \\ 5 & -2 & 1 \\ -1 & 1 & -1 \end{vmatrix},$$

$$|D_y| = \begin{vmatrix} 1 & 2 & 1 \\ 1 & 5 & 1 \\ 2 & -1 & -1 \end{vmatrix}, \text{ and } |D_z| = \begin{vmatrix} 1 & 1 & 2 \\ 1 & -2 & 5 \\ 2 & 1 & -1 \end{vmatrix}.$$

Work through the interactive video to verify that $|A| = 9$, $|D_x| = 9$, $|D_y| = -9$, and $|D_z| = 18$. Therefore,

$$x = \frac{|D_x|}{|A|} = \frac{9}{9} = 1, \ y = \frac{|D_y|}{|A|} = \frac{-9}{9} = -1, \text{ and } z = \frac{|D_z|}{|A|} = \frac{18}{9} = 2.$$

Work exercises 13–18 in this textbook or in the **MyMathLab**® Study Plan.

7.6 Exercises

In Exercises 1–4, use Cramer's rule to solve each linear system.

1. $\begin{array}{r} x + y = 3 \\ -2x + y = 0 \end{array}$

2. $\begin{array}{r} 2x - y = 1 \\ -3x + 3y = 0 \end{array}$

3. $\begin{array}{r} 4x - y = -4 \\ -3x + 3y = -6 \end{array}$

4. $\begin{array}{r} 2x - 3y = 0 \\ 4x + 9y = 5 \end{array}$

In Exercises 5–12, calculate $|A|$.

5. $A = \begin{bmatrix} 2 & 2 & 2 \\ -1 & 5 & -2 \\ 3 & 7 & 4 \end{bmatrix}$

6. $A = \begin{bmatrix} 7 & -3 & 8 \\ 1 & -5 & -2 \\ 3 & 7 & 4 \end{bmatrix}$

7. $A = \begin{bmatrix} 7 & 0 & 4 \\ -1 & 3 & -1 \\ 6 & 2 & 5 \end{bmatrix}$

8. $A = \begin{bmatrix} -1 & 0 & 2 \\ 4 & 0 & -1 \\ 3 & 0 & -4 \end{bmatrix}$

9. $A = \begin{bmatrix} 2 & 0 & 0 \\ 0 & 3 & 0 \\ 0 & 0 & 1 \end{bmatrix}$

10. $A = \begin{bmatrix} -5 & -1 & -2 \\ 8 & 10 & 0 \\ -7 & 5 & 0 \end{bmatrix}$

11. $A = \begin{bmatrix} 1 & 5 & 1 & 9 \\ 2 & -1 & 7 & 3 \\ 9 & 0 & 0 & 1 \\ 12 & -4 & -2 & 8 \end{bmatrix}$

12. $A = \begin{bmatrix} 0 & -1 & 6 & 6 \\ 7 & -3 & 0 & 2 \\ 1 & 0 & 2 & 3 \\ 3 & 0 & 4 & 8 \end{bmatrix}$

In Exercises 13–18, use Cramer's rule to solve each linear system.

13. $\begin{array}{r} x + y - z = 4 \\ 5x - y + 3z = 6 \\ x + y - 5z = 8 \end{array}$

14. $\begin{array}{r} 3x + 3y + z = 6 \\ 4x - y + z = 8 \\ -x + y - 3z = -2 \end{array}$

15. $\begin{array}{r} -2x + 5y + 5z = -4 \\ 5x + 6y + 5z = -23 \\ -3x + 4y + 5z = -3 \end{array}$

16. $\begin{array}{r} x - z = -1 \\ 2y - z = 8 \\ 2x + 2y = 0 \end{array}$

17. $\begin{array}{r} 3x + 5z = 0 \\ 2x + 4y = -14 \\ -y + 2z = -21 \end{array}$

18. $\begin{array}{r} -2w + 6x - 9y + z = 57 \\ -w + x - y + z = 11 \\ w + x + y + z = 1 \\ 3w + 2x + 2y + z = -2 \end{array}$

APPENDIX A

Conic Section Proofs

APPENDIX A CONTENTS

A-1 Parabola Proofs

DERIVATION OF THE EQUATION OF A PARABOLA IN STANDARD FORM WITH A VERTICAL AXIS OF SYMMETRY

To derive the equation of a parabola with a vertical axis of symmetry, start with the graph of a parabola with vertex $V(h, k)$, focus F, directrix D, which passes through the point $P(x, y)$ as shown in Figure 1.

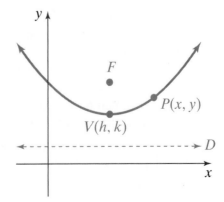

Figure 1

By the geometric definition of the parabola, the distance from any point that lies on the parabola to the focus must be equal to the distance from that point to the directrix. Therefore, the following equation must be true:

$$d(P, F) = d(P, D)$$

Because the vertex lies on the graph of the parabola, it implies that the distance from the vertex to the focus must be equal to the distance from the vertex to the directrix. Let $|p|$ be the distance from the vertex to the focus. Without loss of generality, suppose $p > 0$. Therefore, the coordinates of the focus must be $(h, k + p)$. Similarly, the equation of the directrix must be $y = k - p$. See Figure 2.

Consider the point $D(x, k - p)$ that lies on the directrix. By the geometric definition of the parabola, the distance from P to F must be the same as the distance from P to D. See Figure 3.

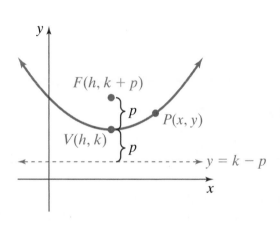

Figure 2   **Figure 3**

By the distance formula, we can rewrite the equation $d(P, F) = d(P, D)$ as

$$\sqrt{(x - h)^2 + (y - (k + p))^2} = \sqrt{(x - x)^2 + (y - (k - p))^2}$$

Squaring both sides of this equation and simplifying we get:

$(x - h)^2 + (y - (k + p))^2 = (0)^2 + (y - (k - p))^2$

$(x - h)^2 + (y - k - p)(y - k - p) = (y - k + p)(y - k + p)$

$(x - h)^2 + y^2 - 2ky - 2py + 2pk + k^2 + p^2 = y^2 - 2ky + 2py - 2pk + k^2 + p^2$

$(x - h)^2 - 2py + 2pk = 2py - 2pk$

$(x - h)^2 = 4py - 4pk$

$(x - h)^2 = 4p(y - k)$

Therefore, the equation of a parabola with a vertical axis of symmetry (and a horizontal directrix) is $(x - h)^2 = 4p(y - k)$. If p is positive, the graph opens upward. If p is negative, the graph opens downward.

Equation of a Parabola in Standard Form with a Vertical Axis of Symmetry

The equation of a parabola with a vertical axis of symmetry is
$(x - h)^2 = 4p(y - k),$

where

the vertex is $V(h, k)$,

$|p|$ = distance from the vertex to the directrix, distance from the vertex to the focus

the focus is $F(h, k + p)$, and

the equation of the directrix is $y = k - p$.

The parabola opens *upward* if $p > 0$ or *downward* if $p < 0$.

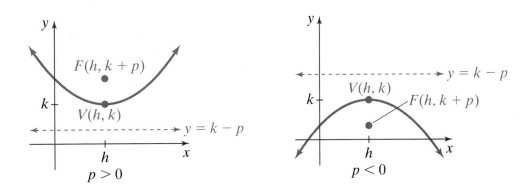

DERIVATION OF THE EQUATION OF A PARABOLA IN STANDARD FORM WITH A HORIZONTAL AXIS OF SYMMETRY

To derive the equation of a parabola with a horizontal axis of symmetry, start with the graph of a parabola with vertex $V(h, k)$, focus F, directrix D, which passes through the point $P(x, y)$ as shown in Figure 4.

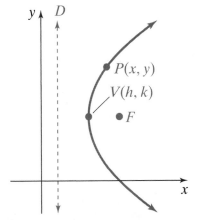

Figure 4

By the geometric definition of the parabola, the distance from any point that lies on the parabola to the focus must be equal to the distance from that point to the directrix. Therefore, the following equation must be true:

$$d(P, F) = d(P, D)$$

Because the vertex lies on the graph of the parabola, it implies that the distance from the vertex to the focus must be equal to the distance from the vertex to the directrix. Let $|p|$ be the distance from the vertex to the focus. Without loss of generality, suppose $p > 0$. Therefore, the coordinates of the focus must be $(h + p, k)$. Similarly, the equation of the directrix must be $x = h - p$. See Figure 5.

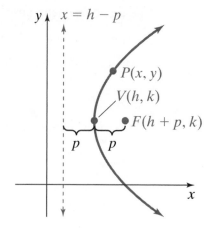

Figure 5

Consider the point $D(h - p, y)$ that lies on the directrix. By the geometric definition of the parabola, the distance from P to F must be the same as the distance from P to D. See Figure 6.

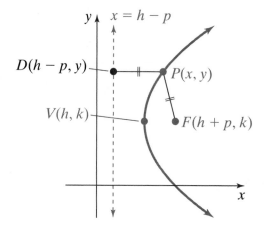

Figure 6

By the distance formula, we can rewrite the equation $d(P, F) = d(P, D)$ as

$$\sqrt{(x - (h + p))^2 + (y - k)^2} = \sqrt{(x - (h - p))^2 + (y - y)^2}$$

Squaring both sides of this equation and simplifying we get:

$(x - (h + p))^2 + (y - k)^2 = (x - (h - p))^2 + (0)^2$

$(y - k)^2 + (x - h - p)(x - h - p) = (x - h + p)(x - h + p)$

$(y - k)^2 + x^2 - 2hx - 2px + 2ph + h^2 + p^2 = x^2 - 2hx + 2px - 2ph + h^2 + p^2$

$(y - k)^2 - 2px + 2ph = 2px - 2ph$

$(y - k)^2 = 4px - 4ph$

$(y - k)^2 = 4p(x - h)$

Therefore, the equation of a parabola with a horizontal axis of symmetry (and a vertical directrix) is $(y - k)^2 = 4p(x - h)$. If p is positive, the graph opens to the right. If p is negative, the graph opens to the left.

> ### Equation of a Parabola in Standard Form with a Horizontal Axis of Symmetry
>
> The equation of a parabola with a horizontal axis of symmetry is
> $$(y - k)^2 = 4p(x - h),$$
> where
>
> > the vertex is $V(h, k)$,
> >
> > $|p|$ = distance from the vertex to the focus = distance from the vertex to the directrix,
> >
> > the focus is $F(h + p, k)$, and
> >
> > the equation of the directrix is $x = h - p$.
>
> The parabola opens *right* if $p > 0$ or *left* if $p < 0$.

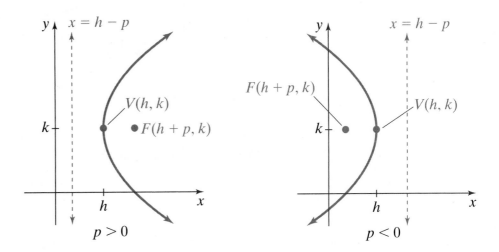

A-2 Ellipse Proofs

FACT 1 FOR ELLIPSES

Given the foci of an ellipse F_1 and F_2 and any point P that lies on the graph of the ellipse, the sum of the distances between P and the foci is equal to the length of the major axis or $d(F_1, P) + d(F_2, P) = 2a$. See Figure 7.

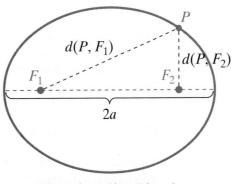

$$d(P, F_1) + d(P, F_2) = 2a$$ **Figure 7**

Proof

To establish Fact 1, note that for some $c > 0$, the foci are located along the major axis at points $F_1(h - c, k)$ and $F_2(h + c, k)$. Let $a > 0$ be the distance from the center of the ellipse to one of the vertices and let $b > 0$ be the distance from the center to an endpoint of the minor axis. See Figure 8. Because the length of the major axis is larger than the length of the minor axis, it follows that $a > b$.

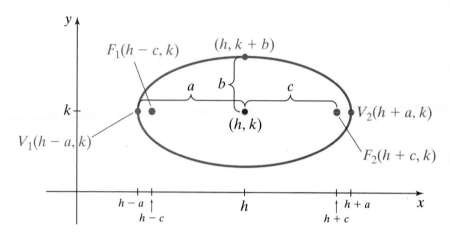

Figure 8

Consider the vertex $V_2(h + a, k)$. The distance from V_2 to F_2 is $(h + a) - (h + c) = a - c$ and the distance from V_2 to F_1 is $(h + a) - (h - c) = a + c$. See Figure 9.

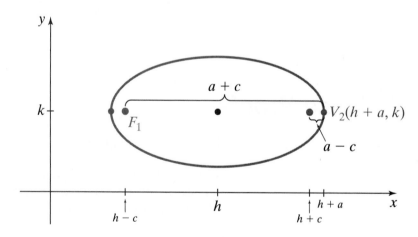

Figure 9

Thus, $d(V_2, F_1) + d(V_2, F_2) = (a + c) + (a - c) = 2a$. By the geometric definition of the ellipse, the sum of the distances from **any** point on the ellipse to the foci is a positive constant. Therefore, this positive constant must be equal to $2a$. This establishes Fact 1.

FACT 2 FOR ELLIPSES

Suppose $a > 0$ is the length from the center of an ellipse to a vertex, $b > 0$ is the length from the center to an endpoint of the minor axis, and $c > 0$ is the length from the center to a focus, then $b^2 = a^2 - c^2$ or equivalently, $c^2 = a^2 - b^2$.

Proof

Suppose that we label the coordinates of one of the endpoints of the minor axis as $(h, k + b)$. See Figure 10.

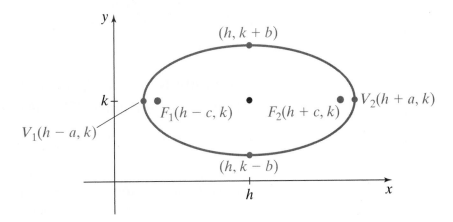

Figure 10

We can construct a triangle by connecting the point $(h, k + b)$ with the two foci as shown in Figure 11.

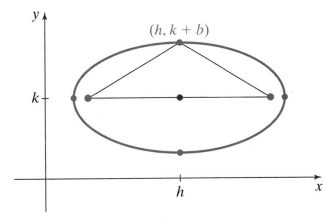

Figure 11

If we draw a line segment connecting the point $(h, k + b)$ with the center of the ellipse, we see that we have constructed two congruent right triangles. See Figure 12. The base of each right triangle has length c and the height of each right

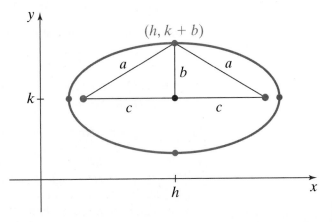

Figure 12

triangle has length b. We know from Fact 1 that the sum of the distances between *any* point on the ellipse and the two foci is $2a$. Therefore, the sum of the distances between the point $(h, k + b)$ and the two foci is also equal to $2a$. Thus, the hypotenuse of each of the right triangles must be equal to a. Therefore, by the Pythagorean theorem we can establish the relationship $b^2 + c^2 = a^2$ or $b^2 = a^2 - c^2$. Thus, we have established Fact 2.

DERIVATION OF THE EQUATION OF AN ELLIPSE IN STANDARD FORM WITH A HORIZONTAL MAJOR AXIS

Suppose we are given the ellipse with horizontal major axis centered at (h, k). See Figure 13. Let $c > 0$, $a > 0$, and $b > 0$ be the distance between the center and a focus, vertex, and endpoint of a minor axis respectively. Because the ellipse has a horizontal major axis, we know that $a > b$. By the definition of the ellipse and by Fact 1, we know that for any point P that lies on the graph of the ellipse, $d(P, F_1) + d(P, F_2) = 2a$.

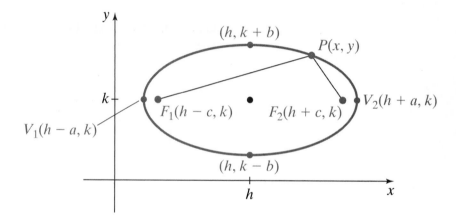

Figure 13

Using the distance formula, we can rewrite the equation $d(P, F_1) + d(P, F_2) = 2a$ as

$$\sqrt{(x - (h - c))^2 + (y - k)^2} + \sqrt{(x - (h + c))^2 + (y - k)^2} = 2a$$

Subtract a radical from both sides.

$$\sqrt{(x - (h - c))^2 + (y - k)^2} = 2a - \sqrt{(x - (h + c))^2 + (y - k)^2}$$

We now square both sides of the equation.

$$\left[\sqrt{(x - (h - c))^2 + (y - k)^2}\right]^2 = \left[2a - \sqrt{(x - (h + c))^2 + (y - k)^2}\right]^2$$
$$(x - (h - c))^2 + (y - k)^2 = 4a^2 - 4a\sqrt{(x - (h + c))^2 + (y - k)^2}$$
$$+ (x - (h + c))^2 + (y - k)^2$$

Subtract $(y - k)^2$ from both sides and simplify.

$$(x - (h - c))^2 = 4a^2 - 4a\sqrt{(x - (h + c))^2 + (y - k)^2} + (x - (h + c))^2$$
$$x^2 - 2xh + 2cx - 2ch + h^2 + c^2 = 4a^2 - 4a\sqrt{(x - (h + c))^2 + (y - k)^2} + x^2$$
$$- 2xh - 2cx + 2ch + h^2 + c^2$$

Combine like terms and subtract $4a^2$ from both sides.

$$4cx - 4ch - 4a^2 = -4a\sqrt{(x - (h + c))^2 + (y - k)^2}$$

Now divide both sides by -4.

$$-cx + ch + a^2 = a\sqrt{(x - (h + c))^2 + (y - k)^2}$$

We now square both sides and simplify.

$$\left[-cx + ch + a^2\right]^2 = \left[a\sqrt{(x - (h + c))^2 + (y - k)^2}\right]^2$$
$$(-cx + ch + a^2)(-cx + ch + a^2) = a^2\left[(x - (h + c))^2 + (y - k)^2\right]$$
$$c^2x^2 - 2c^2xh - 2a^2cx + 2a^2ch + c^2h^2 + a^4$$
$$= a^2\left[x^2 - 2xh - 2cx + 2ch + h^2 + c^2 + (y - k)^2\right]$$
$$c^2x^2 - 2c^2xh - 2a^2cx + 2a^2ch + c^2h^2 + a^4 = a^2x^2 - 2a^2xh - 2a^2cx + 2a^2ch + a^2h^2$$
$$+ a^2c^2 + a^2(y - k)^2$$

Collect the x and y terms on the left side.

$$-a^2x^2 + 2a^2xh - a^2h^2 + c^2x^2 - 2c^2xh + c^2h^2 - a^2(y - k)^2 = -a^4 + a^2c^2$$
$$\underbrace{-a^2x^2 + 2a^2xh - a^2h^2} + \underbrace{c^2x^2 - 2c^2xh + c^2h^2} - a^2(y - k)^2 = -a^4 + a^2c^2$$

Factor $-a^2$ out of the first three terms and factor c^2 out of the next three terms.

$$-a^2(x^2 - 2xh + h^2) + c^2(x^2 - 2xh + h^2) - a^2(y - k)^2 = -a^4 + a^2c^2$$

$$\underbrace{-a^2(x - h)^2 + c^2(x - h)^2} - a^2(y - k)^2 = -a^4 + a^2c^2 \qquad \textit{Factor.}$$

$$(-a^2 + c^2)(x - h)^2 - a^2(y - k)^2 = -a^2(a^2 - c^2) \qquad \textit{Factor out } (x - h)^2 \textit{ and } -a^2.$$

$$-(a^2 - c^2)(x - h)^2 - a^2(y - k)^2 = -a^2(a^2 - c^2) \qquad -a^2 + c^2 = -(a^2 - c^2).$$

RECALL FACT 2: $b^2 = a^2 - c^2$

$$-b^2(x - h)^2 - a^2(y - k)^2 = -a^2b^2 \qquad \textit{Substitute } b^2 = a^2 - c^2.$$

$$\frac{-b^2(x - h)^2}{-a^2b^2} - \frac{a^2(y - k)^2}{-a^2b^2} = \frac{-a^2b^2}{-a^2b^2} \qquad \textit{Divide both sides by } -a^2\,b^2.$$

$$\frac{(x - h)^2}{a^2} + \frac{(y - k)^2}{b^2} = 1.$$

Therefore, the standard equation of an ellipse centered at (h, k) with a horizontal major axis is $\frac{(x - h)^2}{a^2} + \frac{(y - k)^2}{b^2} = 1$, where $a > b$ and $b^2 = a^2 - c^2$. We can derive the standard equation of an ellipse centered at (h, k) with a vertical major axis in a similar fashion to obtain the equation $\frac{(x - h)^2}{b^2} + \frac{(y - k)^2}{a^2} = 1$, where $a > b$ and $b^2 = a^2 - c^2$.

Equation of an Ellipse in Standard Form with Center (h, k)

Horizontal Major Axis

$$\frac{(x - h)^2}{a^2} + \frac{(y - k)^2}{b^2} = 1$$

- $a > b > 0$
- Foci: $F_1(h - c, k)$ and $F_2(h + c, k)$
- Vertices: $V_1(h - a, k)$ and $V_2(h + a, k)$
- Endpoints of minor axis:
  $(h, k - b)$ and $(h, k + b)$
- $c^2 = a^2 - b^2$

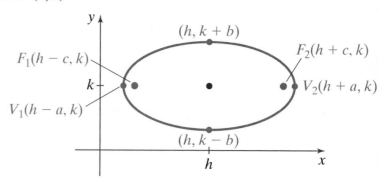

Vertical Major Axis

$$\frac{(x - h)^2}{b^2} + \frac{(y - k)^2}{a^2} = 1$$

- $a > b > 0$
- Foci: $F_1(h, k - c)$ and $F_2(h, k + c)$
- Vertices: $V_1(h, k - a)$ and $V_2(h, k + a)$
- Endpoints of minor axis:
  $(h - b, k)$ and $(h + b, k)$.
- $c^2 = a^2 - b^2$

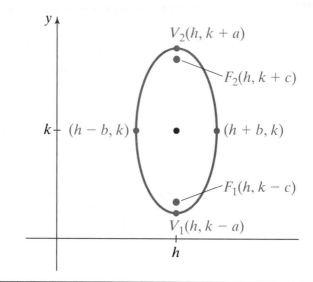

A-3 Hyperbola Proofs

FACT 1 FOR HYPERBOLAS

Given the foci of a hyperbola F_1 and F_2 and any point P that lies on the graph of the hyperbola, the difference of the distances between P and the foci is equal to $2a$. In other words, $\left|d(P, F_1) - d(P, F_2)\right| = 2a$. The constant $2a$ represents the length of the transverse axis.

Proof

Consider the hyperbola centered at (h, k) with vertices $V_1(h - a, k)$ and $V_2(h + a, k)$ and foci $F_1(h - c, k)$ and $F_2(h + c, k)$ as shown in Figure 14.

The distance between the center and either vertex is equal to a, so the distance between V_1 and V_2 is equal to $2a$. The distance between F_2 and V_2 is $\left|(h + c) - (h + a)\right| = \left|c - a\right| = c - a$ because $c > a$. Likewise, the distance between F_1 and V_1 is also equal to $c - a$. See Figure 15.

The distance between V_1 and F_2 is $d(V_1, F_2) = 2a + (c - a) = a + c$. The distance between V_1 and F_1 is $d(V_1, F_1) = c - a$. Therefore, $\left|d(V_1, F_2) - d(V_1, F_1)\right| = \left|a + c - (c - a)\right| = \left|a + c - c + a\right| = 2a$. Thus, by the geometric definition of the hyperbola, the difference between the distances from **any** point P on the hyperbola to the two foci must also be equal to $2a$. That is, $\left|d(P, F_1) - d(P, F_2)\right| = 2a$.

Figure 14

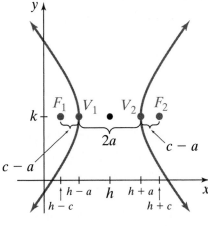

Figure 15

DERIVATION OF THE EQUATION OF A HYPERBOLA IN STANDARD FORM WITH A HORIZONTAL TRANSVERSE AXIS

To derive the equation of the hyperbola with a horizontal transverse axis with center (h, k), use Figure 16. Without loss of generality, suppose that $d(P, F_2) > d(P, F_1)$. We will use Fact 1 for hyperbolas which states that $d(P, F_2) - d(P, F_1) = 2a$ to derive the equation.

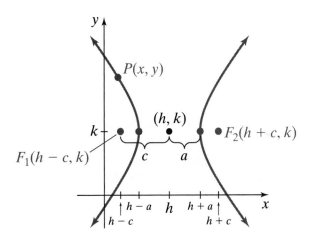

Figure 16

Using Fact 1 for hyperbolas and the distance formula, we get:

$$d(P, F_2) - d(P, F_1) = 2a$$

$$\sqrt{(x - (h + c))^2 + (y - k)^2} - \sqrt{(x - (h - c))^2 + (y - k)^2} = 2a$$

Now add a radical to both sides then square both sides of the equation.

$$\sqrt{(x - (h + c))^2 + (y - k)^2} = 2a + \sqrt{(x - (h - c))^2 + (y - k)^2}$$

$$\left[\sqrt{(x - (h + c))^2 + (y - k)^2}\right]^2 = \left[2a + \sqrt{(x - (h - c))^2 + (y - k)^2}\right]^2$$

$$(x - (h + c))^2 + (y - k)^2 = 4a^2 + 4a\sqrt{(x - (h - c))^2 + (y - k)^2}$$
$$+ (x - (h - c))^2 + (y - k)^2$$

Subtract $(y - k)^2$ from both sides and simplify.

$$(x - (h + c))^2 = 4a^2 + 4a\sqrt{(x - (h - c))^2 + (y - k)^2} + (x - (h - c))^2$$

$$x^2 - 2hx - 2cx + 2ch + h^2 + c^2 = 4a^2 + 4a\sqrt{(x - (h - c))^2 + (y - k)^2}$$
$$+ x^2 - 2hx + 2cx - 2ch + h^2 + c^2$$

$$-4a^2 - 4cx + 4ch = 4a\sqrt{(x - (h - c))^2 + (y - k)^2}$$

Combine like terms and isolate the radical.

$$a^2 + cx - ch = -a\sqrt{(x - (h - c))^2 + (y - k)^2}$$

Divide both sides by −4.

Now square both sides and simplify.

$$\left[a^2 + cx - ch\right]^2 = \left[-a\sqrt{(x - (h - c))^2 + (y - k)^2}\right]^2$$
$$(a^2 + cx - ch)(a^2 + cx - ch) = a^2\left[(x - (h - c))^2 + (y - k)^2\right]$$
$$a^4 + 2a^2cx - 2a^2ch - 2c^2hx + c^2x^2 + c^2h^2 = a^2x^2 - 2a^2hx + 2a^2cx - 2a^2ch$$
$$+ a^2h^2 + a^2c^2 + a^2(y - k)^2$$
$$c^2x^2 - 2c^2hx + c^2h^2 - a^2x^2 + 2a^2hx - a^2h^2 - a^2(y - k)^2 = a^2c^2 - a^4$$
$$\underbrace{c^2x^2 - 2c^2hx + c^2h^2} - \underbrace{a^2x^2 + 2a^2hx - a^2h^2} - a^2(y - k)^2 = a^2c^2 - a^4$$

Factor c^2 out of the first three terms and factor $-a^2$ out of the next three terms.

$$c^2(x^2 - 2xh + h^2) - a^2(x^2 - 2xh + h^2) - a^2(y - k)^2 = a^2c^2 - a^4$$
$$c^2(x - h)^2 - a^2(x - h)^2 - a^2(y - k)^2 = a^2c^2 - a^4$$
$$(c^2 - a^2)(x - h)^2 - a^2(y - k)^2 = a^2(c^2 - a^2)$$

$x^2 - 2xh + h^2 = (x - h)^2$

Factor $(x - h)^2$ out of the first two terms.

$$b^2(x - h)^2 - a^2(y - k)^2 = a^2b^2$$

Let $b^2 = c^2 - a^2$.

$$\frac{b^2(x - h)^2}{a^2b^2} - \frac{a^2(y - k)^2}{a^2b^2} = \frac{a^2b^2}{a^2b^2}$$

Divide both sides by a^2b^2.

$$\frac{(x - h)^2}{a^2} - \frac{(y - k)^2}{b^2} = 1$$

Simplify.

Therefore, the standard equation of a hyperbola centered at (h, k) with a horizontal transverse axis is $\frac{(x - h)^2}{a^2} - \frac{(y - k)^2}{b^2} = 1$, where $b^2 = c^2 - a^2$. We can derive the standard equation of a hyperbola centered at (h, k) with a vertical transverse axis in a similar fashion to obtain the equation $\frac{(y - k)^2}{a^2} - \frac{(x - h)^2}{b^2} = 1$, where $b^2 = c^2 - a^2$.

Equation of a Hyperbola in Standard Form with Center (h, k)

Horizontal Transverse Axis

$$\frac{(x - h)^2}{a^2} - \frac{(y - k)^2}{b^2} = 1$$

- Foci: $F_1(h - c, k)$ and $F_2(h + c, k)$
- Vertices:
  $V_1(h - a, k)$ and $V_2(h + a, k)$
- Endpoints of conjugate axis:
  $(h, k - b)$ and $(h, k + b)$
- $b^2 = c^2 - a^2$
- Asymptotes: $y - k = \pm\frac{b}{a}(x - h)$

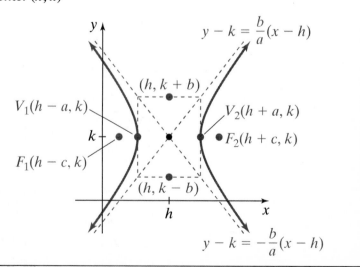

Equation of a Hyperbola in Standard Form with Center (h, k)

Vertical Transverse Axis

$$\frac{(y - k)^2}{a^2} - \frac{(x - h)^2}{b^2} = 1$$

- Foci: $F_1(h, k - c)$ and $F_2(h, k + c)$
- Vertices: $V_1(h, k - a)$ and
  $V_2(h, k + a)$
- Endpoints of conjugate axis:
  $(h - b, k)$ and $(h + b, k)$
- $b^2 = c^2 - a^2$
- Asymptotes: $y - k = \pm \frac{a}{b}(x - h)$

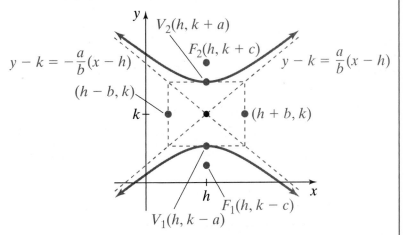

Glossary

Abscissa   The first number of an ordered pair.

Absolute Value   The **absolute value** of a real number a is defined by

$$|a| = \begin{cases} a \text{ if } a \geq 0 \\ a \text{ if } a < 0 \end{cases}$$

and represents the distance between a and 0 on a number line.

Algebraic Expressions   An **algebraic expression** consists of one or more algebraic terms that may include variables, constants, and operating symbols such as $+$ and $-$.

Argument   The **argument** of a function is the variable, term, or expression on which a function operates.

Augmented Matrix   A **matrix** created by combining two matrices. An augmented matrix used to solve a system of linear equations is usually created by combining the coefficient matrix and the constant matrix separated by a vertical line segment.

Axis of Symmetry   An **axis of symmetry** is an invisible line that divides the graph into two equal halves.

Binomial   A **binomial** is a polynomial expression consisting of two terms.

Bisect   Two line segments **bisect** each other if the intersection of the segments divide each segment into two parts of equal length.

Boundary Point   Let $p = a_n x^n + a_{n-1} x^{n-1} + \cdots + a_1 x + a_0$ be a polynomial. A **boundary point** of a polynomial inequality $p < 0$ is a real number for which $p = 0$. The m boundary points divide the real number line into $m + 1$ intervals. *Note:* The symbol $<$ in this definition could be replaced by $>$, $\leq$, or $\geq$.

Carrying Capacity   The **carrying capacity** is the maximum population size that can be regularly sustained by an environment.

Cartesian Plane   Another name for the rectangular coordinate system.

Coefficient   A **coefficient** is a constant of an algebraic expression, equation, or function that stands alone or is multiplied by a variable.

Coefficient Matrix   The matrix created using the variable coefficients of a system of linear equations.

Column Matrix   A **column matrix** is a matrix consisting of m rows and one column.

Common Logarithmic Function   For $x > 0$, the **common logarithmic function** is defined by $y = \log x$ if and only if $x = 10^y$.

Commutative   An operation (such as addition and multiplication) is **commutative** if the ordering of numbers does not matter.

Commutative Property   The **commutative property** is the algebraic property that states that a change in order produces the same result. Addition and multiplication are commutative because for any real numbers a and b, $a + b = b + a$ and $ab = ba$. Subtraction is *not* commutative.

Completely Factored Form   The polynomial function $f(x) = a_n x^n + a_{n-1} x^{n-1} + a_{n-2} x^{n-2} + \cdots + a_1 x + a_0$ where $n \geq 1$ and $a_n \neq 0$ is said to be in **completely factored form** if it is written in the form $f(x) = a_n(x - c_1)(x - c_2)(x - c_3)\cdots(x - c_n)$ where $c_1, c_2, c_3, \ldots c_n$ are complex numbers and not necessarily distinct. Note that in each factor of the form $(x - c)$, the coefficient of x is 1.

Complex Conjugate The **complex conjugate** of a complex number $z = a + bi$ is denoted as $\overline{z} = \overline{a + bi} = a - bi$.

Complex Number A **complex number** is a number that can be written in the form $a + bi$, where a and b are real numbers. We often use the variable z to denote a complex number.

Complex Zero A **complex zero** is a zero of a polynomial $f(x)$ of the form $a + bi$, where a and b are real numbers.

Compound Inequalities Two or more inequalities joined together using the words "and" or "or."

Composite Function Given functions f and g, the composite function, $f \circ g$ (also called the **composition of f and g**) is defined by

$$(f \circ g)(x) = f(g(x))$$

provided $g(x)$ is in the domain of f.

Consistent System A **consistent system** of equations is a system of equations with at least one solution.

Coordinate Plane Another name for the rectangular coordinate system.

Dependent System A system of linear equations is **dependent** if all solutions of one equation are also solutions of the other equations. A dependent system has infinitely many solutions.

Difference of Two Systems If x and y are any two algebraic expressions, then $x^2 - y^2$ is called a **difference of two squares.** The difference of two squares can always be factored as $x^2 - y^2 = (x - y)(x + y)$.

Dependent Variable A **dependent variable** is the variable whose value is determined by the value assumed by an independent variable.

Discriminant Given a quadratic equation of the form $ax^2 + bx + c = 0$, the discriminant is the expression $b^2 - 4ac$.

Element An **element** of a cone is a straight line segment joining the outer edge of the base of the cone with the vertex of the cone. The set of all elements creates the surface of a cone.

Ellipse An **ellipse** is the set of all points in a plane, the sum of whose distances from two fixed points is a positive constant. The two fixed points, F_1 and F_2, are called the *foci.*

Equivalent System Two systems of linear equations are **equivalent** if and only if they have the same solution set.

Even Functions A function f is **even** if for every x in the domain, $f(x) = f(-x)$. Even functions are symmetric about the y-axis. For each point (x, y) on the graph, the point $(-x, y)$ is also on the graph.

Extraneous Solution An **extraneous solution** is a solution obtained through algebraic manipulations that is *not* a solution to the original equation.

Exponential Function An **exponential function** is a function of the form $f(x) = b^x$, where x is any real number and $b > 0$ such that $b \neq 1$. The constant, b, is called the base of the exponential function.

Factor A **factor** is one of two or more expressions that are multiplied together to form a product.

Function A **function** is a relation such that for each element in the domain, there is *exactly one* corresponding element in the range. In other words, a function is a well-defined relation.

Graph of an Equation The **graph of an equation** is the set of all ordered pairs **(x, y)** whose coordinates satisfy the equation.

Greatest Common Factor (GCF) The **greatest common factor (GCF)** is the largest number or algebraic expression that divides two or more algebraic expressions evenly.

Horizontal Asymptote A horizontal line $y = H$ is a **horizontal asymptote** of a function f if the values of $f(x)$ approach some fixed number H as the values of x approach ∞ or $-\infty$.

Horizontal Major Axis An ellipse has a **horizontal major axis** if the horizontal axis of symmetry passes through the two foci and is longer than the vertical axis of symmetry.

Hyperbola A **hyperbola** is the set of all points in a plane, the difference of whose distances from two fixed points is a positive constant. The two fixed points, F_1 and F_2, are called the foci.

Identity Matrix An **identity matrix** is a square matrix where all entries down the diagonal are 1's with zeros everywhere else. We use the notation I_n to denote an $n \times n$ identity matrix.

Imaginary Part Given a complex number of the form $a + bi$, the real number, b, is called the imaginary part of the complex number.

Imaginary Unit The **imaginary unit,** i, is defined by $i = \sqrt{-1}$, or equivalently, $i^2 = -1$.

Independent Variable An **independent variable** is the variable whose value determines the value of other variables.

Index The **index** of a radical expression $\sqrt[n]{b}$ is the integer n.

Integer The set of real numbers that includes the negatives of the natural numbers combined with the whole numbers.

Intercepts The **intercepts** of a graph are points where a graph crosses or touches a coordinate axis.

x-intercepts The **x-intercepts** of a graph are the values of x where the graph crosses or touches the x-axis. The ordered pairs associated with an x-intercept have the form $(x, 0)$.

y-intercepts The **y-intercepts** of a graph are the values of y where the graph crosses or touches the y-axis. The ordered pairs associated with a y-intercept have the form $(0, y)$.

Inconsistent System An inconsistent system of equations is a system of equations that has no solution.

Interval Notation Short-hand notation used to describe an interval of numbers.

Inverse Function Let f be a one-to-one function with domain A and range B. Then f^{-1} is the **inverse function of f** with domain B and range A. Furthermore, if $f(a) = b$, then $f^{-1}(b) = a$.

Irrational Number An **irrational number** is a real number that *cannot* be expressed as the quotient of two integers p/q. Examples of irrational numbers include $\sqrt{7}$, $\sqrt{17}$, $\sqrt[3]{29}$, and π. We can describe the set of irrational numbers using the following set-builder notation: $\mathbb{I} = \{x | x$ is not a rational number$\}$.

Irrational Solution An **irrational solution** is a real number solution to an equation that *cannot* be expressed as the quotient of two integers.

Irrational Zero The value $x = r$ is an **irrational zero** of f if $f(r) = 0$ and if r *cannot* be written as the quotient of two integers.

Leading Coefficient The **leading coefficient** of a polynomial function is the constant that is multiplied by the variable of highest power.

Least Common Denominator The **least common denominator** is the smallest multiple of all denominators in an algebraic expression or equation.

Linear Equation A **linear equation** in one variable is an equation that can be written in the form $ax + b = c$, where a, b, and c are real numbers and $a \neq 0$.

Linear Equation in n Variables A **linear equation** in n variables is an equation that can be written in the form $a_1 x_1 + a_2 x_2 + \cdots + a_n x_n = b$ for variables $x_1, x_2, \ldots x_n$ and real numbers $a_1, a_2, \ldots a_n, b$, where at least one of $a_1, a_2, \ldots a_n$ is nonzero.

Linear Equation in One Variable A **linear equation** in one variable is an equation that can be written in the form $ax + b = c$, where a, b, and c are real numbers and $a \neq 0$.

Linear Function A first degree polynomial function of the form $f(x) = mx + b$ is called a **linear function**.

Linear Inequality A **linear inequality** is an inequality that can be written in the form $ax + b < c$, where a, b, and c are real numbers and $a \neq 0$.

Logarithmic Function For $x > 0$, $b > 0$, and $b \neq 1$, the **logarithmic function** with base b is defined by $y = \log_b x$ if and only if $x = b^y$.

Local Maximum A turning point in which the graph of a function changes from increasing to decreasing. A **local maximum** is also called a relative maximum.

Local Minimum A turning point in which the graph of a function changes from decreasing to increasing. A **local minimum** is also called a relative minimum.

Matrix A **matrix** is a rectangular array of numbers arranged in a fixed number, m, of rows and a fixed number, n, of columns. The size of a matrix is $m \times n$. The numbers in the array are called entries. Each entry is labeled a_{ij}, where i is the ith row of the entry and j is the jth column of the entry. Two matrices A and B are said to be equal ($A = B$) provided both matrices are of the same size and the corresponding entries are the same.

Method of Elimination The **method of elimination** is a method used to solve linear equations by adding a multiple of an equation to a multiple of another equation in the attempt to eliminate one of the variables.

Minor Axis The **minor axis** of an ellipse is the shorter of the two axes of symmetry.

Monomial Function A **monomial function** is a polynomial function consisting of a single term. $f(x) = 6x^9$ and $g(x) = 3$ are examples of monomial functions.

Multiplicative Inverse of a Matrix Let A be a $n \times n$ square matrix. If there exists an $n \times n$ matrix A^{-1} such that $AA^{-1} = A^{-1}A = I_n$ then A^{-1} is called the **multiplicative inverse** of matrix A.

Natural Logarithmic Function For $x > 0$, the **natural logarithmic function** is defined by $y = \ln x$ if and only if $x = e^y$.

Natural Numbers   The set of counting numbers $\{1, 2, 3, \dots\}$.

Negative Reciprocals   For real numbers a and b, the numbers $\frac{a}{b}$ and $-\frac{b}{a}$ are called negative reciprocals of each other.

Noncollinear Points   Three points are said to be **noncollinear** if they do not all lie on the same line.

Odd Functions   A function f is **odd** if for every x in the domain, $-f(x) = f(-x)$. Odd functions are symmetric about the origin. For each point (x, y) on the graph, the point $(-x, -y)$ is also on the graph.

One-to-One Function   A function f is **one-to-one** if for any values $a \neq b$ in the domain of f, $f(a) \neq f(b)$.

Open Interval   An **open interval**, denoted by the symbol (a, b), consists of all real numbers between a and b not including a and b.

Ordered Pair   The collection of two discernible objects listed in a meaningful order. The notation (a, b) is typically used to describe an ordered pair.

Ordinate   The second number of an ordered pair.

Origin   The **origin** is the point in the Cartesian plane where the x-axis and y-axis intersect. The origin has coordinates $(0, 0)$.

Parabola   A **parabola** is the set of all points in a plane equidistant from a fixed point F and a fixed line D. The fixed point is called the **focus,** and the fixed line is called the **directrix.**

Parallel Lines   Two lines having the same slope. **Parallel lines** do not intersect.

Parallelogram   A **parallelogram** is a quadrilateral (four-sided polygon) in which opposite sides are parallel.

Perpendicular Lines   Two lines that intersect at a right angle.

Perfect nth power   A **perfect nth power** is a rational number that can be written as the nth power of another rational number where n is a positive integer.

Polynomial   A **polynomial** in one variable is an expression of the form $a_n x^n + a_{n-1} x^{n-1} + \cdots + a_1 x + a_0$ where $a_0, a_1, \dots, a_n$ are constants and each exponent is an integer greater than or equal to zero.

Polynomial Equation   A **polynomial equation** is an equation that can be written in the form $a_n x^n + a_{n-1} x^{n-1} + \cdots + a_1 x + a_0 = 0$, where $a_0, a_1, \dots, a_n$ are real numbers and each exponent is an integer greater than or equal to 0.

Polynomial Expression   An algebraic expression in which the exponents of all variable factors are nonnegative integers is called a **polynomial expression**.

Polynomial Function   The function $f(x) = a_n x^n + a_{n-1} x^{n-1} + a_{n-2} x^{n-2} + \cdots + a_1 x + a_0$ is a **polynomial function** of degree n, where n is a nonnegative integer and $a_0, a_1, a_2, \dots, a_n$ are real numbers. The domain of every polynomial function is $(-\infty, \infty)$.

Polynomial Inequality   A **polynomial inequality** is an inequality that can be written in the form $a_n x^n + a_{n-1} x^{n-1} + \cdots + a_1 x + a_0 < 0$, where $a_0, a_1, \dots, a_n$ are real numbers, and each exponent is an integer greater than or equal to 0. *Note:* The symbol $<$ in this definition could be replaced by $>, \leq$, or $\geq$.

Polynomial in One Variable   The algebraic expression $a_n x^n + a_{n-1} x^{n-1} + a_{n-2} x^{n-2} + \cdots + a_1 x + a_0$ is a polynomial in one variable of degree n, where n is a nonnegative integer. The numbers $a_0, a_1, a_2, \dots, a_n$ are called the **coefficients** of the polynomial.

Prime Polynomial   A **prime polynomial** is a term that indicates that a polynomial cannot be factored using *integer* coefficients. It is possible for a prime polynomial to be factored using rational or irrational coefficients.

Principal nth Root   For any integer $n \geq 2$, the **principal nth root** of a number b is defined as $\sqrt[n]{b} = a$ if and only if $a^n = b$.

Power Function   Polynomial functions of the form $f(x) = ax^n$ where $n \geq 1$ is an integer.

Quadratic Equation in One Variable   A **quadratic equation in one variable** is an equation that can be written in the form $ax^2 + bx + c = 0$, $a \neq 0$. Quadratic equations in this form are said to be in *standard form.*

Quadratic Equation   A **quadratic equation** is an equation that can be written in the form $ax^2 + bx + c = 0$, where a, b, and c are real numbers and $a \neq 0$.

Quadratic Function   A **quadratic function** is a function of the form $f(x) = ax^2 + bx + c$, where a, b, and c are real numbers with $a \neq 0$. Every quadratic function has a "u-shaped" graph called a *parabola.*

Radicand   A **radicand** is the number, variable, or algebraic expression under the radical.

Rational Equation   A **rational equation** is an equation involving the quotient of two polynomial expressions where the polynomial in the denominator is nonconstant.

Rational Expression   A **rational expression** is the quotient of two polynomials. For example, $\frac{2x - 1}{x + 3}$ is a rational expression.

Rational Functions   A **rational function** is of the form $f(x) = \frac{g(x)}{h(x)}$ where $g(x)$ and $h(x)$ are polynomial functions and $h(x) \neq 0$. The domain of a rational function is the set of all real numbers such that $h(x) \neq 0$.

Rational Inequality   A **rational inequality** is an inequality that can be written in the form $\frac{p}{q} \geq 0$, where p and q are polynomials. *Note:* The symbol $\geq$ in this definition could be replaced by $\leq$, $<$, or $>$.

Rational Number   A **rational number** is a real number that can be expressed as the quotient of two integers $\frac{p}{q}$, where $q \neq 0$. All finite decimals and repeating decimals are rational numbers.

Rational Zero   A **rational zero** is a real zero of a polynomial $f(x)$ of the form $\frac{p}{q}$, where p and q are integers and $q \neq 0$.

Real Part   Given a complex number of the form $a + bi$, the real number, a, is called the real part of the complex number.

Real Zero   A real number $x = c$ is a **real zero** of a function f if $f(c) = 0$. Real zeros are also x-intercepts.

Reciprocal   A **reciprocal** is a number or algebraic expression related to another number or algebraic expression in such a way that when multiplied together their product is 1.

Relation   A **relation** is a correspondence between two sets A and B such that each element of set A corresponds to one or more elements of set B. Set A is called the **domain** of the relation, and set B is called the **range** of the relation.

Reduced Row-Echelon Form   An augmented matrix is said to be in **reduced row-echelon form** when all zeros appear above and below the diagonal *and* 1's appear down the diagonal.

Right Circular Cone   A **right circular cone** is a cone that has a vertex directly above the center of a circular base.

Root Function   The function $f(x) = \sqrt[n]{g(x)}$ is a **root function,** where n is an integer such that $n \geq 2$.

Row-Echelon Form   An augmented matrix is said to be in **row-echelon form** when all zeros appear under the diagonal *and* 1's appear down the diagonal.

Row matrix   A **row matrix** is a matrix consisting of one row and n columns.

Set Builder Notation   Notation used to describe sets or intervals of numbers.

Simple Interest   The formula for **simple interest** is $I = Prt$, where $P =$ principal amount invested, $r =$ interest rate, and $t =$ time (in years).

Slant Asymptote   A line of the form $y = mx + b$, $m \neq 0$ in which the graph of a function approaches.

Slope   If $x_1 \neq x_2$, the **slope** of a nonvertical line passing through distinct points (x_1, y_1) and (x_2, y_2) is given by $m = rise/run =$ change in y/change in $x = y_2 - y_1/x_2 - x_1$.

Square matrix   A **square matrix** is a matrix that contains the same number of rows as columns.

Standard form   A polynomial in one variable is written in **standard form** when it is written in descending order with the term of highest power written first.

Subset   A **subset** is a set contained within a set. A subset can be the set itself. The subset of an interval must contain points within the interval. For example, the interval $[-1, 2]$ is a subset of the interval $[-4, 8]$.

Synthetic Division   A short cut method used to long divide polynomials. The divisor must be of the form $x - c$.

System of Linear Equations in Two Variables   A **system of linear equations in two variables** is the collection of two linear equations in two variables considered simultaneously. The solution to a system of equations in two variables is the set of all ordered pairs for which *both* equations are true.

Test Value   After using the boundary points to divide the number line into intervals, a **test value** is a number chosen from each interval to test whether the inequality is positive or negative on the interval. *Note:* A test value cannot be a boundary point.

The Difference of Two Squares   If x and y are any two algebraic expression then $x^2 - y^2$ is called a **difference of two squares.** The difference of two squares can always be factored as $x^2 - y^2 = (x - y)(x + y)$.

Triangular Form   A system of three linear equations is said to be in **triangular form** when the leading coefficient of the first variable in the second equation is 0 and the coefficients of the first two variables of the third equation are zero. We also say that a system of equations is in triangular form if there are all zeros below the diagonal.

Turning Points   The **turning points** on the graph of a polynomial function are the points at which the graph changes from increasing to decreasing or from decreasing to increasing.

Variable Matrix   A column matrix consisting of the variables of a system of linear equations.

Vertical Asymptote   A vertical line $x = a$ is a **vertical asymptote** of a function f if the values of $f(x)$ approach ∞ or $-\infty$ as the values of x approach a.

Vertex   The highest (or lowest) point on the graph of a parabola.

Vertical Major Axis   An ellipse has a **vertical major axis** if the vertical axis of symmetry passes through the two foci and is longer than the horizontal axis of symmetry.

Whole Numbers   The set of positive integers combined with the number 0.

x-Axis   The horizontal axis in the rectangular coordinate system.

y-Axis   The vertical axis in the rectangular coordinate system.

Zero   A number c is a **zero** of a function f if $f(c) = 0$.